The Student's Companion to Geography

B

The Student's Companion to Geography

Edited by
Alisdair Rogers, Heather Viles
and Andrew Goudie

First published 1992

Blackwell Publishers
108 Cowley Road, Oxford, OX4 1JF, UK

Blackwell Publishers
Three Cambridge Center
Cambridge, Massachusetts 02142, USA

Library of Congress Cataloging in Publication Data

The Student's companion to geography /
edited by Alisdair Rogers,
Heather Viles, and Andrew Goudie.
p. cm.
Includes index.
ISBN 0-631-17088-X : —
ISBN 0-631-17089-8 :
1. Geography. I. Rogers, Alisdair.
II. Viles, Heather A. III. Goudie, Andrew.
G116.S78 1992
910—dc20 91-27557
CIP

British Library Cataloguing in Publication Data

A CIP catalogue record for this book is available from the British Library.

Typeset in 9 on 11 pt Melior
by Photo·graphics, Honiton, Devon
Printed in Great Britain by
T.J. Press Ltd, Padstow, Cornwall

This book is printed on acid-free paper.

Contents

List of Contributors

Abler, Ronald F. Executive Director, Association of American Geographers and Professor of Geography, Pennsylvania State University, USA

Anson, Roger W. Principal Lecturer in Cartography, Oxford Polytechnic, UK

Binns, Tony Lecturer in Geography and PGCE Curriculum Tutor, University of Sussex, UK

Blouet, Brian W. Huby Professor of Geography and International Education, College of William and Mary, Virginia, USA

Bridge, Gary Lecturer in Urban Studies, School of Advanced Urban Studies, Bristol, UK

Burgess, Jacquelin Lecturer in Geography, University College London, UK

Burt, Tim P. Lecturer in Geography, University of Oxford, UK

Chamberlain, David Oxford University Appointments Committee, UK

Cirrincione, Joseph M. Professor of Geography, University of Maryland at College Park, USA

Coones, Paul Fellow in Geography, Hertford College, University of Oxford, UK

Corbridge, Stuart Lecturer in Geography, University of Cambridge, UK

Douglas, Ian Professor of Physical Geography, University of Manchester, UK

Fincher, Ruth Senior Lecturer in Geography, University of Melbourne, Australia

Gale, Fay Vice-Chancellor, University of Western Australia, Australia

Gardner, Nigel Director, Programme on Information and Communication Techniques, University of Oxford, UK

Gold, John R. Senior Lecturer in Geography, Oxford Polytechnic, UK

Goudie, Andrew S. Professor of Geography, University of Oxford, UK

Gregory, Stan Emeritus Professor of Geography, University of Sheffield, UK

Harris, Chauncy D. Professor Emeritus of Geography, University of Chicago, USA

Hope, John Principal Planner in the Policy, Information and Implementation Section, City Planning Officer's Department, Oxford City Council, UK

Jackson, Peter Lecturer in Geography, University College London, UK

Jenkins, Alan Principal Lecturer in Geography, Oxford Polytechnic, UK

Johnston, Ron J. Professor of Geography, University of Sheffield, UK

Kearns, Gerry Lecturer in Geography, University of Liverpool, UK

Kennedy, Barbara A. Lecturer in Geography, University of Oxford, UK

Ley, David Professor of Geography, University of British Columbia, Canada

Livingstone, David N. Reader in the School of Geosciences, The Queen's University, Belfast, UK

Lockwood, John G. School of Geography, University of Leeds, UK

Middleton, Nick J. Lecturer in Geography, Oriel and St Anne's Colleges, University of Oxford, UK

Pawson, Eric Lecturer in Geography, University of Canterbury, New Zealand

Raper, Jonathan Lecturer in Geography, Birkbeck College London, UK

Roberts, Neil Lecturer in Geography, University of Technology, Loughborough, UK

Rogers, Alisdair Lecturer in Geography, Keble College and Lady Margaret Hall, University of Oxford, UK

Sparke, Matt Graduate in Geography, University of British Columbia, Canada

Swyngedouw, Erik Lecturer in Geography, University of Oxford, UK

Talbot, Christopher Computer Assistant/Programmer, Oxford Institute of Retail Management, UK

Taylor, Michael Professor of Geography, University of Western Australia, Australia

Taylor, Peter J. Reader in Political Geography, University of Newcastle-upon-Tyne, UK

Thompson, Derek Associate Professor of Geography, University of Maryland at College Park, USA

Thrift, Nigel Professor of Geography, University of Bristol, UK

Viles, Heather A. Lecturer in Geography at Jesus and St Catherine's Colleges, University of Oxford, UK

Wallis, Helen formerly of the British Library, UK

Whittaker, Robert J. Lecturer in Geography, University of Oxford, UK

Williams, Michael Reader in Geography, University of Oxford, UK

Winser, Shane Expeditions Advisory Centre, Royal Geographical Society, London, UK

Wolman, M. Gordon Professor of Geography, The Johns Hopkins University, Baltimore, USA

Preface

Students who are thinking of doing geography, students who are already doing geography at school and are considering whether to go on and continue the subject at college, university or polytechnic, and students who have just embarked on their courses in higher education frequently want answers to a series of questions. 'Why should I study geography?' 'Why am I studying geography?' 'What do geographers do?' 'What can I do with geography?' 'Who are the key figures in the subject?' 'What should I read?' 'What is fieldwork and how do I do it?' 'How can I go on to do postgraduate work?' 'What are the exciting growth areas in the discipline?'

The purpose of this volume is to try to provide some of the answers to such questions, and to give an indication of the scope, methods, uses, techniques and, above all, the excitement of geography.

We see geography as a diverse discipline, practised by a broad range of individuals, and in that spirit we have tried to persuade our contributors to display their own personalities, enthusiasms, approaches, convictions, and styles. We have not attempted to homogenize and standardize their contributions more than is absolutely essential. We hope you will ponder and debate the diversity and opinions displayed in these pages. We make no apology for them.

AR, HAV and ASG

Acknowledgements

The publisher and authors are grateful to the following for permission to use and for their help in supplying photographs: Actors World Production Limited (VI.5.1); Bodleian Library (III.6.1); By permission of the British Library (V.1.1); Daily Telegraph Colour Library (II.10.1, VI.3.1); Mark Edwards (I.3.1); Environmental Picture Library, V. Miles (I.2.1); © Gad Gross/JB/Katz Pictures 1990 (II.12.1); Library of Congress (III.13.1); Museo Del Prado (IV.1.1); Panos Pictures (II.8.1 Philip Wolmuth, VI.10.1 Jeremy Hartley); Peter Keene (V.7.1, VI.2.1); Royal Geographical Society (V.8.1); Reproduced by permission of the Trustees of the Science Museum (III.4.1); Science Photo Library (II.9.1 David Parker, IV.1.2 Dr Fred Espenak); USDA Soil Conservation Service (II.7.1). All other photographs are by the authors.

Part I
Introductory Essays

Contemporary geography is a flourishing and diverse discipline and so it would be foolhardy to attempt to give a single definition of its range and content. We have instead asked five distinguished geographers to give their personal views on what geography is all about, and why it is worth doing. Gordon Wolman explains how an understanding of the interaction between human activity and the natural world is central to our response to contemporary world issues such as food supply and sustainable development. A historical perspective to these concerns is voiced by Gerry Kearns, while Nigel Thrift argues that we are 'hooked on risk' in our attitudes to the environment and world debt. Two other geographers provide personal accounts of their engagement with geography. For Neil Roberts, it combines the excitement of science with the aesthetic pleasures of working in the field. Fay Gale finds inspiration in the landscapes of Australia and the contrasting perception of Aborigines and Europeans. She adds to the call for a more unified and holistic attitude to life and land, a sensibility which has been, and remains, central to the geographer's imagination. The authors therefore share two themes; that geography is about a practical engagement with the world, a down-to-earth subject; and that a breadth and depth of vision which encapsulates the human and physical environments is the geographer's special quality.

I.1

Contemporary Value of Geography: Applied Physical Geography and the Environmental Sciences

M. Gordon Wolman

The Interests of Physical Geography

The world at large has only recently acknowledged the intimate relationship between human activities and the broader natural world. Physical geographers have for centuries dealt with this interaction. Studying the distribution of land, air and water and the way in which processes in each of these spheres interact, physical geography necessarily enters into the study of the distribution of plants and animals on the planet. Because other activities of human beings are often strongly influenced by the 'natural' environment, and in turn may markedly affect the environment, physical geography is especially concerned with attributes of the natural world which relate to human activites.

Within the last several decades, as observations have revealed the processes within and the interactions between land, water and air more clearly, new scientific approaches have suggested that physical geography can be expressed in terms of interacting systems. On a large scale, these are the atmosphere, geosphere, hydrosphere and ecosphere, interacting systems in which materials and processes in one sphere directly interact with those in the others. Such interactions occur at all scales from the micro to the global level. The ocean circulation, for example, directly affects the behaviour of the atmosphere and the climate as flows in the ocean carry heat and moisture. At the same time, movement of the atmosphere alters the dynamics of the ocean, its temperature and its composition. The water on the surface of the globe cycles from atmosphere to land to ocean, or directly to the ocean, and back to the atmosphere driven by the heat of the sun. The availability of water or moisture in the atmosphere directly affects the distribution of plants and animals and virtually all human activities.

The expression 'interacting systems' is only a few decades old. The idea is not, but the realization that these systems can be described and their interactions specified to provide models of the behaviour of earth processes at many levels is relatively new. Not only has the notion of global interactions, often referred to as 'globalization', of both natural and human phenomena become a guiding light in discovering and understanding how the natural world works, but the development of models of how natural processes work has also become a basis for relating human activities, including social and economic phenomena, to the natural scene. Whether the issue be 'greenhouse gases', acid rain, biodiversity, natural hazards or global transportation, the structure of enquiry and the paths of public policy must involve consideration of the earth system.

Words like 'system' and 'model' convey not only an approach but also a bit of modern jargon. A system can mean anything from the flow field and path of gasoline leaking from an underground storage tank to the carbon cycle in biota, industries, atmosphere and oceans on a global scale. Models of these can be simple or elaborate mathematical formulations describing the processes involved in transporting, changing or storing materials.

Because society has been thrust into a global world, geography and physical geography are coming to have new meaning as individuals and governments discover that activities in one place are inextricably tied to those in places often very far removed from the supposed site of importance. Through the dynamics of earth processes no fields of enquiry are moving faster nor involving tougher questions than those related to the physical features and processes of the earth and the way in which these relate to human affairs.

The close relationship between an understanding of how the earth works and current social problems is illustrated at a variety of spatial and temporal scales in the examples which follow, beginning with the global scale.

Population, Agriculture, Land Use and Climate

Application at a global scale

By now presumably everyone is aware that the population of the globe is growing at an exponential rate. Simply because the total population is already large, approximately 5 billion, a rate of increase of only 1.7 per cent per year, the average growth rate of the world's population, represents a doubling in only forty years. The significance of this enormous number of people is not conveyed by the number itself. Rather, its significance is related to the spatial distribution and activities of the people and the character of the lands on which they dwell. At one extreme, the world's population is becoming rapidly urbanized. By the year 2000, ten metropolitan areas will have populations of over 13 million people, two over 24 million. At the other, many farmers in the developing world farm less than 5 hectares, ekeing out a subsistence in dry lands or on precipitous hillslopes. Urban agglomerations and rural hinterlands alike depend on the land for the food and sustenance on which life and society depend. How do we appraise that land, project its uses and its potential? Does it have a finite capacity to support human society?

Appraisal of the characteristics of the land, the availability of moisture, soil and climate provides the context for subsistence agriculture as well as the potential for enhancement beyond mere subsistence. But the land is not fixed. Its reality and its potential are in the imagination of those who wish to use it and in the tools at their disposal. Thus, the concept of capacity implies both an objective or future and an assumption about society, culture and technology. The physical geographer applies his or her skills within this context.

No more difficult questions confront us today than evaluation of the values of land for different uses, now and in the future. A seemingly dull business of 'land-use classification' is crucial to all efforts at planning in all societies. What land is good for what use, or why is this land better for one use than for another? Whether the requirement is an airport, a rice paddy-field or a high-rise office building, geographers can help to define or differentiate between the attributes that make up the land: gradient, fertility, vegetation, fauna, stability, drainage, moisture, elevation and continuity are often measurable attributes of the land. However, the same criteria are not necessarily appropriate to different potential uses, nor are they intrinsic characteristics sufficient to dictate a given use in the absence of the essential fact of location of other human activities. Development of techniques for handling spatial information has made possible the inventory and storage of information about the land and manipulation of such information through computerized geographic information systems. This new tool, coupled with the ability to observe images of the earth from remote sensors as well as aerial photographs, makes it possible to survey vast areas and to accumulate information about specific, even small, units of the land.

How do we answer the question, can the soils of the globe provide food to feed an estimated 12 billion people for the year 2015? The question can only be answered if we know the distribution of soils of different kinds, their relationship to climatic factors such as the growing season, the technology and fertilizers available for agriculture at a particular site, and the ability of, and incentives for, individuals to make use of these attributes. Present estimates suggest the possibility of feeding three to five times the present world population when looked at from a global perspective, that is, if food surpluses in one place can be transported to another and those who need food have money to buy it. A rough balance over the last several decades suggests increases in agricultural productivity have roughly paralleled world population growth. However, as the spatial scale diminishes from globe to continent to region to nation to district, the apparent balance disappears. Many people starve in different areas of the world while surpluses exist elsewhere. Appraising the land and its capacity at different scales is a first step in seeking to assure the availability of food for all.

A new buzz-word is 'sustainable development', the notion that economic development can be called successful only when it does not take place at the expense of renewable and non-renewable resources essential to sustaining life and civilization on the globe. But how is sustainability to be defined? Consider the soil resource. The physical geographer, concerned with the rate at which erosion takes place on, for example, a steep landscape, is also involved in a broader question. At the observed rate of erosion, what will be the potential loss in agricultural productivity associated with the loss of soil? Or, should erosion proceed, when would the point be reached when the process of degradation would no longer be reversible and loss of productivity permanent? The question requires information about the technology available to promote reversibility, and the economic and psychological motivation of farmers to attempt to reverse the degradation.

We hear much about the degradation of the globe, whether of land, tropical forests or other biota, but little about how to measure this phenomenon and even less about how to reverse it at costs that people are willing to pay. Here physical and human geography join in looking at the whole, not simply at an inseparable part labelled either 'physical' or 'human'.

The climate of the earth over most of earth history has been considerably warmer than it is today. At the same time, within the past several million years, the mean annual temperature of the globe has fluctuated over a range of 6 degrees, several degrees warmer as well as colder than temperatures being experienced today. Today's scene, however, differs markedly from those of the past when the natural landscape was forever altered by the development and spread of agriculture. While the spatial extent of the destruction of natural vegetation has been extensive, the combination of large numbers of people and high technological capacity has made the rate of transformation of both landscape and climate potentially far greater today than at any time in history. The emission of carbon dioxide, methane and other greenhouse gases now exceeds millions of tons per year. These amounts are estimated to be sufficient to promote global warming of the order of several degrees Celsius over the coming decades. Never before in human history has it been possible for human beings to modify the environment to such a global extent. Development of an understanding of the potential impact of industrial effluents is essential to understanding global climate and to influencing climatic change. Moreover, the climate itself is merely the first step; climatic change also involves modification of the hydrologic regime of the world's lands (plate I.1.1) and a myriad of potential changes in landscape processes and biota. These in turn presage potentially significant effects on virtually all human activities.

Applications at smaller scales

A global view does illustrate the relevance of physical geography, a systems perspective, to major human problems of today and tomorrow. The following questions, or vignettes, are intended to illustrate the close relationship between an understanding of natural processes and landscapes and matters of immediate policy decision. They suggest that the solutions to immediate questions of policy hinge directly upon the resolution of questions at the cutting edge of scientific enquiry in physical geography. The issues are posed as questions.

Erosion is rampant on a given landscape. Is it because too many livestock graze there, or because a changing climate provides too little rain in the

Plate I.1.1 Lake Chew Bahir, Southern Ethiopia shows the impact of climatic change. The spit in the foreground, now stranded, was formed at a time of greater humidity when the lake was deep rather than dry.

grass-growing season and too much punishing rain later which erodes the thin cover?

If crops are sprayed with herbicides to control weeds, will the herbicides be degraded by biota, accumulate in groundwater, run off farmland into streams, and accumulate in river muds, ponds, reservoirs and estuaries? Because mud and herbicides from range and farmland pollute the watercourses, environmental policy may dictate controls. But, how, where and with what intensity should they be applied?

Babbling brooks and trout streams are a natural treasure often found in areas covered by forests supporting trees ripe for timber harvest. Will the pools and riffles, critical habitat for fish, be destroyed by extensive roads, fallen trees and slash left behind? What processes in nature can sustain that familiar pool and riffle pattern? Can timber harvesting enhance those processes rather than inhibit them?

Floods frequently over-top the river-banks of channels flowing in broad valleys. Why is the channel not large enough to contain the higher floods – what magnitude and frequency of flow will it contain? Should people be permitted to build houses, highways, parks or warehouses on such nice flat open land? If not at the channel's edge, where? Why not?

Sulphur and nitrous oxides emitted from power plants, factories and automobiles not only foul the air but, chemically transformed in transport in the atmosphere, return to earth as dry and wet acid precipitation. Acid deposition alters the natural rate of weathering and increases the release of aluminium in some soils, which in turn makes its way to rivers and lakes, altering the chemical balance and thus the biota of the water system. Distances from source to deposition may be thousands of kilometres across countless international borders. Can we predict how changing the mix of power plants, factories and automobiles, or the controls on each at different places, will affect the quality of the water environment thousands of kilometres away?

Urban sprawl may house thirty people on each hectare as developments march over the landscape – each individual responsible for several tons of mud produced during construction. The completed surface of housetops, streets and storm sewers becomes the source and transporter of increased local floods and runoff of organic material, dirt, oil and debris comparable with sewage from secondary treatment plants. The water which runs off no longer replenishes the groundwater and small streams in backyards and parks dry up, except during storms when they flood. Do we understand this system of land and water well enough to predict how various designs for urban expansion will change water quality and quantity within the urban area and at the periphery?

Whether the scale is global or local, understanding the interaction of human beings and their environment involves a knowledge not only of the way 'nature' works (humans are, after all, natural) but of the way societies cope as well. The distance between a theory or hypothesis about how water gets into streams, or whether floods are bunched in time, like Joseph's sequences of dry years, and critical policy decisions designed to cope with natural hazards, urban sprawl or the uncertainties of agricultural production is very small. On a crowded planet, growing ever more crowded, policymakers need new information at an accelerating rate about the earth system and its inhabitants. Physical geography is a vital contributor to the science and policy needed to deal with these complex issues.

I.2

Apocalypse Soon, or, Why Human Geography is Worth Doing

Nigel Thrift

A German philosopher once said that by the end of this century we would no longer be talking about how to improve the world but about how to save it. As that time approaches, so this statement rings increasingly true; we are in a race to save the world.

Is this blowing things out of proportion? The answer has to be no. Consider just the four most important global problems that currently sustain the interest of human geographers. Each one of them has serious consquences on its own. But put them together and we are presented with a truly frightening prospect. Not apocalypse now, but apocalypse soon.

The first of these problems is damage to the environment. The natural environment has been treated like a toy for so long that we are still surprised that it can bite back. Now we have to put the damage to rights or pay the consequences like global warming, loss of species and the exhaustion of finite resources. This will be a gargantuan task but it is not an impossible one. After all, nearly all damage to the environment is the result of *social* causes. In developing countries, the uneven distribution of land ownership leads to the massing of poor people on marginal land and to that land's eventual exhaustion; as more and more poor people try to eke a living from the land, so they kill both it and eventually themselves. In countries with planned economies, the result of a remorseless push to industrialize without regard to consequent problems of environmental pollution has been dead rivers and shrinking lakes (plate I.2.1). Developed countries cannot look smugly on. For example, much of the western United States has become a water welfare state. Water has been assumed to be a free and unending resource. The result is that aquifers are drying up and salt deposits are spreading.

The second problem consists of an acceptance of debt as a way of life. Since the 1960s we have grown used to a world dominated by a financial system whose chief purpose is to produce mountains of debt and then recycle them at ever greater speeds, producing more mountains of debt in the process. We tend to assume that this newest, highest and fastest-growing of all mountain ranges is somehow inevitable. But it also represents a future that has been mortgaged. Many of the developing countries found this out in the 1980s, as they battled with debts that were taking up, on average, 60 per cent of gross national product (GNP). Many people, especially children, died so that the interest on these debts could be paid. Some developed countries are now in harm's way too. After a spending spree in the 1980s, the United States has a trillion dollar national debt, and another trillion dollars in liabilities is just beginning to show through. Financing these debts is going to be very difficult in a world of high interest rates caused by Japan's and Germany's increasing concern for problems in their own backyard and a corresponding lack of interest in bailing the United States out. The developing country debt crisis and the United States

Plate I.2.1 Much of the industrial plant in Eastern Europe, exemplified by this factory in Litvinof, Czechoslovakia, is of antiquated design and has created serious environmental problems, including air and water pollution.

debt crisis are only two of many crises that have plagued the new world financial system in the last decade. The concern is that this system is now so inured to crisis that, when the big one arrives, no one will recognize it as such.

The third problem is the 'informational revolution'. We are undoubtedly going through a period of major technological change. There are epic shifts going on in the world of electronics and telecommunications that now rival in scale and importance the original industrial revolution. Instead of the growth of railways, we see the expansion of global digital highways, massive 'smart' networks of fibre-optic cables and satellite links. Instead of the growth of larger and larger industrial combines, we see the rise of 'flexible' firms, networks of subcontracts and alliances tightly co-ordinated by information technology and harnessed to flexible production systems. Instead of the growth of markets attuned first and foremost to the imperatives of production we are witness to the advent of markets where, through information technology, the consumer is truly sovereign. In each case, what defines this new industrial revolution's 'space of flows' is speed – whether it is in the form of a more rapid response to changing market conditions, more and then more truncated product cycles or more rapidly changing markets. Speed is of the essence in this brave new industrial world but it is also, because it brings with it yet higher levels of production and consumption, a problem.

The final problem is one of estrangement. We live in a world in which, through the growth of mass media like television, people are able to know more of what is going on in the world and share more experiences and signs than ever before. But the potential for greater understanding between peoples has hardly been realized: instead it sometimes seems that what are being built are new ways of identifying strangers. Blind nationalisms, ugly racisms, pathetic sexisms, and religious bigotry still persist and thrive for many and various reasons; perhaps because mutual links between people are too often reduced to a visual contact, perhaps because new forms of tolerance of others are a thinly disguised excuse for ignoring them, perhaps because old traditions of disparagement are so engrained. Whatever and wherever the case, it is clear that human difference is still treated with great suspicion.

To summarize, we live in a world that is hooked on risk. There is the risk of environmental catastrophe. There is the risk of falling into a financial abyss. There is the risk that the informational revolution will turn the world into a private playground for a few corporate giants. And there is the risk that no one will care sufficiently about what is going on elsewhere in the world to tackle these other risks. In Hans Magnus Enzensberger's (1990, p. 142) words, what we need is a 'managed retreat' from this world of risks and more risks, away from the old certainties which have only bred uncertainties.

All very dramatic. But what contributions can human geography make to this retreat? More than might be thought, perhaps. First of all, it is these four different but interconnected risks which will form the chief areas of research in human geography over the next decade. A human geography of the environmental crisis is being formed around notions like political ecology and Gaia (see p. 43), one that stresses the interdependence of society and ecology (Blaikie, 1985; Emel and Peet, 1989; O'Riordan, 1989). A human geography of finance and debt is being put together, one that recognizes the monumental importance of money in the modern world, but one that is not mesmerized by the glamour of the dealing room (Thrift and Leyshon, 1987, 1992; Harvey, 1989; Corbridge and Agnew, 1991). A human geography of the informational revolution is also being constructed, one that is as aware of the spatial divisions that are being fostered by this revolution as it is of the way that the world is being drawn into a unified space of corporate flows of information (Massey, 1984; Castells, 1989; Cooke, 1989, 1990). And last but not least, a human geography of estrangement is being forged out of work on subjects like gender, race, culture and the state, one that is alive to the importance of places as a means through which people make the world and are made by it but which also recognizes that this process is no longer purely local (Jackson, 1989; Peet and Thrift, 1989; Wolch and Dear, 1989; Thrift, 1990). In the words of Frederic Jameson, (1988, p. 349) human 'experience no longer coincides with the place in which it takes place'.

Another contribution that human geography can make comes from its very nature as a subject. Human geography has always been a science of difference: if there were no different people and places to describe, explain and appreciate there would be no human geography to write and draw. So human geography comes naturally to stressing the growing interdependence between places and yet, at the same time, valuing all the differences that still exist which give these places their unique worth. Concomitantly, it comes easily to the job of stressing that the differences between the people in these places can unite as well as divide. In other words, the task of communicating about people and places, so that other people in other places realize that they exist as more than just ciphers on a screen, has become more pressing than ever.

A third contribution again comes from the nature of human geography. Human geography has always been an intensely practical project. In the past, at its worst, this made for a rather stodgy subject, frightened to move away from immediate issues and with a tendency to kowtow to authority. But the advent of a proper theoretical tradition (e.g. Gregory and Urry, 1985; Peet and Thrift, 1989) has combined with the old rough-and-readiness to produce a mixture of theoretical foresight and practical analytical skills which is well suited to the times we now live in. Human geography is now down-to-earth in the best possible sense.

To conclude, human geography is a deadly serious pursuit. The state of the world demands it. But it is important to remember that it is not all about gloom and doom. Human geography also takes from its past the sense of optimism that possessed the explorers who helped to found the subject. It is a sense that around the next corner will be something interesting, even redemptive. After all, it is important to believe that there is a way out of the current dilemmas that the world faces. What we need to do now is find the right maps.

References

Blaikie, P. 1985: *The Political Economy of Soil Erosion in Developing Countries.* London: Longman.

Castells, M. 1989: *The Informational City.* Oxford: Basil Blackwell.

Cooke, P. (ed.) 1989: *Localities.* London: Unwin Hyman.

——1990: *Back to the Future.* London: Unwin Hyman.

Corbridge, S. and Agnew, J. 1991: The US trade and budget deficits in global perspective. *Environment and Planning D: Society and Space*, 9, 71–90.

Emel, J. and Peet, R. 1989: Resource management and natural hazards. In R. Peet and N. J. Thrift (eds), *New Models in Geography, Vol. 1*, London: Unwin Hyman, 49–76.

Enzensberger, H.M. 1990: The state of Europe. *Granta*, 30, 136–42.

Gregory, D. and Urry, J. (eds), 1985: *Social Relations and Spatial Structures.* London: Macmillan.

Harvey, D. 1989: *The Condition of Postmodernity.* Oxford: Basil Blackwell.

Jackson, P. 1989: *Maps of Meaning.* London: Unwin Hyman.

Jameson, F. 1988: Cognitive mapping. In G. Nelson and L. Grossberg (eds), *Marxism and the Interpretation of Culture*, Urbana, IL: Macmillan, 347–60.

Massey, D. 1984: *Spatial Divisions of Labour.* London: Macmillan.

Models in Geography, Vol. 2, London: Unwin Hyman, 77–102.
Peet, R. and Thrift, N. J. (eds), 1989: *New Models in Geography*, 2 vols. London: Unwin Hyman.
Thrift, N. J. 1990: Taking aim at the heart of the region. In D. Gregory, R. Martin and G. Smith (eds), *Geography in the Social Sciences*, London: Macmillan.
——and Leyshon, A. 1987: The gambling propensity: bankers, developing country debt exposures and the new international financial system. *Geoforum*, 19, 55–69.
——and Leyshon, A. 1992: *Making Money*. London: Routledge.
Wolch, J. and Dear M. (eds), 1989: *The Power of Geography*, London: Unwin Hyman.

I.3
The Historical Geographical Perspective

Gerry Kearns

Why study geography? Why study historical geography? A lot of what geographers study is on our television screens and in our newspapers every day: green issues and the environment; the plight of the Third World (plate I.3.1). Much of such journalism is ill-informed, bland and often aimed at an attention span of thirty seconds. Geography will continue to interest us all as long as it can do better than this. When geographers study the Third World, they must, as Brookfield (1975) argued some time ago in *Interdependent Development*, pay attention to the links between the rich and poor countries. Poor countries remain poor by keeping rich countries rich. This interdependence, though with a slightly different emphasis, was also the lesson of the two reports of the international commission of expert economists chaired by the former West German chancellor Willy Brandt: *North–South: a programme for survival* (1980), and *Common Crisis, North–South: co-operation for world recovery* (1983). Rich countries need poor countries to develop in order that the expanding markets in the poor countries might stimulate production in the rich. Are there really any solutions as simple as this? There are certainly no simple explanations and both Brookfield and Corbridge (1986, ch. 5) draw attention to the challenge of thinking clearly about exactly how the First and Third Worlds are interdependent. In terms of their analysis, the arguments of the Brandt reports seem no more than wishful thinking.

Another, and perhaps more exciting, example of geographical thinking being taken up by the media is provided by the discussion of green issues. The Brundtland Commission's (1987) report on world environmental problems, *Our Common Future* (produced at the request of the United Nations Secretary-General) is a very interesting attempt to link the questions of economic development raised by Brandt to the new ecological concerns of green politics. Can we leave the world to our children with at least the set of productive possibilities we inherited; is development sustainable? Can we contemplate the economic development of the Third World without being struck by the multiplication of pollution and environmental degradation already seen in the development of the rich world? These are important questions and they will be debated in the classroom, living room and workplace for years to come. The Brundtland Commission drew attention to the need for food security as a prerequisite of human dignity and economic development throughout the world. Indeed, it is around pictures of starvation in Africa that television pundits often spill their most alarming reflections about the unequal struggle between environment and society in the poor countries. In some ways, the question of world hunger crystallizes both the Third World and green perspectives. This is one area where the value of geography should be clear. I want to argue that global issues such as this are at the heart of geography and, furthermore, that we shall fail to

Plate I.3.1 A dust storm in Ethiopia. Such phenomena are manifestations of environmental issues such as desertification. Adverse climatic conditions, including drought, combined with over-exploitation of the land, leads to such degradation.

understand these issues properly if we do not pay attention to the historical processes which shape the decision-making environment for both farmers and consumers. There is a social context to the food question and it is shaped by long-term (hence historical) and regionally variable (hence geographical) transformations of the economy.

The question of food is one of the most important challenges facing the world community; the chronic mismatch between the need for food and the possession of food which reveals itself in the existence of food mountains in the rich countries and malnutrition in the poor is disgusting. Understanding the causes of the food scandal is central to our attempts to discuss the prospect and basis for economic development. It is a test of the moral basis of our world community. Like all important questions it has a number of dimensions.

Economically and politically (see Tarrant, 1980; Grigg, 1981), we see the rich countries subsidizing their farmers to produce surpluses we cannot afford to eat (for fear it will reduce the retail price of the food we already eat) and which we either destroy or 'generously' provide as aid to poor countries unable to feed themselves. This aid, of course, acts as a disincentive to food production in the poor countries and may even be seen as insurance against a strategy of gearing rural economies towards the production of cash crops which can be exported to help repay their debts to the rich countries — the ones with the food mountains. Is it humane to demand repayment from countries forced to deny their own subsistence needs in order to keep faith with these loans?

In cultural terms one might look at the eating habits of the people of poor countries as being related to systems of survival which were appropriate to earlier farming systems but which provide them with much less insurance in the new global market. Or, one might look at the westernization of consumers in poor countries and the cultivation of tastes through advertising which stimulate demand for such products as refined sugar or white bread.

In an ecological sense, we might examine the ecological context of food production. We might look at the energy efficiency of different systems of food production and we would soon discover that the meat-eaters of the world are very wasteful of the sun's bounty whereas wet-paddy rice cultivation converts a very high proportion of the energy reaching the earth's surface into food-calories (Braudel, 1981; Bayliss-Smith, 1982). In ecological terms, we might explore the range of inputs to farming and note the very large amounts of industrial chemicals supplied to the soil in the rich countries and then raise some further questions about the notion of sustainable development. These geographical concerns are at the heart of the report *Our Common Future*, and they will become more rather than less prominent as green issues come even more to the fore in our discussions about society.

To take matters only this far, however, leaves an important part of the geographer's contribution unsaid, for there is a geography to the political, economic, cultural and ecological perspectives developed so far. Obviously, in ecological terms, there will be a geography to the possible systems of food production which will be a function of topography and climate. These relations are immediately wrapped up with considerations of technology and wealth. More simply, how a society *chooses* to deal with the costs of pursuing certain food strategies in particular environments is partly constrained by their wealth and their available technology. There is also a geography to the culture of food which is partly related to the geography of religion and prevailing farming practices but which is also related to the geography of social organization. Food is the very staff of life and social priorities concerning food are also priorities concerning life. These priorities are partly shaped by the way society is organized.

To give a very quick example (from Iliffe, 1987), in subsistence societies, those where the food requirements of the agriculturalist are the first aim of production, the food needs of those unable to work are sometimes met through the family, leaving the orphaned and widowed at greatest risk. During lean times, it is the unattached followed by the old, weak or infirm who find it most difficult to stake a claim to the dwindling supply of food. In capitalist societies, the first aim of the food producer is to meet the needs of the market, their own food needs often being met through purchases. Farmers are not all going to be keen to swap the security of control over land which enabled them to act in a subsistence fashion for the uncertainty of working for wages or on credit to pay rents as in a capitalist system. Indeed, the eviction of the direct producers from the land becomes one of the crucial preconditions of a capitalist economy where most food comes to market. From the point of view of the urban industrialist this eviction removes an awkward bottleneck from the food production system and ensures that urban wages can command available food. From the point of view of the rural producers, it inaugurates a new regime of access to food. The *level* of wages becomes crucial. While family obligations are still recognized, an inadequate wage can plunge whole families, the strong as well as the weak, into desperate straits; and this of course is the way the hungry country continues to feed the high-waged town in a capitalist economy under pressure.

The question of food supply, then, is more than a brute confrontation between nature and society; it concerns how access to resources is socially organized. Indeed, in speaking of the destruction of subsistence economies, we are dealing with a historical process of global dimensions. There is a geography to this transition to capitalism. Furthermore, the transition is a process which took place by means of the connections between places. In thus wedding time to space, the historical geographer can bring the big picture into focus. I want to flesh these claims out a little by returning to the geography of the transition to capitalism and show how access to land and labour in the production of food and how social priorities in the allocation of food were changed by the transition and how the nature of the transition was shaped by the external connections of countries and regions: a geography, then, of the spread of capitalism and a sketch of how a global perspective is essential to the understanding of the transition in any one place (largely based on Wolf, 1982).

When the food producers met their own food needs as a first priority, they were still not acting as completely free agents. All over the world, the development of civilization has entailed the concentration of power based on systems of taxation. Taxation was a sort of cull from the food produced in the countryside. It did not really determine the crop-mix or the number and type of animals in the rural sector. Decisions about farming practices, though, were taken by people operating within the class system of the time and place.

Where the king and nobles largely monopolized military power and where they also controlled the most productive elements of the economy, their world view, as it were, held sway. The crucial factor was the control of land and labour (for a simple summary see the introduction to Kriedte, 1983). They directly controlled large swathes of the countryside but, and more importantly, they had a claim on the labour of others to work it and to a share in the produce of the remaining parts of the countryside. These claims sometimes took the form of serfdom, where the labourer went with the land, so to speak, but at other times and places these claims took the form of custom and practice: this is the way society has been organized since time immemorial, or so it is said. Given a retinue of lawyers, clerks and bailiffs to sustain the world view of the lords, we have, here, a society where the direct food producers are constrained by force and law to pay taxes to and work for the crown and nobility. Furthermore, this power over the food producers is subject to no legitimate check on their part. Such a society is variously called a feudal or a tributary mode of production, depending largely on whether one is adopting a European or a world perspective.

These societies were not all of a muchness, however, and there were important world geographical differences which nudged subsequent transitions in different directions (this comparison of Europe and Asia is drawn from Jones, 1981). The central differences related to power of the exploiting class. In Europe, they were relatively weak. That is, they ensnared a smaller share of total production and this agrarian elite had a weaker monopoly on productive and political power than in Asia. Indeed they were increasingly forced to share power with an urban commercial class. In China, in contrast, the landed elite froze the merchants out of the political arena and they were able to uphold the existing forms of exploitation. This relatively greater monopoly power of the Asian ruling classes allied with a more unstable physical environment had serious consequences for the Asian peasants. The miraculous productivity of wet-rice cultivation merely multiplied misery and provided little incentive for the peasant to invest in new technologies or produce a marketable surplus to stimulate the rest of the economy. Indeed the hazards of the peasants' lives were met by the insurance of high fertility, anticipating the need to replace workers lost in war, famine or epidemic and reflecting the insecurity with which possessions were held.

In other words, even as early as the sixteenth century, there were clear differences in the balance struck between population and resources in Europe and Asia. These differences have been re-established at various times since then but it is important to note that the high-pressure regime of high fertility and high mortality which many see as the starting point of the so-called demographic transition was not characteristic of pre-industrial Europe on the verge of the transition to capitalism. Any theoretical model of the changing relations between population and resources must take account of this fundamental feature of the historical and geographical reality.

From the sixteenth century in western Europe, the feudal nobles, squeezed by their inability to twist the rachet of peasant exploitation even further and by the newly emerging alternative base of economic power in towns and trade, became ever more unsatisfactory to the king as both financial and political ally. Or, to put it another way, the urban commercial classes were able to gain some purchase on political power by means of their blossoming economic power. But now they faced precisely the bottleneck I described above. A subsistence rural sector constrained both the labour and food supplies for the new manufacturing economy. As the urban commercial class made inroads on the feudal lords' political clout, they preached freedom for the economy from the arbitrary, *merely* 'customary', rights of the feudal lords over land and labour. At the same time, they offered the exploited rural classes a sort of freedom from these arbitrary feudal impositions. A coalition of middle-class free marketeers and peasant pressure from below pushed through revolutionary political changes in England and France which set the context for the failure of the state to defend feudal privileges at all by the eighteenth and nineteenth centuries respectively, thus setting the seal on a process which had begun two hundred years before.

Capitalist agriculture, with its tenant farms, waged labour and market orientation proved very responsive to technological innovation and the *labour* productivity of agriculture improved dramatically, thus sustaining a larger manufacturing and trading sector (see Wrigley, 1987). Part of that trading sector was the world market. In the seventeenth century, the taste for stimulants drove the Europeans abroad in search of tea, coffee, sugar and tobacco – like most drugs, high value, low bulk – but lowering transport costs brought more basic goods into world trade and by the nineteenth century coal and manufactures had replaced silver as the spearhead of the European penetration of African, Asian and Latin American societies. By the late nineteenth century labour was being redistributed on a world scale to support the European need for tropical products with plantations sucking labour in from poor countries far and wide. Once again, this involved 'freeing' labour and wrecking subsistence economies. But this time the force for change was as much external as internal. It was the Europeans with their systems of cash taxation who drove Indian peasants to the market and it was the Europeans, with their confiscation of land rights, who proletarianized the African agriculturalists. True, in many cases they worked through local elites, yet it is equally true that in thus involving themselves with domestic political instability they ensured that the resulting internal class struggle was more one-sided than it had been in Europe. This had serious consequences which may partly be grasped by looking at the concessions offered to European labour during their own transition – in particular, state-funded systems of poor relief. Here is a clear illustration of one of the ways that social priorities reflect social organization and social organization reflects contested transitions from one set of power relations to another.

When we take the story on, then, to the present with reports such as *Our Common Future* bemoaning the orientation of Third World economies towards cash crops and away from basic needs and pointing to the impossibility of environmental considerations in the face of the most desperate poverty, we need to see how those social priorities have been framed by a social organization which is the result of a global historical process differentiated from place to place but crucially determined by the links between places. The geography of the transition to capitalism and the subsequent geographies of the different phases of capitalist reorganization is the immediate context of the question of population and resources. This is not to say that there is no real problem of feeding the world; rather it is to suggest that, considered in its historical and geographical context, this is an issue about the way options are opened and closed for different groups in various places during the development of a capitalist world economy. Historical geography can help to bring the 'big picture' into frame (for a discussion of global perspectives see Kearns, 1988). As Third World and green issues assume greater importance in geography this will become ever more necessary if we are to bring analytical rigour to these questions of general concern.

References

Bayliss-Smith, T. 1982: *The Ecology of Agricultural Systems*. Cambridge: Cambridge University Press.

Brandt, W. (Chairman of the Brandt Commission) 1980: *North–South: a programme for survival*. London: Pan.

— — 1983: *Common Crisis, North–South: Co-operation for world recovery*. London: Pan.

Braudel, F. 1981: *The Structures of Everyday Life: the limits of the possible*. London: Collins.

Brookfield, H.C. 1975: *Interdependent Development*. London: Methuen.

Brundtland, G.W. (Chairman of the World Commission on Environment and Development) 1987: *Our Common Future*. Oxford: Oxford University Press.

Corbridge, S. 1986: *Capitalist World Development: a critique of radical development geography*. London: Macmillan.

Grigg, D. 1981: The historiography of hunger: changing views on the world food problem 1945–1980. *Transactions, Institute of British Geographers, New Series*, 6, 279–92.

Iliffe, J. 1987: *The African Poor: a history*. Cambridge: Cambridge University Press.

Jones, E. L. 1981: *The European Miracle: environments, economies and geopolitics in the history of Europe and Asia*. Cambridge: Cambridge University Press.

Kearns, G. 1988: History, geography and world-systems theory. *Journal of Historical Geography*, 14, 281–92.

Kriedte, P. 1983: *Peasants, Landlords and Merchant Capitalists: Europe and the world economy, 1500–1800*. Leamington Spa: Berg.

Tarrant, J. 1980: The geography of food aid. *Transactions, Institute of British Geographers, New Series*, 5, 125–40.

Wolf, E. 1982: *Europe and the People without History*. Berkeley, CA: University of California Press.

Wrigley, E. A. 1987: *People, Cities and Wealth: the transformation of traditional society*. Oxford: Basil Blackwell.

I.4

Not So Remote Sensing of our Environment

Neil Roberts

Malham – April 1970 A blustery, showery, spring day in the Yorkshire Dales, northern England. I'm walking from Malham Tarn on a sixth-form field excursion with C. M. G. The stream we are following disappears down a sink hole to leave a dry bourne. Downvalley lies the great natural amphitheatre of Malham Cove with its limestone pavements and Vauclusian spring resurgence. Despite the topographic connection, the re-emerging stream is not the same one that disappeared into the earth some miles upvalley. Was the cove once Yorkshire's Niagara Falls, with torrents of water cascading over the cliff top? And exactly what combination of processes – tectonic, climatic, erosional – have sculpted and created this magnificent landscape? Fifteen months later I obtained my Geography A-level, grade B.

Kara Dağ, Turkey – May 1978 A black volcanic mountain rises out of the flat Konya plain, dry enough today for sand dunes to form. We drive up a rough track past stands of wild einkorn, distant ancestor of our domestic wheat. From a vertical kilometre above the plain, and despite the dust haze, the panorama is stunning. It is also geographically revealing (I believe in getting to the highest and best vantage point of a field area, which usually means a mountain top and a lot of climbing). It is possible to trace a line of low sand ridges and eroded bluffs, marking the edge of a vast former lake. I try to imagine how this same landscape would have looked 20,000 years ago; the plain transformed to water and Kara Dağ almost an island. Was this lake the product of a wet 'pluvial' climate or a cold dry one, and was it surrounded by woodland or by steppe? (It turns out to be the latter; the Tibetan Plateau might be a good place to look for the modern equivalent of palaeo-lake Konya.) On the descent we skid and all but overturn the landrover; D. A. and M. K. are not sure whether to curse my careless driving or praise our good fortune. Two years later I submitted my PhD thesis on the physical geography of the Konya Basin.

Bumburwi dambo, Zimbabwe – August 1984 We drive out from Harare for an hour before turning off into Chiota, one of Zimbabwe's many former native reserves, now renamed Communal Areas. The tarmac road immediately changes to potholes and dirt. The leafy landscape is abruptly replaced by khaki-coloured treeless terrain, almost every square metre used for peasant farming. Only the dambos (small valley wetlands) remain green, and this despite a three-year drought. We stop and R. W. leads us down into Bumburwi dambo, its centre meshed by an extensive gully network. How fast has this been expanding? A villager tells us that when he was young the gully head was next to his brick kiln. I measure out the distance that it has expanded since then. (Study of old air photographs later proves the villager right.) Apart from a few fenced-off areas, the grass cover on Bumburwi dambo is seriously degraded through cattle grazing. Are other forms of dambo land use also associated with accelerated gully erosion? (No, it transpires from subsequent work, pointing the blame for desertification at overgrazing.) Three years later a fat report on environmentally sustainable forms of dambo resource use was submitted to Britain's Overseas Development Aid Agency (ODA).

Geographers need to have use of laboratories (so do chemists); they work in archives (as do historians); and they use computers (who doesn't?). But ultimately, geography concerns the real world, and the only way to experience and understand the complexity of the real world is in the field. The field can be in Yorkshire, in Turkey or in Africa, and it may be in city or in forest, but without direct sensing of it there can be no true geography. Geographers, of course, also rely on remote forms of sensing, maps, air photographs, satellite imagery, geographic information systems – even looking down from mountain tops is remote sensing of a sort. These are vital tools of the trade, especially in identifying broad-scale spatial patterns not visible from the ground. But they are not a substitute for field observation and verification.

I was attracted to geography not only because I loved to travel and wanted to see places, but also because of the challenge of trying to understand what I found when I got there. Zapping across the world, dropping in for a taste of Tahiti or a soupçon of San Francisco may appear glamorous but, like the snacks at a cocktail party, it leaves me, at least, hungry for more. Geography is, in this sense, as much about stopping as about travelling; or at least about travelling slowly. You see little of the landscape from a car or a coach, even from a Landrover. Trains and planes are better, but best of all are the tried and tested methods of perambulating on two legs or in some cases those of a four-legged friend! Travelling slowly, stopping often, allows one to look at the landscape, to interrogate it, search for clues which help explain it. There is a real challenge and pleasure in unravelling and decoding landscape. For me, a physical geographer, this is a scientific process, one of searching for regularities and causal connections, for I believe nature to be fundamentally orderly and well organized. There is a logic to the existence of tree lines or to the orientation of sand dunes, even if it is is not apparent to us at first sight. Understanding the logic of the natural world generates the same sort of excitement as it does for all those involved in making scientific discoveries – whether Crick and Watson's search for the double helix, or Vine and Matthews' explanation of sea floor spreading. Physical geographers are fortunate indeed, because not only do we share the excitement of science, but we also enjoy the aesthetic pleasures that can come with working in the field. Lack of sleep and damp boots notwithstanding, few scientists experience a desert sunrise, the scent of aromatic mediterranean herbs, or flamingos rising from an East African salt lake as part of their studies. And these pleasures are by no means restricted to exotic faraway places. Getting off the beaten track, as geographers have to, can be equally revealing close to home. Above all, when a geographer works somewhere, he or she needs to get to know it from close quarters and to know it well.

I.5

A View of the World Through the Eyes of a Cultural Geographer

Fay Gale

Geography, for me, is about how we view the world, how we see people in places. As primarily a cultural geographer, I think of geography as a study not so much of people in the environment but of how people visualize and use that environment. It is not so much what is there as what we believe to be there. A geographical training enables us to understand not only the relationship between people and their environment but also how people's reactions to that environment are influenced by their cultural conditioning. It is this broad interrelated view of human relationships in the environment that makes geographers such employable graduates in a very wide range of fields. It is also why geographers predominate, well in excess of their relative numbers, in senior management positions.

I was attracted to geography as a young student because it was so broad. Its field of study was the whole world and all the people in it. Unlike many other disciplines it offered me a vast array of choices, and a great deal of freedom. No set road had to be taken, no line of inquiry was prohibited. Geography not only allowed, it encouraged free thought, and a creative use of intelligence. Not being a subject of a particular specialist field had its difficulties. Narrow-minded friends would say, 'Why are you doing geography? It doesn't lead anywhere', or 'What's geography anyway? It's just bits of everything'. Only with the passing of time did I come to realize just how valuable it was to be trained in a discipline that was not specific and did not lead down any one particular path.

Geography has never had great theory or scientific laws. To the uninitiated this is seen as a decided disadvantage but to geographers it is a very real advantage. We do not have to follow the set rules or, in an intellectual sense, worship the gurus or acclaimed theorists. We are free to think for ourselves. It is a great subject for lateral thinkers. There are no boundaries, and going sideways is just as acceptable as going forward.

For me, growing up at a time when 'a woman's place was in the home' and a professional university education was 'wasted on a girl', the breadth of geographical thinking and intellectual exploration were crucial. I am sure that, had I gone down a set scientific path, the professional doors would have closed in my face, as they did for so many of my friends. But because geography 'did not go anywhere' I could not be stopped. I could study it just for fun. After all, one day I would stop and get married and have a family, so it did not matter what I did in the meantime. I did get married, and have a family, but I did not stop being a geographer. Partly because it was not seen as a definite career, there were not the same pressures to make me stop. The completion of my PhD was described by my supervisor as a photo-finish with the birth of my daughter!

For me also the strong field work basis of geography was absolutely crucial. One of the great strengths of geography, indeed its essential core, is that we study real people in real landscapes. It is the ability to interrelate the two, and understand the interaction of the one on the other, that makes geographers good analysts and managers. The essential 'feet on the ground' approach of geography has also ensured that geographers saw what was happening to the world around them. The search for scientific theory did not blind them to the reality of life.

Geography spawned environmental science largely because of its broad brush approach combined with its acquisition of knowledge through basic field work. In the 1920s and 1930s very active geography schools like those at Berkeley, California, began to study the impact of 'man' on the environment. Indeed W. L. Thomas's 1956 volume *Man's Role in Changing the Face of the Earth* (University of Chicago Press, Chicago, IL) grew out of his concern for the future of the globe. The ideas that led to the environmental movement, to a large degree, came out of that 'non-specialized' work in geography. By not narrowing our discipline to specific areas of study we were able to see the emerging problems of the whole world. Our fieldwork origins were crucial in this. 'Dirty-boots' geographers saw more of what was really happening to our globe than did the 'armchair' theoreticians.

Growing up in Australia was also important to my development as a cultural geographer. The cultural geographer can scarcely travel anywhere in Australia without stumbling over example after example of the contrasting ways in which Aboriginal Australians understood and used the land differently from European Australians. British colonists came to Australia during the great scientific revolution of the eighteenth and nineteenth centuries. During this period Western scientific thought separated 'man' from 'nature'; it dichotomized 'man' and the 'environment'. To Aboriginal people the land, and the human beings who occupied it, remained one interrelated whole, each influencing the other. Only now, with the advent of the environmental movement, has Western scientific thought begun to realize that a unified view of the earth, and our place within it, rather than superior to it, is crucial for our survival.

Studying cultural geography in Aboriginal Australia taught me that the landscape I had thought to be so real, so concrete, was actually a cultural construction. It was what we made it, not a physical fact separate from human action. Humans could nurture the land because they believed themselves to be part of it as in Aboriginal thought, or they could change it, indeed destroy it, because they believed themselves to be separate from, and therefore superior to, nature. The twenty-first century may yet prove that the Aboriginal view gives more long-term hope to human survival. Western education in Australia taught me that people live *on* the land, but studying geography amongst Aboriginal people taught me that humans live *in* the land and cannot dissociate themselves from it.

Australian history is littered with the bodies of Europeans who failed to understand this essential truth. Thus Burke and Wills on the Cooper Creek in the northeastern corner of South Australia and Kennedy on Cape York in Queensland are famous Australian explorers who died of starvation in areas of very rich food resources because they did not understand the land, and felt too superior to Aboriginal people to be able to learn from them. To the early British explorers and settlers, the Australian environment was viewed as harsh and barren (plate I.5.1). The saving power of technology was seen as essential to 'tame' it. But to Aboriginal people the Australian landscape, whether desert or tropical coast, was a lush rich country provided that one lived in a respectful relationship with nature.

I came to learn, through Aboriginal eyes, that the landscape is not, as my initial science education had taught me, a matter-of-fact reality. It is what we

Plate I.5.1 The central portions of Australia provided a harsh environment for early explorers such as Burke and Wills. Camels were used extensively because of their hardiness, but even they could be overcome by extreme sandstorms.

make it to be; it can be a friend or an enemy – the reality lies in what we believe it to be. Thus Burke and Wills died alongside a rich water course because they believed central Australia to be a desert, unfriendly, barren and remote. By contrast the large Aboriginal population which inhabited this region viewed it as very hospitable because for them there was plenty of water, and an enormous variety of food sources. Thus cultural geography taught me, through graphic field work experiences, that the value of any part of Australia, like any country, depended on what the cultural attitudes of the occupants created out of the landscape.

In the greatly varied Australian landscape, geography also gave me the essential spatial models that ensured an interrelated, and interacting, vision of the countryside.

Occasionally very heavy rains in central Queensland cause the rivers to flow and convert the usually salt-encrusted Lake Eyre into a large freshwater lake. Fish and birds suddenly abound. Pelicans fly in from coastal areas hundreds of kilometres away. Eggs are laid and young pelicans are hatched in their thousands. In Aboriginal times people came in, like the birds, from huge distances, to enjoy great social gatherings, and feast on the rapid appearance of large quantities of food. Pelicans provided eggs for food and feathers for ceremonial activities. European pastoralists destroyed the Aboriginal people so that none now live a traditional life-style in this area. But each time when the lake fills and pelicans hatch, and nowadays die in their thousands instead of providing food for people, popular science television programmes bemoan the death of the pelicans, and biologists try to explain the reasons for the apparent suicides. This narrow scientific view of the environment fails to appreciate the human interaction in the landscape scene. For me the broader view of geography, and cultural geography in particular, can explain the apparent paradox.

The world is an interrelated whole, people and land, past and present. The realism of this holistic perspective of geography is one of the subject's great strengths.

Part II
What is Geography? Past, Present and Future

Most students of geography approach the subject through the study of special topics; it is often only at the end of a course that the interrelationships between them become fully apparent. In this section of the volume we provide eleven reviews of the current state and future prospects of some of the main specialisms that students will encounter. Each review identifies the central ideas and preoccupations of geographers in the various areas of the discipline, covering both physical and human geography. The choice of specialisms is made on the assumption that, wherever the department and whatever the nature of the individual course, there are certain common topics that are taught. Goudie (chapter V.8) lists the study groups of the Association of American Geographers and the Institute of British Geographers. From these it is clear that we have not addressed the full range of special topics, which would probably require a volume by itself! Good reviews of other topics can often be found in two journals, *Progress in Physical Geography* and *Progress in Human Geography*.

The section begins with a historical essay by David Livingstone which describes the major dialogues within the discipline that have engaged geographers from the ancient Greeks to the present day.

II.1 A Brief History of Geography

David N. Livingstone

Geography has meant, and still means, different things to different people. For some it conjures up images of far-away places and intrepid explorers going where none has gone before. For others, the geographer is regarded as the person with an encyclopaedic knowledge of the longest rivers, the highest mountains, the largest cities and so on – a sort of talking atlas invaluable for television quiz-shows but useful for little else. For still others geography is the subject that deals in charts and globes; history is about chaps, it is said, geography about maps. In all likelihood today's professional geographers would reject all these commonplace notions as definitions of their discipline and provide their own explanation of just what geography is all about.

I do not intend to adjudicate these disparate claims. All of them – and doubtless many others – are valid interpretations of geography to one degree or another. Instead my intention is to look at what people *have taken geography to be* over the years and to trace the evolution of what I like to call 'the geographical tradition'. Accordingly I have no desire to defend any particular definition of geography, as many historians of the subject have done; rather I just want to consider some of the different ways people have thought about it down through the ages.

In order to confront this task I propose to identify some ten different discourses – conversations, if you will – in which geography has been engaged. Certainly my list is not exhaustive. Nor does it have to be. For the heart of my argument is simply that geography changes as society changes, and that the best way to understand the tradition to which geographers belong is to get a handle on the different social and intellectual environments within which geography has been practised. Some of the topics which we shall explore will undoubtedly seem bizarre, or exotic, or quaint to modern eyes; but if we are to take history seriously we will have to learn to understand past geographies in *their own contexts* without subjecting them to twentieth-century judgements.

To the Ends of the Earth

The story of geography, like the history of many sciences, has frequently been traced back to the Greek and Roman worlds, to figures like Thales, Anaximander, Herodotus, Strabo, Ptolemy and a dozen or more others. Their contributions – frequently of a mathematical character – did much to advance geographical theory. But it was through the explorations of Muslim scholar-travellers like Ibn-Batuta and Ibn-Khaldun, and the voyages of the Scandinavians, the Chinese and medieval Christian adventurers that first-hand knowledge of the world began to contribute to geographical lore. Eventually the European explorers of the fifteenth and sixteenth centuries helped to transform these earlier fragmentary gleanings into a more or less coherent body of knowledge about the terrestrial globe.

Indeed it could be argued that the voyages of discovery, so called, made a vital contribution to the development of science in the West. Many of these seafarers, for example, saw themselves as involved in world-scale experiments to test the accuracy of Renaissance concepts inherited from the ancient classical world. This is not to say, of course, that they all thought of themselves as proto-scientists; many were just lustful for adventure on the high seas or greedy for the untold riches of exotic kingdoms. But the information they gathered helped challenge the scholarly authorities of the day by demonstrating that people *did* inhabit the southern hemisphere or that there were varieties of plant and animal that just did not fit into Aristotle's taxonomy. Besides all this, the whole business of navigation required sophisticated technological and scientific skills to determine a ship's position at sea and, more important, to chart the way back to safe havens. So it is not surprising that the navigational institute that Prince Henry the Navigator established at Sagres in the early fifteenth century – and which drew together experts in cartography, astronomy and nautical instrumentation – has been seen as a crucial early move in the development of Western science. The names of Diego Cão, Bartholemew Dias, Vasco da Gama, the Cabots, Christopher Columbus, Francis Drake and Ferdinand Magellan – to name but a few – thus all occupy as important a niche in the early annals of modern geography as the re-publication of Ptolemy's *Geography* in 1410.

Of course geography's engagement with exploration did not come to an end in the fifteenth century. Voyages of reconnaissance continued to expand geographical knowledge of the globe throughout later centuries and special mention might be made of the eighteenth-century journeys of James Cooke and Joseph Banks into the South Pacific and the nineteenth-century circumnavigations of such naturalists as Charles Darwin and Thomas Henry Huxley. At the same time the significance of scientific travel was being championed by men like Alexander von Humboldt (plate II.1.1), Henry Walter Bates and Alfred Russel Wallace through their own explorations of the Far East and South America. Indeed the Royal Geographical Society, which did so much to advance overseas exploration in the Victorian era, continues to sponsor expeditions of this sort right up to the present day. Moreover geographers have continued to speak of expeditions in other contexts: expeditions into the urban jungle, ethnic ghettoes and other such 'threatening' environments. The vocabulary of exploration thus continues to capture the spirit of certain aspects of the geographical tradition. My argument here is simple: geography has always been closely associated with the exploring instinct.

Geography is Magic!

Even while new geographical knowledge was challenging accepted scholarly traditions there were ways in which geographical lore continued to confirm long-held beliefs. Thus, just as other nascent sciences were deeply implicated in various magical practices, so too was geography. This is plain for example in the early development of modern astronomy. Much interest in the stars was stimulated by astrological concerns and among the earliest Copernicans there is evidence of a continuing interest in that enterprise. Kepler, for example, cast his own horoscope every day, and in so doing he was far from unique. Aside

Plate II.1.1 Alexander von Humboldt, who died in the same year (1859) that Charles Darwin's *The Origin of Species* was published, made important contributions to many areas of natural history. His influential book *Cosmos: A sketch of a physical description of the universe* was published from 1845 onwards.

from this the belief that various plants possessed hidden occult powers that could be harnessed for medicinal powers led to important pharmacological and chemical findings. Moreover the writings of such giants of the scientific revolution as Bacon and Newton reveal a substantial interest in such seemingly arcane practices.

Geography, of course, was no less identified with astrology and natural magic than these other fields of discourse. Numerous early writers on geography, like William Cunningham, Thomas Blundeville, John Dee, and Thomas and Leonard Digges involved themselves in various aspects of magic. For some, like Dee, the key lay in the mystical significance of number – the celestial and terrestrial worlds were held together in certain mathematical relationships in such a way that changes in one directly influenced the other. For others, like the Digges, astrology was of first importance and their early meteorological efforts were all of a piece with astrological knowledge; to them weather forecasting required acquaintance with the significance of celestial changes in the moon, the stars and the planets. For still others, notably Jean Bodin and Cunningham, the diversity of the world's peoples and cultures was closely bound up with which sign of the zodiac governed the particular region they inhabited.

No doubt this chapter in the history of geography will seem utterly bizarre to modern eyes. But it would be mistaken to ignore it, or suppress it, as historians of geography have all too frequently done, because it demonstrates the role of apparently non-rational discourse in the evolution of the discipline. Moreover, this geographical interest in the mystical has continued to manifest itself right up to the twentieth century. Recent work has shown various mystical elements in the history of the modern conservation movement – in late nineteenth-century and early twentieth-century figures like

Francis Younghusband and Vaughan Cornish, for example – and that strain of thought which spiritualizes, even divinizes nature, continues to be with us to the present day.

A Paper World

The knowledge explosion occasioned by the European voyages of exploration soon brought new cartographic challenges and accomplishments. To be sure, the science of cartography was not born in the sixteenth century. Portolano sea charts had been circulating for long enough around the Mediterranean and of course there already existed numerous symbolic depictions of the world in the form of various Mappaemundi. But now whole new worlds had to be reduced to paper and that brought new challenges. Gerard Mercator solved some of the mathematical problems associated with transferring a sphere to a flat surface with his famous map projection. Soon a series of Dutch and Belgian cartographers such as De Jode, Jodocus Hondius and Petrus Plancius splendidly mapped the progress of overseas discovery. Closely associated with these accomplishments was the development of surveying skills and instruments, and so instrument making was frequently one of the craft competences of the early cartographer.

Map making, of course, was as artistic a practice as it was scientific. Frequently maps were elaborately decorated and skilfully executed, so much so that they frequently became *objets d'art* in their own right. Besides, the whole cartographic impulse in painting is nowhere more clearly manifest than in the Dutch art of the seventeenth century. And this serves to remind us of the early associations between geography and humanistic endeavours.

In the following centuries geography's links with cartography have continued to be maintained. The progress of the Ordnance Survey's work in nineteenth-century Britain was regularly reported at the Royal Geographical Society; geographers frequently involved themselves in the thematic mapping of drift geology, soils, disease, populations and so on; now in our own day geographers maintain this tradition when they turn to remote sensing and computer mapping. The mapping drive has thus always been strong in geography; so much so that Carl Sauer believed that, if a geographer was not fascinated by maps to the extent of always needing to be surrounded by them, then that was a clue that he or she had chosen the wrong profession.

A Clockwork Universe

In the wake of the Mechanical philosophy that came to dominate science in the seventeenth century, there were numerous efforts to retain the integrity of religious discourse in the face of the apparently naturalistic implications of a mechanistic world picture. One of the most common strategies, defended by men like Newton and Boyle, was to argue that the world was essentially like a grand clock, comparable with that at Strasbourg, and that by investigating the world machine scientists were interrogating the very mind of the Great Designer. This logistic move was to play a key role in the evolution of the geographical tradition. Numerous writers during the period of the Enlightenment developed a style of natural history called Physico-theology. Regarding the world as teleologically designed and providentially controlled they interpreted the world environment as a functioning revelation of divine purpose. In the writings of Thomas Burnet, John Ray, John Woodward, William Derham, as later in the works of William Paley, the world's geography

– its physical and organic forms – was seen as pointing beyond itself to nature's God.

Of course these practitioners of natural theology differed frequently among themselves on both detail and strategy; but between them they delivered to history a vision of nature as a holistic system, a sort of ecological picture, that emphasized the interrelationships and interdependences among organisms and environment. Here the image of a warfare between science and religion turns out to be something of a historical fiction. Indeed there were geographers like Bartholomaus Keckermann in Germany (author of *Systema Geographicum*) and Nathanael Carpenter in England (author of *Geography Delineated Forth*) whose commitment to the theology of the Reformation encouraged them to *reject* ecclesiastical authority in matters of science and to argue for the liberation of science from scholastic censure.

This particular intellectual trajectory continued to inform geographical thought over the next centuries. In the nineteenth century Karl Ritter exemplified the same stance, and the Ritterian vision was propagated in the United States by his disciple-devotee Arnold Guyot. Besides these there is much evidence of teleological thinking in the works of Mary Somerville and David Thomas Ansted in England, and Matthew Fontaine Maury and Daniel Coit Gilman in the United States. Indeed H. R. Mill, writing in 1901, was entirely correct when he noted that teleological modes of reasoning were 'tacitly accepted or explicitly avowed by almost every writer on the theory of geography'. Even more recently the self-same teleological vision comes through in the writings of the Dutch geographer De Jong. Here geography continues to operate as the handmaiden to theology.

On Active Service

If, as we have just seen, geography could subserve theological ends, its services to external interests did not stop there. Throughout the nineteenth century it was frequently cast as the *aide de camp* to militarism, imperialism, racism and doubtless a host of other 'isms'. Maps, it was long known, were as vital implements of warmongering as gunnery, and it is no surprise that institutional geography first flourished in military schools. Indeed the prehistory of the Ordnance Survey can be traced back to military needs during the Jacobean era, while in the twentieth century geographers like Isaiah Bowman played their parts in America's involvement with post-war European reconstruction.

By the same token British expansion overseas aroused a renewed interest in geography for its functional purposes. At the inaugural meeting of the Royal Geographical Society of London in the early 1830s the need for such a society was defended on the grounds that geography was vital to the imperial success of Britain as a maritime nation. Accordingly there was, and continued to be, considerable debate in British – not to mention German and American – geography on the subject of acclimatization because the question of white adaptation to the tropical and subtropical worlds was of pressing international significance. Here geographers worked closely with medical experts to delineate the signifiance of climatic factors. Indeed in so doing they kept alive an ancient tradition, rejuvenated by Montesquieu, that explained the cultural in terms of the natural.

Besides this there were certain aspects of geographical theory ripe for manipulation. Environmental determinism – a doctrine emphasizing the moulding power of physical conditions – could be used for a range of purposes. Some found in it justification for a racial ideology;

indeed racial questions were commonplace in geography texts around the turn of the century and in some cases long after that. Others saw in it a doctrine with strategic potential. Halford Mackinder, for example, outlined a theory of world political power that crucially depended on the control of a particular piece of territorial space in the Old World. Friedrich Ratzel in Germany erected an organic theory of the state on his notion of *Lebensraum*, urging that the character and destiny of a *Volk* was umbilically tied to a definite area or *Raum*. In the United States the Ratzelian viewpoint was propagated by Ellen Semple who used it to chart the necessitarian course of American history, while Ellsworth Huntington turned to climate as the great mainspring of civilization. In all of these, as in the stop-and-go determinism of Griffith Taylor, the constitutive links between geographical theory and social outlook are clearly displayed. This is not to say, of course, that geographical determinism as a precept was *just* social ideology writ large. But it *is* to recognize that there is a *social* history of geographical ideas as well as a purely *cognitive* one.

The Regionalizing Ritual

Even while environmental determinism in one form or another was spreading like wildfire among professionalizing geographers, there were those who insisted on the capacity of human culture to transform its natural milieu rather than remaining in nature's deterministic grip. In Britain H. J. Fleure emphasized the importance of human agency in modifying environment and thus turned away from the conventional concentration on natural regions towards the significance of transitional zones of culture contact down through history. Moreover, even those like A. J. Herbertson, in whose geography the concept of natural region occupied a strategic place, nevertheless recognized the subtle interplay of environment, heredity and consciousness in producing the geographical patterns of human diversity across the face of the globe. For both the idealist strain in Lamarckian evolution – an evolutionary model stressing the significance of life-force and will – was of crucial importance.

Another strain of environmentalist critique was forthcoming from a different, though related, conceptual source around the turn of the century, namely the vibrant tradition of French cultural geography associated with Vidal de la Blache. For Vidal and the Vidalians environment was to be seen, not as a determinative force, but rather as a limiting factor setting limits on cultural possibilities. Possibilism, as this doctrine was styled, also emphasized the science of human regions because it was in specific physical milieux that distinctive *genres de vie* – modes of life – found expression.

A third strand of determinist criticism emanated from Carl Sauer and the Berkeley school of cultural geography in the United States. Here inspiration was derived less from evolutionary biology than from cultural anthropology and can be traced back to the seminal influence of the anthropologist Franz Boas. Boas had begun his academic career as a physical geographer but turned to anthropology when his work among the Inuit led him to question environmental determinism. The mild cultural relativism that he came to espouse was mediated to Sauer through anthropological colleagues at Berkeley and Sauer built on these foundations as he emphasized the importance of residual material culture as historical artefacts of cultural diversity.

Whatever the differences in approach, all these geographers shared a conception of geography as a study of regions. And this brand of geography received its benediction in Richard Hartshorne's influential monograph *The Nature of Geography* in which he argued his apologetic case from a partisan review of historical – and in particular German – sources. Thus the notion of geography as the 'regionalizing ritual', provided a paradigm that still governs much geographical work, whether in the qualitative contributions of writers of regional personality or in the more quantitative emphasis of the practitioners of regional science.

The Go-between

Alongside these efforts to delineate for geography a piece of cognitive territory – a sector of conceptual space in the academic scheme of things – there were those who were rather more inclined to stress its functional role. Frequently the case was made that geography was the integrating discipline *par excellence* that kept the study of nature *and* culture under one disciplinary umbrella. W. M. Davis, for example, otherwise remembered for his elucidation of the cycle of erosion, nevertheless felt that physical geography was incomplete without ontography, its human counterpart. This go-between function was valuable in a number of contexts. For one thing it was appealed to to justify geography as a coherent and independent academic discipline both in Britain and the United States. Indeed Halford Mackinder in Britain found this to be the only foundation on which geography as a causal science could be built. In the United States Isaiah Bowman championed the same view.

Besides this, geography's bridging role between nature and humanity frequently took the form of a strenuous engagement with questions of resources. In America the roots of this geographical tradition go back to such figures as Nathaniel Southgate Shaler and George Perkins Marsh, and later J. Russell Smith, whose contributions were resurrected by early twentieth-century geographers seeking the recovery of a tradition of environmental sensitivity. For some this emphasis led to a historical reassessment of 'man's role in changing the face of the earth'; for others the needs of the future fostered an engagement with environmental systems analysis or with ecological energetics in the attempt to model the changing human–nature interface. In our own day, as the environmental crisis has bitten even more deeply, geographers like Timothy O'Riordan and Andrew Goudie have done much to keep this tradition at the forefront of geographical discourse. Moreover, so far as institutional identity is concerned it is noteworthy that university and college geography is not infrequently housed in schools of environmental studies.

A Science of Space

If some identified geography's essence in its focus on regional integration, there were those who found the emphasis on the particularity of places lacking in methodological rigour. To them, all the talk of bridging the gulf between the sciences and the humanities seemed little more than academic–political rhetoric, and the idea of regional personality frankly unscientific. Fred Schaefer spearheaded the attack with his article on 'Exceptionalism in geography' published in the *Annals of the Association of American Geographers* in 1953. Schaefer's critique was designed to transform geography into a true science by urging that it become a law-seeking explanatory discipline concerned with universal laws, not regional specifics, or, as he put it, 'exceptions'. Schaefer's paper, it is commonly believed, heralded the introduction of logical positivism into the discipline and its curriculum was defended in William Bunge's *Theoretical Geography* of 1962 and David Harvey's *Explanation in Geography* published at the end of the decade. And thus was born the idea of geography as a science of spatial distribution – locational analysis as it was frequently styled – and soon various theorems seeking to explain the location of economic behaviour were introduced to geography by figures such as W. L. Garrison in America and Peter Haggett in Britain. In particular the earlier economic theorizing of Von Thünen, Alfred Weber, Walter Christaller and August Lösch soon began to receive an airing in the discipline.

Along with this definition of geography as spatial science came the paraphernalia of scientific know-how, and thus geography received its newest initiation into scientific method and statistical technique. Not of course that geography had been utterly innocent of quantification hitherto. The roots of geography as a mathematical practice can be traced back at least to the period of the scientific revolution in the seventeenth century, and doubtless before that. Nor does it mean that all geography was quantified; plainly many areas of the tradition remained statistically immune. Still, positivism did make substantial inroads into geographical theory and practice from the 1950s and a variety of reasons for geography's relatively late baptism in positivist philosophy have been put forward. A Marxist-converted Harvey believes it represented, at least in America, a strategic attempt by geographers to escape the political suspicion falling on social science in the post-McCarthy era by retreating into the safety of number-crunching. At least as compelling, I think, would be an explanation that takes seriously the perceived need for geographers to accrue to themselves a set of craft competences which bolstered their professional vested interests in creating a spatial *science*.

Statistics Don't Bleed

Whatever the causes of geographical quantification may have been, recent decades have witnessed a sequence of attacks on positivism from different perspectives. From the radical side comes the complaint that the whole quantitative procedure is ideologically laden from the start. The argument here is that by keeping geography just a sort of spatial calculus, a geometric technique for depicting distributions, fundamental questions of justice and political involvement are simply – and too comfortably – ruled out of court. Accordingly various contemporary radical geographers see themselves in a geographical lineage stretching back to figures such as Elisée Réclus, Peter Kropotkin and Karl Wittfogel who strenuously advocated social engagement. In this scenario, and it has to be admitted that it is far from unified, there is something of an emphasis on the determinative role of economic structure. Whether investigating the significance of residential segregation, the vicissitudes of the world economic system or the historical change from feudalism to capitalism, this same *motif* regularly reasserts itself.

From another perspective, there are those humanistic geographers who insist that the quantitative tabulation of economic data and other activities has dehumanized geography by ignoring, not to say suppressing, human agency. Statistics are simply not made of flesh and blood. Whole acres of human experience – fear, imagination, emotion – are left out of the picture. And these geographers have seen it as their task to keep the geographical world open to the artistic side of its history by their interrogation of literary texts and their championing of subjectivism in the subject. Yi-Fu Tuan's meditations on 'topophilia' and 'topophobia', David Ley's excursion into the mind of the inner city ghetto, and

Leonard Guelke's turning to Collingwood's idealist philosophy of history are just some of the currents to have swept through the discipline recently. Again partisans are quick to point out that this is not a wholly new departure: some claim that the earlier behavioural geography of J. K. Wright, David Lowenthal and William Kirk accorded a key role to subjective experience, while others – ignoring Vidal's natural science aspirations for *géographie humaine* – speak of the revivification of the Vidalian tradition.

Everything in its Place

These respective emphases on the role of social structure and human agency in accounts of geographical phenomena have most recently led some to wonder whether explanatory privilege ought to be accorded to either side of the equation. In the attempt to find a way out of the impasse, some geographers have turned to the theory of 'structuration' advanced by the Cambridge sociologist Anthony Giddens. This account of social formation and transformation highlights the interplay of both forces: human beings find themselves in structural circumstances not of their choosing, but through the exercise of their own agency can do something to bring about change. The never-ending ebb and flow of agent–structure intercourse provides the engine power of social transformation. Where geography enters the picture is in the need to 'earth' this general model of historical change. Just how the interplay of social structure and human agency falls out is evidently different from place to place and depends crucially on the particular arena of encounter. Hence geographers – arguing for the prime significance of locale – increasingly call for the geographizing of social theory.

What has given further encouragement to this renewed emphasis on the significance of place is a whole series of philosophical and social developments. The details need not concern us, save to note that the idea of cultural and epistemological pluralism now seems inevitable. Fragmentation of knowledge, social differentiation, and the questioning of scientific rationality have all coalesced to reaffirm the importance of the particular, the specific, the local. And in this social and cognitive environment a geography stressing the centrality of place is seen as having great potential. Once again the constitutive nature of the relationship between geography's internal domain and external context is clearly evident.

Geographical Conversations

Little needs to be said in conclusion. My argument throughout has been that the geographical tradition, like a species, has evolved as it has adapted to different social and intellectual environments. Geography, as was noted at the beginning, has meant different things to different people at different times and in different places. It has employed different vocabularies to suit different purposes – from magic and theology to science and art. Sometimes these discourses have been in conflict; at other times they have been mutually reinforcing. Sometimes the conversations have admitted a range of geographers; sometimes only a select group were allowed to take part. Either way what is important is that in telling the story of the tradition to which geographers belong there needs to be a recognition of the integrity of each of these diverse discourses in their own terms. Otherwise the history and future of geography will be enslaved to partisan apologists who wish to monopolize – even hijack – the conversation in order to serve their own sectarian interests.

II.2

Geomorphology: Today and Tomorrow

Andrew S. Goudie

Geomorphology, both as a term and as a discipline, became established in the 1880s, largely through the efforts of a group of American scientists which included Major J. W. Powell, G. K. Gilbert, WJ McGee and W. M. Davis. For long it was concerned with elucidating the history of earth surface development, and in particular with establishing the progress and sequence of erosional events through time. Such an evolutionary slant to geomorphology, which was manifested in the denudation chronology of British geomorphologists (e.g. Wooldridge and Linton, 1955) and the climato-genetic geomorphology of some German workers (e.g. Büdel, 1982), was dominant until recent decades, though on the continent of Europe the importance of climatic differences in determining landform type was often stressed (e.g. Tricart and Cailleux, 1972).

More recently, geomorphologists in many countries have become less concerned with landform history (and, indeed, with form itself) and more concerned with the nature and rate of geomorphological processes, and with understanding the way in which geomorphological systems operate. Thus they have tended to look at geomorphology in terms of systems, have tried to model geomorphological phenomena (see Anderson, 1988) and have attempted to quantify force, resistance and form. The impetus for such a new approach has often been attributed to an engineer, Robert E. Horton (who died in 1945), to various of his disciples (including A. N. Strahler, S. A. Schumm and R. J. Chorley) and to various members of the hydraulic geometry 'school' of the U.S. Geological Survey, notably Luna B. Leopold. Such work has often involved studies of small areas over short time spans.

Like all disciplines geomorphology is varied in scope and constantly evolving. Today, the quantitative process-oriented approach is still healthy and vibrant, but there are various new developments that are worthy of note, and which may assume a greater relative importance in the future. Of these we select for further consideration the importance of global tectonics, long-term climatic change, human actions and applied work, the quantification of rock properties and their relationship to landform type, and organic influences (biogeomorphology).

Global Tectonics

The discovery, in the 1960s, that the earth's surface is composed of a number of moving plates amounted to a revolution in the earth sciences. This revolution has prompted some geomorphologists to shift their researches to mega-scale phenomena (e.g. mountain ranges, the global distribution of deltas, and global coastal types) but the full implications of the new global tectonics have yet to be fully explored. Nonetheless, the realization of the importance of ongoing tectonic mobility is starting to have a profound effect on thinking. How, for example, can one understand the evolution of the western coastline of Canada and Alaska without reference to the movement of 'exotic terranes'? How can one understand the development of the landforms of the British Isles without reference to the opening of the North Atlantic and the rifting and subsidence of the North Sea Basin? Ollier's *Tectonics and Landforms* (1980) gives an indication of the scope of this field.

Long-term Climatic Changes

Approximately contemporaneous with the plate tectonics revolution was a revolution in our appreciation of environmental changes on a variety of time-scales. In particular, the analysis of deep sea cores revealed the duration, frequency and degree of change over the last few millions of years, and new dating techniques (e.g. radiocarbon) permitted regional correlations to be made with some degree of confidence (Bowen, 1978). It became apparent that present-day processes may be atypical of those that have persisted for much of the past. It also became apparent that low latitudes, including deserts and rain-forests, had not remained stable in the Quaternary. This has created a renewed interest in historical geomorphology, an interest that is being strengthened by the desire to use past climatic scenarios (e.g. the mid-Holocene, the last interglacial, or the Pliocene) as analogues for future climatic changes produced by green-house warming.

Human Actions

Until the 1960s many geomorphologists consciously tried to select field areas where human interference was at a minimum. However, in the last two decades there has been a growing realiz-ation of the importance of human activities in creating landforms and modifying the rate of operation of geomorphological processes (Goudie, 1990). This concern is currently being accelerated as a result of the increasing significance of global environmental changes promoted by greenhouse warming. Geomorphologists are starting to seek answers to various major ques-tions: how quickly will polar ice sheets and glaciers melt; what will happen if permafrost starts to thaw; which areas will be inundated or eroded as a result of sea-level rise; how will mangrove swamps, coral reefs and salt marshes cope with a rising sea level; what will be the geomorphological consequences of increased hurricane activity; and what will happen to geomorphological processes if major rainfall and vegetation belts shift?

Applied Work

A variety of influences have caused many professional geomorphologists to become concerned with the application of their work for the solution of practical problems (Cooke and Doornkamp, 1990). These influences include pressures from government and society for academics to justify their existence, and the attrac-tions of coming up with answers that are useful and satisfactory to clients (plate II.2.1). Armed with traditional field skills (e.g. the ability to recognize potentially dangerous landslide scars or avalanche tracks), mapping and remote sensing techniques, and various hard-ware devices for monitoring processes, geomorphologists have found that they have the ability to map resources (e.g. soils and gravel), to identify and locate actual or potential hazards (e.g. land-slides, karstic collapse, dune encroachment), to determine rates of change (e.g. of retreating cliffs) by means of instrumental monitoring or by sequen-tial maps and remote sensing images, and to predict the environmental impacts of human activities (e.g. the effect of

Plate II.2.1 Many physical geographers, including geomorphologists, are concerned with natural hazards. The Karakoram Mountains of Pakistan display many such hazards: unstable screes, flood-prone rivers, mudflows and boulder falls.

groyne construction on beach erosion and deposition).

Rock Properties and Landform Type

It is self-evident that rock properties are a vital control of landform. However, it can be argued that, of the Davisian trilogy of structure, process and stage, it is the first of these that has been most neglected by geomorphologists. Indeed, the quantification of rock properties and their correlation with particular forms has proved to be one of the most difficult problems in geomorphology. However, the development of rapid analytical techniques for chemical and mineral analysis of rock (e.g. X-ray fluorescence) and the application of theories of rock mechanics and slope stability may make this a major growth point in the discipline. Gerrard (1988) provides a comprehensive, although rather qualitative, review of progress to date.

Biogeomorphology

One further field which has been seriously under-researched in geomorphology is the study of biological influences (plate II.2.2) – a subdivision of the subject which has recently been termed biogeomorphology (Viles, 1988). However, organisms (e.g. termites) move large quantities of material and expose it to the influence of other agencies (e.g. rain splash); bacteria, algae, lichens and their ilk are capable of mobilizing elements and participating in weathering reactions; and the nature of vegetation cover controls the operation of many geomorphological, hydrological and pedological processes. In the past there

Plate II.2.2 Recent years have seen an increasing appreciation of the role of organisms in affecting the nature and rate of geomorphological processes. In the tropics, termites play a fundamental role in sediment distribution.

has been a tendency to regard biological influences in geomorphology as being unusual and producing bizarre microforms (e.g. gopher mounds or solutional rills on limestone by rock wallaby urine), whereas in reality biological processes are pervasive (Thornes, 1990).

Organization

With the growth of institutes of higher education in the post-war era, the number of practising geomorphologists increased, and various organizations were established to encourage research and facilitate communication. A major example of this was the establishment, in 1961, of the British Geomorphological Research Group (BGRG). In 1985 the BGRG organized the First International Geomorphological Congress in Manchester, England, the proceedings of which have been published at length (Gardiner, 1987). These provide a good guide to the state of the discipline in the 1980s. Four years later, at Frankfurt, Germany, a second congress took place, and this saw the establishment of an International Association of Geomorphologists.

References

Anderson, M. G. 1988: *Modelling Geomorphological Systems*. Chichester: Wiley.

Bowen, D. Q. 1978: *Quaternary Geology*. Oxford: Pergamon.

Büdel, J. 1982: *Climatic Geomorphology*. Princeton, NJ: Princeton University Press.

Cooke, R. U. and Doornkamp, J. C. 1990: *Geomorphology in Environmental Management*, 2nd edn. Oxford: Oxford University Press.

Gardiner, V. (ed.) 1987: *International Geomorphology 1986*, 2 vols. Chichester: Wiley.

Gerrard, J. 1988: *Rocks and Landforms*. London: Unwin Hyman.

Goudie, A. S. 1990: *The Human Impact*, 3rd edn. Oxford: Basil Blackwell.

Ollier, C. D. 1980: *Tectonics and Landforms*. London: Longman.

Thornes, J. (ed.) 1990: *Vegetation and Erosion*. Chichester: Wiley.

Tricard, J. and Cailleux, A. 1972: *Introduction to Climatic Geomorphology*. London: Longman.

Viles, H. A. (ed.) 1988: *Biogeomorphology*. Oxford: Basil Blackwell.

Wooldridge, S. W. and Linton, D. L. 1955: *Structure, Surface and Drainage in South-east England*, 2nd edn. London: George Philip.

II.3
Biogeography and Ecology

Robert J. Whittaker

Ecology is the study of environmental relationships: geography is the study of space relationships . . . what is not clear is where the one stops and the other starts.

J. L. Davies (1961, p. 415)

Biogeography and ecology are typical of many branches of geography in being the concerns not only of geographers but also of specialists in other disciplines, notably geologists, zoologists and botanists (Brown and Gibson, 1983; Taylor, 1984; Begon et al. 1986). The purpose of this review is to establish why geographers should be interested in these subjects and to hint at ways in which geographers can make a distinctive contribution to them.

Biogeography can be defined as the study of the spatial distribution of plants (plate II.3.1) and animals, but this is a rather bland definition of a very broad, dynamic and, at least at times, useful branch of geography. Applications range from helping the oil industry find oil-bearing strata (through the study of fossil pollen and spores) to specifications for the layout and management of nature reserves (Spellerberg, 1981).

Ecology can be defined as the study of the relationships of plants and animals to each other and to their environment. The subject is of increasing interest, particularly to geographers concerned with the degradation of the natural environment (see chapter II.6).

The distinction between biogeography and ecology can be difficult to pin down, each has as many definitions as there are textbooks on the subjects, and it is simplest to accept that there is considerable overlap between the two traditions and both can contribute to solving a problem such as 'why are there more species in place x than place y?'

It is not possible in such a brief review as this to provide more than a taster of the subject, and so the following is very much a personal sample of ideas and themes.

Organizing Concepts

Scale

'Everything is relative' may sound a rather 1960s sort of statement, but you cannot get far in understanding biogeography without sorting out the scale of your phenomena. As an illustration, most geographers are familiar with regular, random and clustered distributions. A plant population may exhibit each of these patterns depending on the scale of the study. On a small scale the pattern may be dependent on the morphology of the plant and its reproductive habits, whereas on a larger scale it may correspond to particular micro-habitats randomly scattered through an area, thus giving rise to a clustered pattern on a micro-scale and a random pattern on a larger scale (Kershaw and Looney, 1985). A statement that plant x is randomly distributed is thus incomplete; the scale must be specified too. This applies equally to community studies. The Cape region of Southern Africa is famous for its species-rich heathlands, called 'fynbos', yet compared on an equal-area basis, at a

Plate II.3.1 A major component of physical geography is biogeography, that part of the discipline concerned with the patterns of life on the face of the earth. This plate shows one major biome, the boreal forest of Colorado, United States.

scale of 20,000 km^2, it is no richer than is predicted as a function of its climate. Taken at that scale, what it does have is a high proportion of endemics (species found nowhere else). The question 'why is it so diverse?' must thus be replaced by 'why is it so rich in endemics?'

Hierarchy theory

Biogeographical and ecological phenomena can be studied over time-scales from days to millennia and spatial scales from continents down to the number of insects on the flower heads of individual thistles. In discussing debates that focus on different scales, the question arises as to the linkages between these phenomena. Recently Blondel (1987) and others have put forward an organizational framework – hierarchy theory – based on the notion that not only are there links, but it is rarely possible to arrive at a complete understanding of biogeographic phenomena without bringing in considerations on more than one scale. I shall return to this theme later.

Big Ideas in Ecology

The ecosystem concept

The ecosystem concept really is a big idea. It has perhaps been *the* organizing principle of twentieth-century ecology. The essence of it is that within any given area (such as a lake) the organisms interact with each other and with the abiotic (physical) environment and thus that the ecosystem consists of both biotic and abiotic parts (Tansley, 1935). Ecosystems function in the sense that there is a constant cycling of nutrients and a flow of energy through the system. By the laws of thermodynamics we know that energy is neither created nor destroyed, but that entropy increases as it is passed through the system, i.e. that energy is lost to the system in the form of heat, so that at each 'trophic level' less is available for use than at the level before. The implications of this are numerous (Elton, 1927; Lindeman, 1942). For example, as Colinvaux (1980) points out, it explains why large fierce animals are rare. The argument in its simplest form is that about 0.1 per cent of the radiant energy is captured by the photosynthesizers (plants = primary producers), subsequently about 0.015 per cent is captured by the herbivores (primary consumers), 0.0003 per cent by carnivores (secondary consumers), and so only about 0.00004 per cent is available for the big fierce tertiary consumers (lions, tigers etc.). At one time, it appeared that much of ecology could be reduced to the operation of biogeochemical systems (e.g. Odum, 1969), but, if this is so, we do not yet know enough about them to explain all ecological phenomena. One weakness of ecosystem analysis is that the ecosystem, as a spatial unit, is an artificial imposition. Its boundaries are not fixed. Another problem is that organisms at one trophic level are not necessarily interchangeable and the substitution of one by another can alter flows of energy and nutrients within the system (Stoddart, 1986).

Gaia

Gaia is another big idea, formulated by Lovelock (1988). Whether it is strictly scientific or not is questionable. The basis of Gaia appears to be the application of the ecosystem concept to the entire biosphere (i.e. the living planet). In particular it recognizes that not only does the environment tend to regulate organisms, but organisms tend to regulate the environment. It is envisaged that as in any ecosystem there are feedbacks (both positive and negative) between the abiotic and biotic components of the biosphere, and that up to a point the negative feedbacks of the biosphere are capable of damping down changes, i.e. of self-regulation. The proponents of Gaia argue, however, that humans are endangering this natural balance, by threatening to tip the system into a phase where positive feedbacks dominate, throwing the biosphere out of its present equilibrium. Whatever else may be said about it, Gaia has focused attention on the large-scale functioning of the biosphere (e.g. Myers, 1985).

Succession

Succession is an old idea in ecology, formalized by F. E. Clements (1916). In its simplest form, succession is a straightforward and comprehensible phenomenon, observable universally. The fun starts when you try to build in a bit of realistic complexity or to derive a predictive model. Succession is the term given to directional community development, such as takes place on a new volcanic substrate or an abandoned field. It is usually used in reference to plant communities, although it can be applied to animals. From an initial sterile surface (= primary succession) or from a heavily disturbed area (= secondary succession), plants colonize, grow, reproduce, die and are replaced by other species of plants, which in their turn are replaced by others, and so on, until a stable end-point is reached. The study of succession has been bound up with several other big debates. Two of particular interest are the following: first, are there biotic *communities* as such, or is it every man – or plant – for himself?; second, the notion of *climatic climax* – is there such a thing as an 'end-point' at which a stable balance is reached?

Clements considered the community

as a sort of superorganism, with succession its developmental process, one set of organisms facilitating the take-over of the site by the next set. If the process were interrupted, it should resume along the same path as before, a course determined by the prevailing climate and by edaphic controls. In contrast, the 'individualistic' view, associated with Gleason (1926), is that each species should be considered separately, that succession is not predictable and that processes other than facilitation may be important. This debate is still not settled, although the evidence appears to be on the side of the individualistic view, and there certainly appears to be a large stochastic (chance) element to many successions, with alternative pathways being followed after interruption rather than always a resumption of the previous course (e.g. Whittaker et al., 1989). As for the question of the stable end-point, the overwhelming evidence from palaeoecology is that communities are ever-changing through time, and that although species may not always travel alone in response to climatic change, neither do they travel as discrete assemblages (see Sauer, 1988). The ideal of a balance between climate and community is something that there will always be a tendency towards, but the time-scales of distributional adjustment and of climatic change are such that a truly stable balance may rarely if ever be met.

Succession, of course, undoubtedly occurs; but it is like Tantalus, always striving but never arriving.
(Flenley, 1984, p. 95)

Nonetheless, in the absence of an equally neat concept to the 'climatic climax', most textbooks still use it as the end-point to succession. For further reading on succession see Golley (1977) or Miles (1979).

Population dynamics

Population ecology is a discipline in its own right. Its concern is the interactions between organisms, either of the same or of different species (Begon and Mortimer, 1986). A useful illustration of the latter is provided by the 'pest pressure hypothesis', which is based on the idea that predation prevents competitive exclusion amongst prey species and therefore increases community diversity (Paine, 1966). This approach has recently been extended to the problem of explaining rainforest diversity. Here the idea is that insect pests and pathogens, favoured in comparison with their temperate latitude cousins by year-round bounty, will disproportionately affect tree species when they are common. This will be most important in the early stages of life, causing a high rate of mortality of seedlings growing near to parent trees, where such pests and pathogens are of course concentrated. This process should thus allow in other, perhaps less competitive, species, enhancing the local diversity of the forest (Connell, 1979).

Population interactions underlie *changes* in communities too, and so, for example, the nitty-gritty of testing competing hypotheses of successional change can come down to studying the recruitment, performance and mortality of individuals in a population (Colinvaux, 1986, ch. 23).

Big Ideas in Biogeography

Vicariance versus dispersalism

This debate is generally at a very large scale – continental distributions and millions of years – and it concerns the means by which fossil or extant (living) groups of organisms arrived at their particular distributions. Historical explanations for disjunct (i.e. nonoverlapping) distributions generally fall into two categories, a *dispersal explanation*, in which organisms are assumed to have migrated across pre-existing barriers, and *vicariance explanations* in which the formation of barriers fragmented once continuous taxa. In the latter case, once fragmented, the initially similar populations may differentiate evolutionarily, giving rise to different species in the separate areas. In the dispersalist view a variant evolves in one region and then disperses and gradually replaces the earlier form throughout much of the range, with subsequent new forms then displacing it, and so on. Huge methodological arguments have been generated around this simple dichotomy, with authors attempting to discredit one approach or the other, but as Stoddart (1986) and Blondel (1987) have pointed out, there is room for both explanations and both methodologies (see also Myers and Giller, 1988).

Speciation and extinctions

The study of long-term biogeographic change inevitably involves the appearance of new lineages and the disappearance of old. However, as Darwin showed in *The Origin of Species* (1859), evolution can be studied in the living as well as in the fossil record. Questions of continuing interest in this field include the following: (a) Does speciation occur at the centres or the peripheries of ranges? (b) How important is geographical isolation in enhancing speciation rates? (c) Do new forms arise at a steady rate, or are there bursts of evolutionary change? (d) How and where do extinctions take place? (e) Is there evidence for regular extinction spasms? (f) How important is environmental change in enhancing extinction rates? Some clues on these themes are scattered through the next section.

Refugia theory

Refugia theories abound in relation particularly to the events of the Quaternary period (i.e. around the last 2 million years). One of the most interesting theories concerns the Amazon Basin. The traditional view of the tropics as climatically stable through the glacial periods has been rejected for some time (see Flenley, 1979). Based on studies of the patterns of endemicity of various groups within the Amazon (e.g. Haffer, 1969, on birds), a theory was developed that in times of aridity (linked to the patterns of glaciation in the temperate latitudes), the rainforests of Amazonia contracted to disjunct patches. In these 'refugia', evolutionary changes took place, so that, on re-expansion in periods of climatic optima, forms from different refugia remained distinct, reproductively isolated from their relatives elsewhere. This was a neat idea and has even been suggested as providing a basis for the conservation of species within the Amazon, i.e. save the refugia and you save most of the species (see also the next section). However, as more groups are studied more intensively, the pattern of refugia becomes less clear. Furthermore, there is as yet very little palaeoecological data (i.e. from the fossil record) to support the theory, and small but increasing amounts of data that appear to refute it. Watch this space!

Island biogeography

Island biogeography really began with the work of Darwin and Wallace in the nineteenth century and a great deal has been learnt about evolutionary processes from the study of groups of organisms within groups of islands. Islands are in addition excellent places to do your field work, especially if they are fringed with coconut palms and surrounded by warm reef-bearing seas. They provide a sort of massive natural laboratory in which numerous factors can be varied or controlled by careful selection of particular combinations of islands. The students of island biogeography have been much taken with this feature of islands and the largest body of work over recent decades has concentrated around the 'Equilibrium theory of island biogeography' proposed by MacArthur and Wilson (1967). Their work stemmed from observations that there was a predictable increase in species number with increasing island area (see the review by Williamson, 1981). They postulated that this pattern might result from a *dynamic* balance between continually occurring immigration to the island and local extinction on it (due to competitive interactions). A newly created or sterilized island would initially have a high immigration rate to it, as any propagules landing would be new colonists, combined with a low extinction rate, as there would be little competition. In time, the immigration rate would decline and the extinction rate would rise owing to increasing competition for limited resources, and this would result in a gradually flattening curve of species number versus time until equilibrium was reached. The principal variables that have been invoked in setting the form of these curves and the eventual equilibrium point include island area, isolation, biotic richness of the surrounding area, and various measures of habitat diversity (e.g. altitude). A great deal of productive research output has been generated, yet despite this interest the theory has rarely been satisfactorily tested (Gilbert, 1980). Those who have proposed the use of the theory in designing the size and layout of protected areas in threatened regions (e.g. the rapidly disappearing forests of Amazonia) may well be working from false premises. Zimmerman and Bierregaard (1986) illustrate the limitations of this approach in a study of Amazonian frogs, demonstrating that a knowledge of the species autecology, in this case principally relationships to the environment at different stages of the life cycle, is critical to a successful conservation strategy and arguing that it is at this level that research effort should be concentrated.

Hierarchies of Scale: a Case Study

Returning to the theme of hierarchy theory, its advocates argue that many apparently mutually exclusive paradigms could probably be seen to be compatible if scale factors were addressed more carefully. The problem, Blondel (1987) argues, is that biogeography (and ecology) has lacked an integrating theoretical base, because of a lack of bridges between the different specialist disciplines and the different scales of study, each of which has their own frameworks, methods and literature (see also Stott, 1984). Blondel provides an illustration of how interactive changes in the time-scale can be employed to solve biogeographical problems which would be inexplicable by ecology alone.

If you examine bird communities along ecological successions in the Mediterranean, there is a puzzle: the more complex the vegetation, as the late successional stage of mature forest is reached, the less the birds are of Mediterranean origin. Moreover, multivariate analysis showed convergence among the bird communities of older mature forests for sites from Poland, Corsica, Provence and Burgundy. No more than 2 per cent of the birds of the Mediterranean forests actually evolved in Mediterranean habitats. So why should there have been such a poor speciation rate in birds, in contrast with the pattern in the flora in which, for example, 50 per cent of the region's flowering plants are endemic? The answer seems to lie in the Pleistocene history of the Mediterranean. Palynological analyses indicate that forest belts were more or less continuous in space during the Pleistocene. Only their extent varied. There was thus a lack of geographic isolation until recently, which may explain why little speciation took place. The history of the mattoral (scrub vegetation, characterized by low shrubs with tough thick evergreen leaves) is different. It was formerly discontinuous and of varying sized patches as climate varied; only since Neolithic times has the area of this habitat been greatly increased by human pressure. The patterns of isolation, with phases of contraction and expansion in range, allowed evolutionary divergence and the most conspicuous cases of Mediterranean speciation (among the avifauna) are amongst the species of open, or semi-open, habitats (i.e. mattoral), e.g. warblers of the genus *Sylvia*. The same is true elsewhere in Europe and therefore the divergence of avifauna is more pronounced in the first stages of vegetation successions than in later stages.

Concluding Remarks: the Ecological Geographer

This has not been intended as a comprehensive review, but I have attempted to illustrate both the diversity of the subject matter and the connecting strands that unite the various themes together under the labels biogeography and ecology. Geographers can contribute to debates in these fields in many ways, but principally, I believe, by bringing an interdisciplinary approach to problems that have no respect for the boundaries of disciplines, by an appreciation of the importance of scale and by a keen interest in spatial patterns and processes, fundamental to most issues in the subject.

References

Begon, M. and Mortimer, M. 1986: *Population Ecology: a unified study of animals and plants*, 2nd edn. Oxford: Blackwell Scientific.
——Harper, J. L. and Townsend, C. R. 1986: *Ecology: individuals, populations and communities*. Oxford: Blackwell Scientific.
Blondel, J. 1987: From biogeography to life history theory: a multithematic approach illustrated by the biogeography of vertebrates. *Journal of Biogeography*, 14, 405–22.
Brown, J. H. and Gibson, C. 1983: *Biogeography*. St Louis, MO: Mosby.
Clements, F. E. 1916: *Plant succession: an analysis of the development of vegetation*, Publications of the Carnegie Institute 242. Washington, DC: Carnegie Institute.
Colinvaux, P. 1980: *Why Big Fierce Animals are Rare*. London: Penguin.
——1986: *Ecology*. New York: Wiley.
Connell, J. H. 1979: Tropical rain forests and coral reefs as open non-equilibrium systems. In R. M. Anderson, B. D. Turner and L. R. Taylor (eds), *Population Dynamics*, Oxford: Blackwell Scientific, 141–63.
Darwin, C. 1859: *The Origin of Species by Means of Natural Selection*. London: John Murray.
Davies, J. L. 1961: Aim and method in zoogeography. *Geographical Review*, 51, 412–17.
Flenley, J. R. 1979: *The Equatorial Rain Forest: a geological history*. London: Butterworths.
——1984: Time scales in biogeography. In J. A. Taylor (ed.), *Themes in Biogeography*, London: Croom Helm, 63–105.
Gilbert, F. S. 1980: The equilibrium theory of island biogeography: fact or fiction? *Journal of Biogeography*, 7, 209–35.
Gleason, H. A. 1926: The individualistic concept of the plant association. *Bulletin of the Torrey Botanical Club*, 53, 7–26.
Golley, F. B. (ed.) 1977: *Ecological Succession*, Benchmark Papers in Ecology, vol. 5. Stroudsburg, PA: Dowden, Hutchinson & Ross.
Haffer, J. 1969: Speciation in Amazonian forest birds. *Science*, 165, 131–7.
Kershaw, K. A. and Looney, J. H. H. 1985: *Quantitative and Dynamic Plant Ecology*, 3rd. edn. London: Edward Arnold.
Lindeman, R. L. 1942: The trophic dynamic aspects of ecology. *Ecology*, 23, 399–418.
Lovelock, J. E. 1988: *Gaia: a new look at life on earth*, 2nd edn. Oxford: Oxford University Press.
MacArthur, R. H. and Wilson, E. O. 1967: *The Theory of Island Biogeography*. Princeton, NJ: Princeton University Press.
Miles, J. 1979: *Vegetation Dynamics*. London: Chapman & Hall.
Myers, A. A. and Giller, P. S. 1988: *Analytical Biogeography: an integrated approach to the study*

of animal and plant distributions. London: Chapman & Hall.
Myers, N. (ed.) 1985: *The Gaia Atlas of Planet Management*. London: Pan.
Odum, E. P. 1969: The strategy of ecosystem development. *Science*, 164, 262–70.
Paine, R. T. 1966: Food web complexity and species diversity. *American Naturalist*, 100, 65–75.
Sauer, J. D. 1988: *Plant Migration: the dynamics of geographic patterning in seed plant species*. Berkeley, CA: University of California Press.
Spellerberg, I. F. 1981: *Ecological Evaluation for Conservation*, Studies in Biology 133. London: Edward Arnold.
Stoddart, D. R. 1986: *On Geography*. Oxford: Basil Blackwell (see chs. 12, 13).
Stott, P. 1984: History of biogeography. In J. A. Taylor (ed.), *Themes in Biogeography*, London: Croom Helm, 1–24.
Tansley, A. G. 1935: The use and abuse of vegetational concepts and terms. *Ecology*, 8, 118–49.
Taylor, J. A. (ed.) 1984: *Themes in Biogeography*. London: Croom Helm.
Whittaker, R. J., Bush, M. B. and Richards, K. 1989: Plant recolonization and vegetation succession on the Krakatau Islands, Indonesia. *Ecological Monographs*, 59, 59–123.
Williamson, M. 1981: *Island Populations*. Oxford: Oxford University Press.
Zimmerman, B. L. and Bierregaard, R. O. 1986: Relevance of the equilibrium theory of island biogeography and species–area relationships to conservation: with a case from Amazonia. *Journal of Biogeography*, 13, 133–43.

II.4 Climatology

John G. Lockwood

Introduction

Not very long ago climatology was considered to be unbearably dull and not particularly relevant to the rest of geography (plate II.4.1). Typical climatology textbooks contained some statistics on the regional distribution of temperature, rainfall, sunshine, number of frosts etc. and perhaps a few energy and water balance diagrams. Climatic change would consist of a few words about ice ages, often with the comment that we had no clear understanding about their causes. Over the last two decades there has been a revolution in the study of climatology which has taken it to a position where climatic changes are seen as a threat to human society and are frequently on the agenda at meetings of world leaders, and most political parties have a policy on atmospheric matters.

The Climatological Revolution

Many factors have contributed to this revolution. The study of sedimentary evidence from ocean cores, coupled with a new sophistication in theoretical investigation, has revealed a completely unexpected richly varied tapestry of past climatic variability. In the 1970s the analysis of new sedimentary cores from the ocean floor produced a completely new picture of the extent and causes of the ice ages. Space-based observing systems have allowed scientists to observe the global atmosphere and surface in unprecedented scope and detail. The voyages of space craft to the other planets of the solar system have provided much information on other planetary atmospheres. Many of these planetary atmospheres are completely different from that of the earth, but nevertheless they have provided insights into how our atmosphere works. Powerful electronic computers and advanced mathematics have allowed meteorologists to simulate the various global atmospheric processes numerically and to reproduce the global climate from the basic laws of physics. These so-called atmospheric general circulation models or global climate models enable meteorologists to undertake numerical experiments in which one aspect of the climatic system is varied, forests are replaced by grassland for example, and the results in the climate model are observed. Numerical experiments of this type have greatly enhanced our understanding of the causes of both climates and variations in climate.

The advances in technology over the last two decades have completely changed our perception of the global atmosphere. In particular they have produced a picture of an atmosphere in which many climatic changes, probably including ice ages, are the result of variations in the so-called greenhouse gases. In the greenhouse effect, incident short-wave radiation from the sun warms the earth during sunlit hours. This heating is balanced by outgoing long-wave infrared blackbody radiation from the top of the atmosphere 24 hours a day. Nearly 80 per cent of the infrared

Plate II.4.1 The so-called 'tablecloth' of cloud on Table Mountain, Cape Province, South Africa. Geographers have made major contributions to the study of weather and climate.

radiation emitted by the earth's surface in the wavelength region from 7 to 13 μm escapes to space because the pre-industrial atmosphere is transparent in this region. Outside this so-called 'window' region, most of the infrared radiation from the earth's surface is absorbed by the various atmospheric trace gases (table II.4.1) and then reradiated to space at the much colder atmospheric temperatures. These atmospheric trace gases also emit downward infrared radiation towards the earth's surface; this downward flux warms the earth and is known as greenhouse radiation because the earth's surface temperature tends to increase in response to this flux. The natural background greenhouse warming at the earth's surface is some 30 K, since the total amount of infrared radiation to space from the earth–atmosphere sysem is such that it corresponds to a mean equivalent blackbody temperature of about 254 K, while the observed mean surface temperature is around 286 K.

The important radiatively active gases in the present atmosphere are H_2O, CO_2 and O_3. Atmospheric CO_2 has undergone wide variations in the recent past, from about 200 ppmv at the end of the last glaciation (18,000 BP) to 342 ppmv in 1983. Indications are that concentrations of atmospheric CO_2 are partly controlled by biological activity in the ocean surfaces, and that there may be a close link via atmospheric CO_2 between ocean biological activity and the growth and decay of temperate latitude ice sheets. If the ocean were lifeless – just a big puddle of water sitting beneath the atmosphere – CO_2 would distribute itself between water and air solely on the basis of the gas's chemical solubility. That would require the ocean to give up much of its present reservoir of dissolved CO_2 to the atmosphere, tripling its concentration there. Foraminifera in the ocean have a marked effect on atmospheric CO_2; the CO_2 content of surface water is lower than that of deep water as a result of the continual removal of carbon from the surface by photosynthesis. Carbonate skeletons and other organic matter formed by photosynthesis sink irretrievably into the deep ocean. Once in the deep ocean, much of the carbonate dissolves and the organic matter is oxidized to dissolved carbonate, which cannot mix with most surface waters because of sharp temperature and density differences between the surface and the deep waters. Life in the ocean acts to pump carbon from the atmosphere into the deep ocean, and the faster the biological pump works, the less CO_2 remains in the atmosphere.

One of the most exciting developments of the last two decades has been the recognition of the long-term control of the atmosphere and the global climate by the biosphere. Many textbooks discuss the influence of climate on natural vegetation, and this has been known for many years. That there could be a two-way interaction between the atmosphere and the biosphere is an exciting new idea (Lovelock, 1979; Reynolds and Thompson, 1988; Rosswall et al., 1988).

Table II.4.1 Trace gas concentrations and trends

	Concentrations		*Observed trends for 1975–1985*
Gas	*Pre-1850*	*1985*	*(+%)*
CO_2	275 ppmv	345 ppmv	4.6
CH_4	0.7 ppmv	1.7 ppmv	11.0
N_2O	0.285 ppmv	0.304 ppmv	3.5
CFC-11	0	0.22 ppbv	103.0
CFC-12	0	0.38 ppbv	101.0
CH_3CCl_3	0	0.13 ppbv	155.0
CCl_4	0	0.12 ppbv	24.0

ppmv, parts per million by volume; ppbv, parts per billion by volume.
Source: After Ramanathan, 1988

Increasing Greenhouse Gases

Many synthetic trace gases (table II.4.1) have been produced during the last decade, the most important of which are the chlorofluorocarbons CFC-11 and CFC-12 (Ramanathan, 1988). Not only do these gases destroy stratospheric ozone but they also absorb strongly in the infrared window region and have the effect of 'dirtying' the atmospheric window. Therefore the synthetic gases are extremely effective greenhouse gases compared with CO_2, and the addition of one molecule of CFC-11 or CFC-12 can have the same greenhouse effect as the addition of 10^4 molecules of CO_2 to the present atmosphere.

Measurements taken at Mauna Loa, Hawaii, show that the atmospheric CO_2 concentration has increased from 316 ppmv in 1959 to 342 ppmv in 1983, an 8 per cent increase in 24 years. A variety of direct atmospheric CO_2 measurements and indirect reconstructions indicates that the pre-industrial CO_2 concentration during the period 1800–50 was 270 ± 10 ppmv. Much of this recent increase in atmospheric CO_2 is due to the use of fossil fuels (gas, oil, coal), and projections suggest that it will continue into the future.

There is growing evidence that increasing concentrations of CO_2 and of other radiatively active trace gases in the atmosphere may be having a long-term effect on global climate. The earth is at present experiencing a long time-scale climatic warming, with global mean temperatures increasing by 0.3–0.7 K over the last 100 years. Although this observed increase cannot yet be ascribed in a statistically rigorous manner to the increasing concentration of CO_2 and other trace gases, its direction and magnitude lie within the predicted range of their effects. Recent assessments suggest that increases in CO_2 and other trace gases will probably result in the equivalent of a doubling of the atmospheric CO_2 concentration probably between 2050 and 2100.

Virtually all computer models of the climatic effects of doubling CO_2 and other radiatively active trace gases in the atmosphere indicate significant increases in global temperature of between 1.3 and 4.2 K. However, a change in temperature is not the only consequence of increasing CO_2 levels. Major changes in the hydrological cycle (e.g. in evaporation and precipitation rates) are likely to have regional and global consequences (Lockwood, 1989). Unfortunately, general circulation models which have been applied to the problem of the effect of increasing greenhouse gas levels on climate do not give an unequivocal picture of how precipitation or soil moisture might be expected to change. Nevertheless some conclusions are reasonably well established. In general, precipitation increases are predicted poleward of 30 °–35 °N and 30 °–35 °S and in the immediate vicinity of the equator (5 °N to 5 °S). In the intervening zone, the results are more variable, but a tendency for a decrease in precipitation rate in one or more seasons is generally apparent (Bradley et al., 1987).

Climatological Research

Over at least the last million years the earth's climate has been characterized by an alternation of glacial and interglacial episodes, marked in the northern hemisphere by the waxing and waning of continental ice sheets, in both hemispheres by periods of rising and falling temperatures, and in the tropics by hyperarid and pluvial phases (Lockwood, 1985). It has already been commented that one key link in driving these climatic changes is variations in atmospheric CO_2. Therefore a better understanding of the functioning of the climate system may be obtained by a study of past climatic conditions when the influence of man upon the climate system was negligible. Modelling past climate fluctuations that led to extreme climates such as ice ages, for which typical parameters are known from proxy data, should help to assess the power of mathematical models to simulate the real climate under known conditions (Street-Perrott et al., 1983).

A study of proxy data for past climates suggests that major climatic changes can take place in relatively short time periods of several decades to several hundred years. Further, during such transitional times, the essential external parameters and boundary conditions like the insolation pattern, sea level, and continental ice-cover probably changed in a very limited range. It therefore seems possible that for a given set of boundary conditions the climatic system has more than one mode of operation. Proxy data for the last major glacial oscillation (Würm/Wisconsin) suggest the existence of a bi-stable climate system during this period, one data set describing the warm and the other the cold climate state. After a period of one or two thousand years, the stabilizing mechanisms of one mode of operation seem to become ineffective and the system organizes itself in a new mode. The transitional periods are often masked by a series of rapid oscillations between modes before the climatic system settles into its new mode. An example of this is provided by the sudden warmings and coolings in Europe during the melting of the ice sheets at the end of the last glaciation. Such research is relevant to the present greenhouse warming, and suggests that the transition to a warmer greenhouse world may not be smooth, but may take the form of a series of sudden unexpected warmings.

Global climatic models are one of the main tools used to investigate the possible impact of increasing atmospheric greenhouse gases. Studies with global climate models have already indicated considerable sensitivity of their climate simulations to changes in surface properties such as soil moisture, albedo, surface roughness, evaporation and sensible heat transfer (Henderson-Sellers and McGuffie, 1987). Thus there is a need for a deeper understanding of the various land-surface processes and in particular how to represent them adequately in global climate models. Over most of the land surface the effective interface between the atmosphere and the soil is vegetation. Therefore much research effort in recent years has concerned the study of the explicit effects of vegetation on the climate system and the development of micrometeorological and hydrological models of typical vegetation types (Sellers et al., 1986).

A related research area is the creation of reliable global data sets of geographically (and seasonally) varying surface characteristics such as soil type, vegetative covering, albedo, soil moisture and surface roughness, and the investigation of their use in climate models (Wilson and Henderson-Sellers, 1985). In this respect, valuable new information should be available in the growing archives of remotely sensed data, particularly from satellite-borne instruments.

The interactions of climates, ecosystems and societies have received increasing attention from both natural and social scientists in recent years (Parry et al., 1988). Two obvious and fundamental weaknesses generally accompany any assessment of the impact of possible future climatic changes on ecosystems and society (Parry, 1985). First, we have inaccurate information on their present-day sensitivity to climatic variability. Second, we are uncertain what changes of climate will occur in the future. In most global climate models, the increase in CO_2 concentration is modelled as an abrupt change from one concentration to another, not as a gradual change through time. In practice we are experiencing climatic changes as the climate system responds to slowly increasing greenhouse gas concentrations. Therefore among the most crucial questions concerning the impacts of climatic changes are, for example, how intrinsically adaptable are ecosystems and farming systems to different rates of climatic change? Where system adaptability is inadequate to absorb the climate impact, what can be done to mitigate the resulting shocks to the system? To be useful, climate impact analyses should be about particular ecosystems and farming systems in real locations, e.g. Canadian prairie wheat production, moorland ecosystems in upland Scotland.

Final Comments

The consensus among climatologists is that the atmosphere, and indeed the entire global environment, is changing, and that humans are today the principal agents of change. The prime human intervention is by releasing pollutant gases, many of which have a strong greenhouse effect, into the atmosphere. Other human activities of climatic importance include tropical deforestation (also extra-tropical afforestation), desertification of marginal desert regions, pollution of the oceans etc. The climates of all the world's nations, rich and poor alike, will change, and in some cases the changes will be towards a markedly more hostile environment. In nearly all regions the adjustments to the climatic changes are likely to be both costly and difficult. Climatologists have an important role in both predicting the direction of climatic changes in particular regions and assessing their likely impacts.

References

Bradley, R. S., Diaz, H. F., Eischeid, J. K., Jones, P. D., Kelly, P. M. and Goodess, C. M. 1987: Precipitation fluctuations over northern hemisphere land areas since the mid-19th century. *Science*, 237, 171–5.

Henderson-Sellers, A. and McGuffie, K. 1987: *A Climate Modelling Primer*. Chichester: Wiley.

Lockwood, J. G. 1985: *World Climatic Systems*. London: Edward Arnold.

——1989: Hydrometeorological changes due to increasing atmospheric CO_2 and associated trace gases. *Progress in Physical Geography*, 13, 115–27.

Lovelock, J. E. 1979: *Gaia. A New Look at Life on Earth*. Oxford: Oxford University Press.

Parry, M. C. 1985: Estimating the sensitivity of natural ecosystems and agriculture to climatic changes. Guest Editorial. *Climatic Change*, 7, 1–3.

——Carter, T. R. and Konijn, N. T. 1988: *The Impact of Climatic Variations on Agriculture*: vol. 1, *Assessments in Cool, Temperate and Cold Regions*; vol. 2, *In Semi-arid Regions*. Dordrecht: Kluwer Academic.

Ramanathan, V. 1988: The greenhouse theory of climate change: a test by an inadvertent global experiment. *Science*, 240, 293–9.

Reynolds, E. R. C. and Thompson, F. B. 1988: *Forests, Climate, and Hydrology: Regional Impacts*. Tokoya: The United Nations University.

Rosswall, T., Woodmansee, R. G. and Risser, P. G. 1988: *Scales and Global Change*. Chichester: Wiley.

Sellers, P. J., Mintz, Y., Sud, Y. C. and Dalcher, A. 1986: A simple biosphere model (SiB) for use in general circulation models. *Journal of Atmospheric Science*, 43, 505–31.

Street-Perrot, A., Beran, M. and Ratcliffe, R. 1983: *Variations in the Global Water Budget*. Dordrecht: Reidel.

Wilson, M. F. and Henderson-Sellers, A. 1985: Land cover and soils data sets for use in general circulation climate models. *Journal of Climatology*, 5, 119–43.

II.5 Hydrology

Tim P. Burt

Introduction

The science of hydrology deals with the occurrence, distribution, movement and properties of water in its different forms over, on and under the surface of the earth (plate II.5.1). The scope of hydrology is extremely broad and it is linked by common content to many other environmental sciences, particularly agriculture, botany, climatology, ecology, geology, geomorphology, meteorology and soil science. Traditionally, hydrology was regarded as a part of civil engineering. Engineers seek practical numerical solutions to specific hydrological problems using the most reliable methods available; often they need little or no knowledge of physical reality in order to achieve their aim. By contrast, the study of water within the environmental sciences listed above essentially demands knowledge and understanding of the physical system involved. Usually, the hydrological cycle (figure II.5.1) provides the focus of attention and often the drainage basin is the fundamental spatial unit of interest; study is made of the processes controlling the input, storage, transfer and output of water for the area in question. Although this interest in hydrological processes is not exclusive to geographers, it is nevertheless fair to say that 'geographical hydrology' and 'physical hydrology' are identical subjects. Given this outlook, it is not surprising that, as noted below, a good number of today's practising 'hydrologists' started life as 'geographers'!

Hydrology Within Physical Geography

Hydrology represents an important part of teaching in physical geography and it is seen as a central component of the discipline; in addition, it has important applications in human geography, particularly in resource management. Elements of the subject are taught in some form in most geography degree courses, and often there are specialist options available later on in the course. In addition, knowledge of hydrology is vital to other specialist subdisciplines of physical geography, notably fluvial geomorphology and ecology.

This pivotal position of hydrology within physical geography, however, is a relatively recent development and arose through a growing interest in process studies from the 1960s onwards. Within geography itself, fluvial geomorphology provided the main reason for an interest in hydrology. The publication in 1964 of Leopold, Wolman and Miller's *Fluvial Processes in Geomorphology* showed that some facets of hydrology were of particular importance to geomorphological processes and stimulated much research into hillslope and channel forms. The value of integrating these interests is well illustrated by Gregory and Walling (1973) for example. Gradually some aspects of this geomorphological research evolved into a specific concern with physical hydrology. For example, research by members of the Geography Department

Plate II.5.1 A major branch of physical geography is hydrology, much of which is concerned with trying to understand the nature of stream flow. This is the Zambezi River at Victoria Falls, Zimbabwe.

of Bristol University on the karst landforms of Mendip stimulated interest into the behaviour of streams feeding the cave systems; Darrell Weyman's research on the generation of subsurface runoff on hillslopes was a direct result. By the same token, the textbook *Hillslope Hydrology* (Kirkby, 1978) includes chapters by a number of authors whose interests straddle this undefined boundary between the two subdisciplines of hydrology and fluvial geomorphology. At the same time, other geographers were becoming interested in the application of hydrology to environmental problems such as soil erosion, flood control, pollution and water supply. Much of this work was stimulated by the programme of the International Hydrological Decade (1965–74). This led in the United Kingdom to texts such as *Floods* (Ward, 1978); in the United States a series of hazard research studies by White, Burton, Kates and others also showed the value of hydrological studies. The growing importance of hydrology within geography in the late 1960s is indicated in particular by the publication of two texts, one by Ward (1967, the first systematic textbook treatment of the subject by a geographer) and the other by Chorley (1969). Outside geography, there was also a growing interest in hydrological processes at this time, most notably by forest hydrologists. Hewlett's (1961) definition of the variable source area underpins much modern research in hillslope hydrology. Catchment experiments in both the United States and the United Kingdom indicated that forested basins yield less water than unforested ones; this has led to a large number of studies concerned with the hydrological effects of changes in land use.

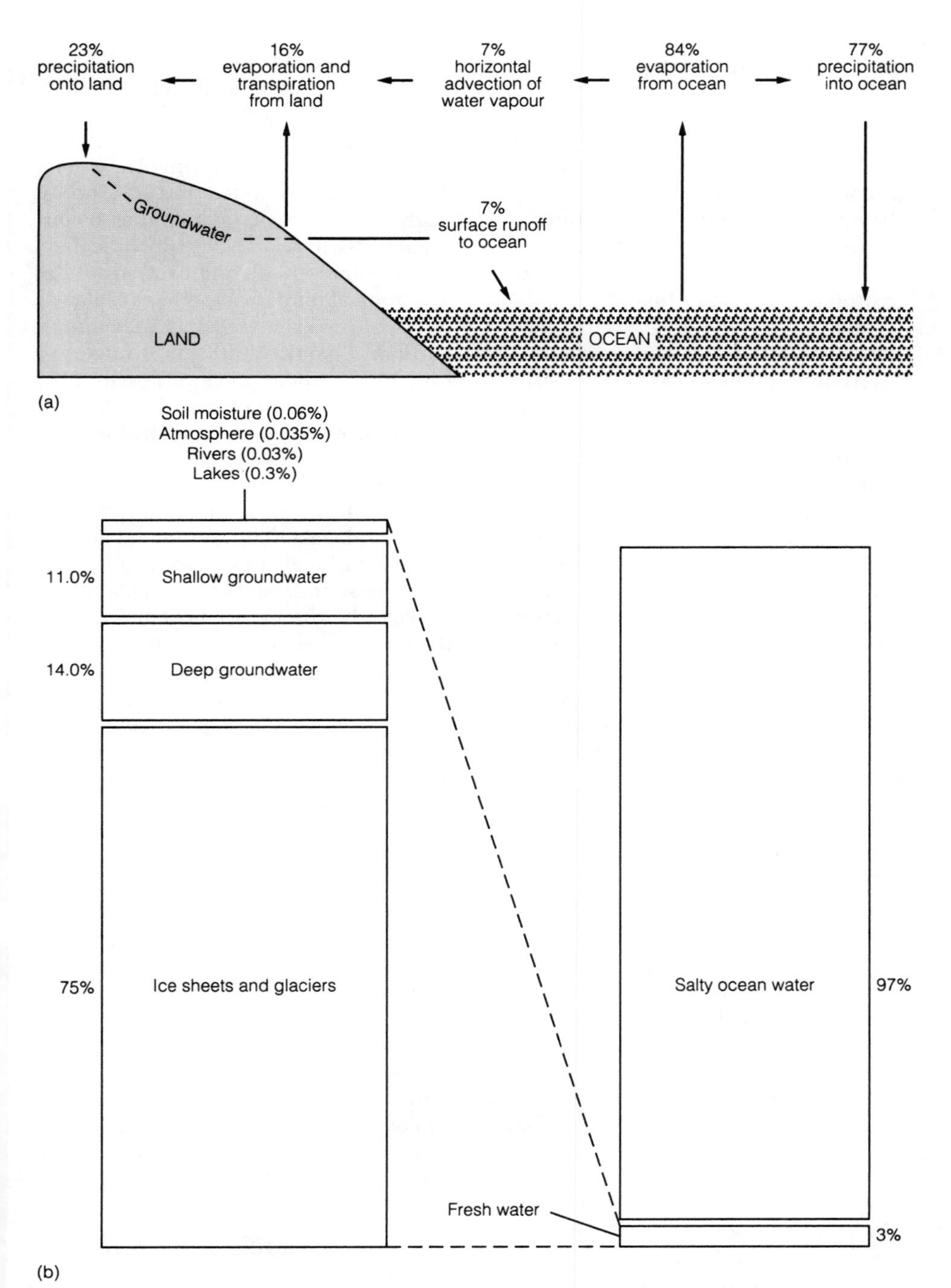

Figure II.5.1 The world hydrological cycle: (a) the main components; (b) fresh-water and saltwater components.

The Nature of Hydrology Courses

An introduction to hydrology, comprising typically about 20 per cent of the physical geography course, is provided in the first year of most geography degree courses. Lectures provide the main foundation and a number of courses focus on the global hydrological cycle at this stage, comparing the hydrology of the major physiographic zones, and linking up neatly with other aspects of physical systems at the global scale as a result. Many courses offer some introductory statistical exercises: such work might deal with water balance computations, for example, or flood hydrograph analysis. A few courses even get first-year students into the laboratory doing simple experiments (e.g. measurement of hydraulic conductivity) and writing reports. Most courses seem to include hydrology as part of their first-year residential field course: simple techniques for measuring runoff dynamics and water quality may be introduced, often organized around a simple theme such as storm runoff generation in a small catchment. In the second year, hydrology is often included as part of a more extensive course concerned with the physical environment. Quite often, the link is with fluvial geomorphology, but increasingly hydrology forms the basis for a course in environmental resource management, with the drainage basin as the basic unit of interest.

Where individual courses in hydrology are offered in the second year, the focus tends to switch to physical hydrology and catchment studies and may well include aspects of water quality such as nitrate pollution and sediment yields. The emphasis is usually on detailed investigations of hydrological processes at the hillslope or catchment scale, including measurement techniques and methods of analysis. Such courses often provide a foundation for the individual project undertaken by undergraduates during the vacation at the end of their second year (most courses have at least one such project). Hydrological studies are popular and usually involve a good deal of fieldwork, often abroad. Topics might include meltwater studies below glaciers (Switzerland, Norway or Iceland could be the venue), interception studies in a tropical rainforest (perhaps Malaysia or Brazil) or karst (e.g. Yugoslavia or northern Spain). Within the United Kingdom a whole range of possibilities spring to mind – increased flooding due to land-use change in Lincolnshire; the turbidity of Cotswold streams; patterns of nitrate leaching in a south Devon catchment; the sedimentation record of a Midlands lake; infiltration and erosion studies on blanket peat, to name just a few! These independent projects often prove to be the most enjoyable part of the course and, as in other areas of geography, a hydrology project often encourages students to consider a career which includes some aspect of hydrology (see below). Third-year courses continue to focus on physical hydrology but often include applications in river management. Themes might include flood prediction, river pollution, water resource planning and the design of hydrometric networks. If not introduced before, third-year courses often provide an opportunity for students to use computer simulation models for flood forecasting. Such courses usually include a fair amount of fieldwork and laboratory work, field visits to water company installations such as sewage works or water treatment plants, and may include residential fieldwork, often based at a research catchment.

Current Research: a Personal View

It seems appropriate to give a little more detail about some of the hydrological topics which geographers are studying at the moment, and which undergraduates would be introduced to as part of their studies. It is perhaps safest to give a personal account: though biased, of course, this avoids causing offence by mentioning some people and not others (everyone objects this way!). I should add that much of this work is joint research with research students and/or other colleagues: typically hydrological studies involve teamwork rather than being a lone pursuit.

My main interest is in hillslope hydrology which is concerned primarily with flow processes within the soil and over its surface. Water follows different pathways which delay and lessen the flow to different extents. A knowledge of the relevant flow processes is relevant both for an understanding of runoff generation and because different hydrological pathways affect the amount of sediment and dissolved material carried by the water. Much of this work involves field study although computer modelling of hillslope hydrology is also one of my ongoing interests. One of my research students has been studying infiltration processes on a landslide near Daventry (central England), examining the way in which recharge depends on water moving down large voids in the clay soil. This will help us to understand how and why landslides develop. Another project at Slapton in south Devon (southwest England) involves studying storm runoff processes on different types of agricultural land and measuring the amounts of soil, nitrogen and phosphorus lost from each area. This is significant in terms of understanding eutrophication, fertilizer loss, and those areas which are especially sensitive to generating pollution. One research student is looking at the possible use of buffer zones on floodplains as a barrier between runoff from intensively farmed land and rivers where pollutants can cause eutrophication and substances like nitrate can impair the quality of drinking water supplies. Excess nitrates may cause health problems, especially for young babies, and some fears have been expressed that they can also increase cancer incidence. Another research student is examining the influence of hydrological pathways on the leaching of aluminium associated with acid rain. Aluminium becomes soluble in acid water and is toxic to fish; at Loch Dee in Galloway (Scotland) where the research is based, the fish population has diminished markedly in recent years. Once again there is also a fear that high levels of aluminium can impair human health. Hillslope studies are also relevant at the larger catchment scale where the spatial and temporal patterns of runoff and water quality can be related to runoff derived from different hillslope source areas. Within the upper Thames basin, and on the River Windrush in particular, we are concerned to identify sources of nitrogen pollution in its various forms (especially nitrate and ammonium) and the reasons why some rivers are turbid (because of a large suspended load) and others are not. Some of this work is funded by the National Rivers Authority, Thames Region and has clear implications for river basin management. Much of this work involves sophisticated laboratory techniques and computer simulation as well as field experiment. Further afield, I have recently been involved in studying soil erosion and lake sedimentation in Mexico, and the effects of deforestation on streamflow in the Coweeta experimental basin in North Carolina, United States.

This sort of catalogue of research interests is by no means unusual; nor is the combination of theoretical and practical work. Many colleagues in other geography departments could provide much more impressive and varied lists of research interests. The important point is that, as in other specialist areas, undergraduates can expect to become acquainted with some of these projects as their courses unfold. In hydrology, this means that experience can be gained in field and laboratory techniques, and in computer applications, for a wide range of environmental issues which are of great public concern at the present time.

Career Possibilities

Having done all this work, some geographers decide that a career in hydrology would suit them. What prospects exist? It is important not to paint too rosy a picture about career opportunities – many career openings require a degree in civil engineering or a postgraduate qualification in hydrology and geographers with only a first degree stand little chance of obtaining posts, although some do. Usually, a student must take a Master's degree in a topic such as water resources or engineering hydrology before a Water Authority or a branch of the National Rivers Authority would consider employing them. A number of geographers do gain places on such courses each year. An alternative career route is to become a postgraduate student researching some aspect of hydrology. Again, a number of geographers with Master's or PhD degrees in this general area go on each year to jobs with Water Authorities or to organizations such as the Institute of Hydrology. Thus, there are quite a few cases where geographers have been successful in gaining employment in the field, and in some cases whole sections (e.g. the Water Resources Planning Unit of the National Rivers Authority, Thames Region) seem to be staffed entirely by geographers!

References

Chorley, R. J. (ed.) 1969: *Water, Earth and Man.* London: Methuen.

Gregory, K. J. and Walling, D. 1973: *Drainage Basin, Form and Process.* London: Edward Arnold.

Hewlett, J. 1961: Soil moisture as a source of base flow from steep mountain watersheds. Paper 132, *US Department of Agriculture Forest Service, S.E. Forest Experimental Station.*

Kirkby, M. J. (ed.) 1978: *Hillslope Hydrology.* Chichester: Wiley.

Leopold, L. B., Wolman, M. G. and Miller, J. P. 1964: *Fluvial Processes in Geomorphology.* San Francisco, CA: Freeman.

Ward, R. C. 1967: *Principles of Hydrology.* London: McGraw-Hill (2nd edn. 1975).

——1978: *Floods: a geographical perspective.* London: Macmillan.

II.6
The Human Impact
Ian Douglas

The relationship between people and their environment is probably the most important core theme of geography. Originally the discipline looked at how environments influenced, restricted or even determined what people could do. Today, the enquiries are much more into how people have modified environments to make them more amenable to economic and leisure activities, and in doing so how accidental, unsuspected and sometimes unwanted and harmful side effects have been produced. Geographers are now increasingly concerned with such human impacts at all scales from soil deterioration and local beach erosion to global climatic modification. Many of these impacts are mentioned in other chapters of this book, e.g. a discussion of research into acid rain impacts in the chapter on hydrology, but this section will consider them in an integrated manner. Apart from the most desolate ice-covered wastes and arid deserts and a few still undisturbed high mountains and virtually uninhabited forests, people have altered the entire land surface of the globe. Many of the moorlands of upland Britain (Edwards and Ralston, 1984; Goudie, 1989) and the woodlands of New England developed after forest clearance and agricultural land use. People drain swamps and create artificial lakes, introduce alien species – including pests and weeds – which transform ecosystems, and even landscapes. The blackberry is a noxious weed in Australia, where perhaps a quarter of the plant species have been introduced since European settlement and where some animal introductions, such as the Queensland cane toad, are so extending their range that they are a serious menace to native species (Adamson and Fox, 1982). The impact of people thus extends beyond the simple direct on-site effects, such as the deliberate clearing of forest or the development of new housing estates, to the more subtle, even unexpected, spread of alien plants and animals and the downsteam off-site consequences affecting other ecosystems and communities, coastal areas, the seas, the groundwater, and even the atmosphere. An idea of the relationships of these impacts to people can be obtained from a simple matrix which plots deliberate and inadvertent, or unwanted, impacts against on-site and off-site receipt of impact (figure II.6.1). More specifically, the impacts of particular human activities can be tabulated, e.g. the effects of European settlement in Australia (table II.6.1) or of urbanization (table II.6.2).

Type of action	Location of effect: On site	Location of effect: Off site downstream
Deliberate	1	2
Inadvertent	3	4

Figure II.6.1 Types of on-site and off-site environmental impact (after Douglas, 1989).

Table II.6.1 Effects of European settlement on vegetation in Australia

Consequences of European settlement	*Effects and comments*
Displacement of Aborigines	Change in long-established fire regimes
Cultivation	Vegetation clearance, introduction of plants, use of chemicals, accelerated erosion, salinity build-up
Irrigation	Salinization, flooding behind dams, changed river discharges, weed problems etc.
Grazing	Vegetation removal and alteration, grazing pressure around watering places, trampling effects
Forestry	Exotic forest species introduced. Native forests managed to produce uniform-aged stands of fast-growing eucalypts
Plant introductions	Many naturalized introductions were originally accidental contaminants of seed etc. or imported as ornamental plants. One-fifth of flora in Sydney region is introduced
Animal introductions	Many animals introduced for draught, hunting, as pets, for food or accidentally have become feral. Examples include donkeys, camels, rabbits and buffalo. Many have significant direct or indirect effects on native vegetation

Source: Adamson and Fox, 1982

People–Environment Interaction in Geography Courses

So fundamental is this relationship between human activity and human surroundings that discussion of the human impact pervades many geography courses. Many introductory courses contain elements of the basic physical and biochemical processes that are involved in physical geography, but the better physical geography texts (Strahler and Strahler, 1977; Muller and Oberlander, 1978; White et al., 1984; Briggs and Smithson, 1985, among others) all contain examples of how these processes are modified by human activities. In some physical geography courses specific topics related to human activities such as the greenhouse effect and the ozone holes in polar regions are discussed. Where historical geography forms an important part of first-year teaching, the stages of land settlement, forest clearance, draining of wetlands, reclamation of wastes, enclosure and agricultural improvement often form a major theme. Many first-year courses contain some instruction on the problems of development in the poorer countries of Latin America, Africa and Asia and inevitably discuss resources, soils and climatic factors affecting agricultural activity and the provision of adequate food supplies. The impact of high rates of population growth in some of these countries, especially in Africa, is often seen in terms of overgrazing and soil deterioration. Introductory human geography courses often discuss human impacts in terms of living conditions in urban areas or recreational impacts on scenic resources, e.g. footpath erosion in British National Parks or coastal erosion in the dunes of the east coast of the United States caused by off-road vehicles. Some first-year courses place special emphasis on the people–environment theme, devoting whole lecture series and practical sessions to consideration of the human transformation of the face of the earth and the contemporary environmental issues of pollution, waste disposal and exploitation of non-renewable mineral and soil resources. Other institutes give particular attention to their local area or region. For example, a college in Hamburg might have a special course on the geography of Hamburg, including

Table II.6.2 Interaction between urban form, flows and processes at four different scales

Scale	*Micro-scale*	*Meso-scale*	*Macro-scale*	*Mega-scale*
Approximate areal magnitude	10–1000 m^2	1000 m^2–1 km^2	1–10 km^2	10–1000 + km^2
Type of area	In and around individual buildings and structures	Census tracts Housing estates Industrial zones	Topographic units in urban areas, valleys, ridges, escarpments	Urban stress as a whole Conurbations Metropolitan regions
Energy flows	Effects of structures on winds, insolation, reflection Thermal comfort of different house styles	Effects of land-use density and ground cover on heat emissions, albedo and radiation balance	Influence of aspect on solar radiation receipt and of water bodies on winds and heat island development	Urban heat islands and wind regimes
Water flows	Runoff from roofs, driveways, yards; individual wells and groundwater pumping	Stormwater drainage including grassed waterways, separate or combined sewers	Channelization and stormwater detention reservoirs, multiple use of river valleys	Urban-induced rainfall Modification of regional groundwater levels; downstream effects of urban runoff and interbasin transfers on major rivers
Materials flows				
(a) Airborne pollutants	Emissions from domestic grates, industrial chimneys, individual vehicles	Concentration of exhaust fumes along major roads, impact of factory emissions on local communities Dust from construction and demolition sites	Effects of topography on concentration and movement of pollutants and scouring effect of winds	Urban dust domes; emission of oxides of nitrogen and sulphur producing downwind, transfrontier and even intercontinental acid rain
(b) Waterborne pollutants	Point-source emissions from individual premises	Pollutants from local septic tanks and aeration ponds; sediment in roadside and construction site runoff	Pollution of urban lakes by atmospheric fallout (e.g. lead) and siltation of lakes and streams	Impact of urban stormwater runoff, including heavy metals, and sewage treatment effluent on major rivers Oceanic discharge of inadequately treated sewage

Table II.6.2 (*cont'd*) Interaction between urban form, flows and processes at four different scales

Scale	*Micro-scale*	*Meso-scale*	*Macro-scale*	*Mega-scale*
(c) Solid wastes	Rubbish disposal at factories, warehouses and retail outlets Illicit dumping of domestic refuse	Community rubbish dumps and garbage depots	Site problems of noxious waste disposal: dangers of groundwater contamination from landfill	Regional waste disposal management such as the use of gravel pits near Didcot for London's rubbish; ocean dumping of sludge from municipal sewage
Geomorphic effects	Subsidence of individual buildings over old mines, quarries and landfill sites	Foundation problems of glacial deposits and buried karst Swelling clay problems Elimination of minor channels	Mass movements on unstable hill slopes Floodplain delimitation and flood risk estimation	Regional subsidence due to removal of liquids, such as water, brine, or petroleum (as at Houston, TX); relationship of urban areas to sea-level change
Biological effects	Insects and vermin in individual buildings Bacterial contamination Character of individual gardens	Effect of housing conditions on health and well being; influence of working environment on human and other life forms	Plant and animal invasions of derelict urban land Planned valley reclamation schemes, urban parks	Adaptation of birds and animals to urban living Evolution of strains of disease vectors resistant to pesticides Urban stress-related diseases

Source: After Douglas, 1989

consideration of the urban heat island, the recently rising groundwater levels, the flood problems, the improved river water quality and the waste disposal problems of the city, in addition to the character of its built form and the spatial distribution of human activities within it.

Many introduction courses have integral field classes which provide an immediate direct contact with major transformations of the earth's surface. A day class in the local area will reveal modified river channels, perhaps with flood protection works, or areas of landfill, wetland drainage or modification of hillslopes for civil engineering works, each of which can engender a discussion of the on-site and off-site consequences of the changes. A class in a different environment, perhaps on the coast or lakeshore, may enable remedial works designed to prevent environmental damage, such as breakwaters and groynes, to be examined and evaluated. Longer field classes away from the home environment, in a different climate or tectonic setting, bring to life the limitations imposed by cold, aridity, steepness of slope or irregularity of rainfall. Many British universities, for example, take students to southeast Spain where the effects of rare floods and gully erosion in Europe's most arid region can be seen.

At second- and third-year level,

geography courses contain a range of systematic options, many of which are devoted to environmental topics dealing with human impacts. Some are oriented towards resource utilization, others towards environmental planning and control, paying special attention to the wealth of policies and techniques now used by governments to assess and regulate environmental impacts. The names and titles of these courses vary, but usually some concentrate on specific resource issues, such as water or coastal resources, while others look at problems of human attitudes and behaviour in the face of environmental hazards, and a third group concentrates on the policies and technology, examining how projects are evaluated, the assessment of economic benefits and environmental costs, multi-objective planning, ecological analysis in land-use planning and environmental impact assessment. In a few geography schools, much of this material is found in courses on applied geography which specifically consider the areas where modern geographers are active in consultancy and advising governments at all levels. Elsewhere, as a survey of third-year courses in northern England (Douglas, 1986) showed, regional geography options remain popular. Nevertheless, they too often contain material on the human impact, particularly in terms of major projects that alter the country's economic structure, such as major water resource developments in the western United States or in the USSR. As there are an increasing number of academic geographers engaged in applied environmental research, it is inevitable that their specialist teaching reflects their research and practical experience.

The learning experience at degree level is not confined to listening to lectures, and the practical and small-group teaching sessions provide splendid opportunities to get involved in debating and investigating environmental issues. Indeed, bright determined students can persuade class teachers and tutors to answer their questions on the relevance of geography to current events, be they local issues such as proposals to build a new freeway or shopping centre or international issues such as the logging of tropical rainforests. Teachers may well respond by organizing the class into a simulation exercise, giving each member a specific role to play, such as the highway planner, the long-distance trucking operator, the local environmental lobby group organizer, the women's action group, the local chamber of commerce, the town mayor and so on. Each student then prepares an argument from the standpoint of the assigned role and puts her or his case in a debate or discussion chaired by the teacher. In another class activity, a small group of students might be asked to work together to prepare a project report or desk study of an environmental issue, looking at all the evidence on particular types of activity or individual development projects.

At this level, fieldwork is often related to specific courses and will involve practical skills of the type used by professional environmental scientists. A water quality study will probably involve developing a field sampling scheme to take account of variations in river flow and possible times of pollutant discharge; considerations of sample deterioration in storage and transport; and appropriate laboratory and statistical analyses. A questionnaire survey of people's attitudes to or awareness of environmental issues or hazards would involve questionnaire design, sampling and subsequent statistical analysis. Alongside fieldwork, students might use one or more of the large environmental databases now available, such as the records of air pollution levels (plate II.6.1) and water quality held by national authorities, or they might explore the archives held in their local public libraries to discover the campaigns waged by public health officers to reduce the impacts on air and water quality fifty or even a hundred years ago. Much of the modern data on air quality for some cities are based on such dense networks that they can be incorporated into small-scale geographical information systems (GISs) (see chapter III.7), enabling the changes from one date to another to be portrayed readily on the computer screen.

Opportunities for Individual Projects and Dissertations

Armed with ideas from the courses and skills from fieldwork and practical work, a student is ready to tackle the test of designing a research exercise and carrying out her or his own investigation into a particular environmental issue or type of human impact. The problems investigated range in excitement and originality from the ever-popular studies of footpath erosion in recreational areas and water quality changes above and below effluent discharges from industrial premises to comparing surface soil loss under logged and unlogged tropical rain-forest, investigating air pollution in a Chinese industrial city, examining the heavy metal pollution of soils, possibly linked to discharges from a British Peak District cement works, and an evaluation of salinity in soils of a California irrigation area.

Plate II.6.1 Tree decline is a widespread problem in Europe and North America. The causes are complex and include climatic change, bad management of woodland and various types of pollution including acid deposition.

With the large number of major practical research programmes being conducted by geographers, students often have the opportunity to carry out their projects or dissertations in association with more experienced scientists. Sometimes this provides the chance to visit new environments or to use extra specialized equipment. Other students group themselves into expeditions with a special purpose, either as a team of geographers or with students from other disciplines. Such endeavours not only enhance the learning experience, but if carefully prepared over a long enough period, can produce results of considerable merit. No such expeditionary work should be undertaken in another country without adequate consultation with the appropriate national authorities and a report on the outcome should always go to the authorities and the local university or other higher educational institution.

Prospects for Further Study

Many students find that they need further intensive specialized training after graduation before they can find work in environmental consultancies, planning authorities or resource management agencies. The traditional approach of training in planning has decreased in popularity compared with the uptake of Master's degree courses in environmental science, pollution and impact assessment. Such courses have varied emphases: some are strongly biological, others more chemical or engineering in scope. Many demand basic science skills which are likely to be possessed by some but not all geography graduates. A number of the impact courses are concerned with social as well as environmental impact and may be particularly attractive to students with a good basis in all aspects of geography. Such courses are often specifically geared towards the environmental legislation of the country in which they are taught and overseas students should carefully consider whether they meet their special needs.

Apart from the specialist environmental courses, many other coursework higher degree programmes relating to the practical aspects of people–environment interactions exist, often oriented towards improving design and construction. Programmes in water resource management, recreation planning, landscape design, housing and urban studies often have, and all ought to have, a major emphasis on ecologically sustainable, environmentally friendly resource use and land management.

Effective environmental assessment and management requires good information. A major part in the information flow is now played by remote sensing and GISs. A further training in one or both of these fields will be of great help to a person seeking professional work in environmental management. Many agencies are now setting up their own or are using established GISs for land information. The ability to update inventories of forest resources, areas affected by soil erosion (plate II.6.2), land subject to flooding, locations of new buildings, utility pipelines and cableways and the extension of the road network is essential if proposed new developments are to be set into their context and their impacts fully evaluated. GISs can be used to demonstrate how, for example, water flows or traffic movements will be altered by a new development, and to show alternative ways of planning the development. Remote sensing enables seasonal changes in ecological conditions to be monitored and can be used in conjunction with GISs to record and evaluate changes in land use.

Many graduates learn more about the human impact by engaging in research into environmental issues or by working as graduate assistants on research projects investigating particular environmental problems. In an urban context such projects might be looking at the water quality of old industrial canals being redeveloped for recreational use, examining the impact of storm sewer overflows on rivers, studying the distribution of heavy metals in soils, especially in inner cities where new houses with gardens are being built on cleared and reclaimed industrial or dockland sites, or evaluating sources of atmospheric acidity in relation to changing industrial, domestic and traffic emissions.

Outside the urban area, the field of research is equally wide, but particular importance in recent years has been attached to evaluating the effects of past human impacts in order to set them against the effects of climatic change. Reconstructing past land use changes by examining the records of lake sediments and correlating them with documentary evidence provides good data on soil erosion rates. Tying these changes to the climatic and harvest records enables the interplay of natural, economic and political conditions to be unravelled and the climate–environment relationship to be more firmly established. From such deductions better forecasts of the conse-

Plate II.6.2 Severe badland development in Swaziland, southern Africa. Soil erosion is one of the most serious of environmental issues.

quences of possible climatic changes can be made.

Equally significant is all the research into ways of reusing unwanted inadvertent human impacts such as safer waste disposal, increased recycling, less dependence on fossil fuels, improved and cheaper public transport systems, more sustainable agriculture and forestry, less wasteful water use and greater control of the release of contaminants to the atmosphere, hydrosphere, biosphere and lithosphere. These are fields to which many disciplines contribute, but in all of them geography graduates have found research opportunities and employment.

Individual Responsibility

Knowledge is precious and should be used wisely. Understanding the complexity of people–environment relationships enables people to advise others on issues that affect their everyday lives. No person is immune from the effects of the way others use resources or dispose of wastes. If a new incinerator or factory chimney is planned in one's home locality, the responsibility of the geographer might be to help local people understand both the possible dangers and the effectiveness of proposed safety measures. The understanding gained from degree-level study should be passed on at the appropriate moment to enable other people to obtain their goals, and maintain or enhance the quality of their lives. Joining local pressure groups or societies, working for national or international non-governmental organizations, from Friends of the Earth and Greenpeace to the Sierra Club or the National Trust, or even lobbying for a political party with a sound environmental policy is a legitimate outlet for skills acquired through studying people–environment relationships in a geography degree. A goal for all people with such skills ought to be to pass on to all the people of the world a sensitivity and awareness of the environmental consequences of their individual and collective actions, thus getting everyone to be more ready to think globally and act locally.

References

Adamson, D. M. and Fox, M. D. 1982: Change in Australasian vegetation since European settlement. In J. M. B. Smith (ed.), *A History of Australasian Vegetation*, Sydney: McGraw-Hill, 109–49.

Briggs, D. J. and Smithson, P. A. 1985: *Physical Geography*. London: Longman.

Douglas, I. 1986: Too many third year tutorials? *Journal of Geography in Higher Education*, 10, 191–5.

——1989: The rain on the roof: a geography of the urban environment. In D. Gregory and R. Walford (eds), *Horizons in Human Geography*, Basingstoke: Macmillan, 217–38.

Edwards, K. J. and Ralston, I. 1984: Postglacial hunter-gatherers and vegetational history in Scotland. *Proceedings of the Society of Antiquaries, Scotland*, 114, 15–34.

Goudie, A. 1989: The changing human impact. In L. Friday and R. Laskey (eds), *The Fragile Environment*, Cambridge: Cambridge University Press, 1–21.

Muller, R. A. and Oberlander, T. M. 1978: *Physical Geography Today*, 2nd edn. New York: Random House.

Strahler, A. N. and Strahler, A. H. 1977: *Geography and Man's Environment*. New York: Wiley.

White, I. D., Mottershed, D. N. and Harrison, S. J. 1984: *Environmental Systems: an introductory text*. London: Allen & Unwin.

II.7
Landscape Geography
Paul Coones

The Landscape Tradition in Geography

The study of landscape constitutes one of geography's principal traditions (plate II.7.1), albeit a diverse and problematic one. The term 'landscape' is itself imprecise, variously implying the generalized or composite visible and visual scene (stemming from topographical description, exploration and travel), an actual scenic view (its currency among painters) and a tract of land, owned and inhabited (the Anglo-Saxon usage). It has been associated with such concepts as environment, region, habitat and 'locale' (an erroneous but well-established English form of the French word *local*). In embodying the qualities of 'place', the interactions of society and nature as expressed on the ground or in the land, and the characteristics of particular environments – especially those altered through human agency and occupancy – the idea of landscape is directly tied to several of geography's basic concepts and long-established themes of enquiry. Consequently, it has featured in sundry guises in the work of geographers of diverse interests and persuasions. Furthermore, in methodological terms, landscape has played a significant role as one of the few unifying foci for the subject, holding the promise of bridging the human–physical divide, weaving the different threads together and creating the geographical synthesis. Indeed, for some, the study of landscapes has been regarded as an end in itself and as the discipline's chief concern and purpose; landscape provided geography with its (otherwise elusive) object of study, defined its individuality and served to delimit its scope.

Most British geographers today would not endorse such a view, contending that landscape can neither embrace all the enterprises and approaches of geography nor contain within itself the key to the processes which need to be understood in order to produce explanations of its characteristics. The landscape has also been criticized as an elusive, vague and intangible object and mode of study – one that is lacking in rigour, theoretically weak and concerned with 'mere' description of superficial form rather than with analysis of underlying process. The charge has also been made that as a pleasant means of passing the time it appeals to old-fashioned generalist geographers unwilling or unable to address new specialist ideas, methods and techniques, and who prefer an escapist, antiquarian, uncontentious pursuit in which diligence can act as a substitute for intelligence! Its practitioners naturally reject these charges, and indeed in recent years the very subjectivity of landscape has been hailed as an asset for constructing a geography in which human values, symbols and aesthetic concerns replace the quantifiable, value-free, abstract and unreal worlds of the spatial scientists and their allegedly objective appraisals of 'rational man' (*sic*). (For general reviews of the place of landscape in geography see Hartshorne, 1939; Wagner and Mikesell, 1962; Mikesell, 1968; English and Mayfield, 1972; Appleton, 1975.)

Plate II.7.1 The rural landscape has been a major area of geographical research. The landscape illustrated here is the agricultural belt of the mid-west of the USA in Wisconsin.

There are, in any case, many conceptions, genres and schools of landscape study, involving a number of individual themes. Particularly influential in the English-speaking world have been the writings of Carl Sauer (1889–1975) and the members of the Berkeley School in California (Leighly, 1978; Williams, 1983; Entrikin, 1984; Solot, 1986). Sauer's seminal paper on the subject was 'The morphology of landscape' (Sauer, 1925):

The cultural landscape is the geographic area in the final meaning. . . . Its forms are all the works of man that characterize the landscape. . . . The cultural landscape is fashioned out of a natural landscape by a culture group. Culture is the agent, the natural area is the medium, the cultural landscape is the result.

Sauer's argument involved several interrelated ideas: the separation of landscapes into 'natural' and 'cultural', and the transformation of the former into the latter by human communities; the emphasis upon human culture as the dynamic factor in this process; the identification of episodes of change, and the visualizing of the transfiguration in terms of temporal stages; the resultant tendency to conceive of geography as 'culture history'; and, ultimately, the use of cultural landscape studies as an implied denial of environmental determinism and as an affirmation of the power of human agency in altering nature, culture being the shaping force. Sauer spoke, in Davisian terms, of cycles of development and the rejuvenation of the cultural landscape that follows the introduction of a different or alien culture. Whittlesey (1929) appealed directly to evolutionary biology and human ecology in his concept of 'sequent occupance', which concentrated upon sequences of stages stemming from the proposition that 'human occupance of area, like other biotic phenomena, carries within itself the seed of its own transformation'. Whittlesey's successors, however, were less cautious than he in their use of this analogy as a basis of the sequent occupance model (Mikesell, 1976), although Broek's (1932) study of the Santa Clara Valley, California, stands out as a classic.

It is not difficult to appreciate the appeal of these precepts in an American context. Ideas about landscape not unnaturally reflect the societal experiences and of course the actual landscapes of particular times and places. The various phases of settlement which characterized North America's history of colonization, immigration and frontier are manifested in successive superimposed cultural landscapes such as Higbee's (1952) 'three earths of New England' ('Indian', 'colonial' and 'modern'). Elements of this tradition are still in evidence (see Conzen, 1990). Ideas of stages of progress and the conquest of nature also figure. Many of the intellectual roots, like those of Sauer himself, lay in European (especially German) soil – and landscape. In Britain, cultural landscape studies have tended to be dominated by historical approaches, notably through the work of historians (most famously, W. G. Hoskins's *The Making of the English Landscape* of 1955 (Meinig, 1979)), archaeologists (see Roberts, 1987) and historical geographers such as H. C. Darby (Williams, 1989), who in 1951 set out an agenda of 'progress' themes ('clearing the wood', 'draining the marsh' and 'reclaiming the heath'). To some, these generally anti-environmentalist, historicist, developmental, stage-model approaches to sequences of human settlement seem to represent all that the landscape is perceived to involve and signify (see for example Aston, 1983). Again, the prevailing view is that 'Man made the Land' (Baker and Harley, 1973), which is to a large degree understandable in a country with such a long and rich history recorded in its diverse landscapes.

A comparable intricacy and particularism, especially of a rural and pre-industrial nature, is evident in France. It is manifested in the small individualistic districts (*pays*), conceived as the creations of integrated modes of livelihood (*genres de vie*). In the work of the most famous French proponent of holistic *géographie humaine*, Vidal de la Blache (1845–1918), the unity of society and nature was upheld; landscape (*paysage*) expressed the symbiotic relationship between communities and environments which in particular milieux moulded the characters of regions (see Buttimer, 1971). The link with the regional concept was particularly strong in Germany, where the complex concept of *Landschaft* implies far more than 'landscape'; indeed, the web woven around it is so imbued with mystical meanings, fraught with etymological convolutions and productive of philosophical connotations that even protracted investigation (Hartshorne, 1939) has failed to delimit *Landschaftskunde* and draw accessible conclusions from the typically German terminological and classificatory field day which it afforded. Associated in the earlier years of the century with scholars such as Schlüter and Passarge, *Landschaft* studies have continued and developed (Leser, 1976), variously involving or recalling several different concepts and elements, notably geographic area, region, a field of vision as seen in perspective, the concrete material objects of a piece of territory, *all* sensually perceptible features, and a harmony of interrelated parts.

The Russians adopted the German word in *landshaftovedeniye* (inadequately translated as 'landscape study'), and thereby preserved the sense of region. In contrast with the small-scale variety of Europe, the landscapes of Russia are generally uniform, monotonous and colossal; a powerful environment is manifested in great natural complexes of soil, climate and vegetation which are expressed in the broad latitudinal landscape zones made famous by V. V. Dokuchayev (1846–1903). Landscape study in the hands of Russian and Soviet scholars such as L. S. Berg (1876–1950) has exhibited a strong environmental (bioclimatic) component and a concomitant integration (Berg, 1950; Isachenko, 1968, 1977). Landscapes are conceived as unified, functioning, interdependent natural systems of great practical relevance, which society can only modify, not create: 'anthropogenic landscapes' are deemed a misnomer by Isachenko (1974); 'unfortunately, many geographers still believe that they are investigating the dynamics of landscape when they simply trace historical changes of land use or the development of a territory' (Isachenko, 1975).

There are some important potential lessons here. The first and perhaps principal one is that, whatever approach he or she adopts, the geographer is capable of treating landscape in an integrated fashion, the unity of the subject reflecting the unity of the subject matter studied. Specialists, notably in the historical disciplines such as landscape history and archaeology – which focus upon individual aspects of the landscape – are apt to violate this synthesis. People notice and select what interests them, and the study of typologies of artefacts, systematic themes of change or the development of society as a supranatural category (independent of environmental elements and forces, constraints and resources) appeals to those who see the landscape essentially as the anthropogenic creation of human history. Yet the landscape 'is in truth nothing less than the complex, interrelated and unified material product of the geographical environment, a seamless totality in which the immemorial processes of nature and the much more recent activities of mankind interpenetrate' (Coones, 1985). There is a crucial difference between the analysis of bits and pieces *in* the landscape and the embracing of the landscape itself.

Second, landscape studies reflect the difficulties that geographers have created for themselves by their widespread adoption of morphogenetic methods allied with the mutual separation of physical and human geography in a misguided attempt to escape the charge of environmental determinism. In the British concentration upon sequentially evolving historical remnants in the landscape, the parallel between the denudation chronology precepts of a genetic geomorphology (see chapter II.2) and a historical geography based upon 'vertical themes' of temporal change was clearly evident in the 1950s (see Wooldridge and East, 1951). Paradoxically, this apparent methodological twinning of geomorphology and historical geography (Darby, 1953) actually served, in the main, to weaken and diminish the interaction of the two spheres of the discipline. It has fragmented rather than integrated landscape studies, divorcing them furthermore both from broadly conceived physical geography (free from domination by geomorphology) on the one hand and from the general traditions of human geography on the other.

In addition, geographical studies conducted within such a framework run the risk of becoming economic history; at worst, they lapse into mere narrative, concerning themselves with the evolution of individual elements at the expense of the geographical context, overemphasizing stages of development and, in logical terms, falling prey to historicism, the genetic fallacy and the confusion of sequence with consequence – the assumption that temporal succession is evidence of causal relation (Coones, 1979).

Third, landscape geography, in attempting to explain why landscapes are as they are, is of direct relevance with respect to certain fundamental problems experienced within geography as a whole. It provides an excellent demonstration of the principle that geography is an 'outlook' subject, whose enquiries are best directed by questions relating to specific problems and issues rather than being defined in terms of subject matter. In this way, geography retains its identity as a distinct, coherent and unified discipline, addressing itself to the interaction of communities and environments as manifested in the character of places. Specialization by subject matter, on the other hand, has led to fragmentation, the proliferation of branches studying the 'geographical aspects' of other subjects, and the view – widespread among members of the academic community and the general public alike – that geography is incapable of a meaningful independent existence.

Applications

A corollary of the interconnected propositions outlined above is that landscape studies are important not only for their own sake but also in their potential applications – conceptual, empirical and practical. It has already been suggested that landscape offers *one* possible bridge over the great divide, thereby making a contribution to the construction of an integrated geography. Such a geography does not claim to be the ultimate all-embracing synthesis; it has a definite, not an infinite, scope. It is concerned with the appraisal of communities within particular milieux; moulded around human values, natural resources and constraints, real and perceived worlds of the present and past, it is tied at every stage to the nature of place as the product of the interaction of societies and environments. Severe difficulties may well be encountered in the fulfilment of this demanding task, but geography is inherently a difficult subject.

Much of the empirical work conducted within the sphere of landscape studies (discussed further in chapter III.12) is of direct practical value. Applied landscape research has made its mark most obviously in landscape evaluation and related methods employed in the management and conservation of historic and valued landscapes and in planning generally (Appleton et al., 1975; Brandon and Millman, 1980). Britain has been bedevilled by piecemeal planning, myopic concentration upon particular features rather than the whole fabric of the landscape, and by insidious intrusions which progressively erode the quality of the visual scene. In order to plan wisely, integrated appraisals of landscapes need to be interpreted in connection with balanced assessments of the nature and speed of change and the conflicting pressures generated by modern society and economy, accompanied by an appreciation of the apparent intangibles of taste and perception.

A consideration of values is of the essence at every stage in such procedures. Who is to define a 'valued' landscape, and upon what grounds and by which criteria? What constitutes acceptable alteration or justifies preservation? Views on these matters are subject to fashion and vary with social class, standard of living and the political persuasion of individuals. At base they reflect the evolution of society itself. The controversy surrounding the 'heritage industry' (Hewison, 1987) is a case in point. Landscape is being treated increasingly, by post-industrial urban society, as a resource – an aesthetic and spiritual resource as well as a commodity which, in an age of decline, can be marketed as a celebration of former glory, status and achievement. Assessed in this way, it can be a valuable asset to a society, sections at least of which are affluent, leisured and mobile to an extent hitherto unknown. Perceptions of the past, of scenic beauty and of landscapes as embodiments of certain beliefs and nostalgic images can be moulded, promoted and exploited. In an era of alienation, the significance of landscape as a context, a means of escape or an indulgence is considerable. There exists a concomitant danger that landscapes will be conceived merely as subjectively selected fragments of the open air scene which, in satisfying such needs experientially, are deemed worth 'protecting'. In this way, as well as by reason of its philosophical associations, the term landscape becomes value laden to the point of being as much a 'visual ideology' (Cosgrove and Daniels, 1988) as a material object.

In view of modern society's potential for transforming the landscape more rapidly and comprehensively than ever before, the fundamental issues concerning the ethics, rationale and control of landscape change and conservation (Lowenthal and Binney, 1981; Penning-Rowsell and Lowenthal, 1986) are consequently very much to the fore. The geographer, able to gather together the diverse elements involved, is in a strong position to contribute an informed appraisal. Few subjects are so innately geographical in their content, significance and ramifications as the study of landscape, involving the physical and the human, the past and the present, the reality of the environment and the realm of ideas, and, not least, in this period of concern with 'relevance', the pure and the applied.

References

Appleton, J. 1975: *The Experience of Landscape.* London: Wiley.

——and others, 1975: Landscape evaluation. *Transactions of the Institute of British Geographers*, 66, 119–61.

Aston, M. 1983: The making of the English landscape – the next 25 years. *The Local Historian*, 15, 323–32.

Baker, A. R. H. and Harley, J. B. (eds) 1973: *Man Made the Land: essays in English historical geography.* Newton Abbot: David & Charles.

Berg, L. S. 1950: *Natural Regions of the U.S.S.R.*, trans. by O. A. Titelbaum and edited by J. A. Morrison and C. C. Nikiforoff. New York: Macmillan.

Brandon, P. and Millman, R. (eds) 1980: *Recording Historic Landscapes: principles and practice*, Occasional Publication. London: Polytechnic of North London, Department of Geography, Historic Landscapes Steering Group.

Broek, J. O. M. 1932: *The Santa Clara Valley, California: a study in landscape changes.* Utrecht: Oosthoek.

Buttimer, A. 1971: *Society and Milieu in the French Geographic Tradition.* Chicago, IL: Rand McNally for the Association of American Geographers.

Conzen, M. P. (ed.) 1990: *The Making of the American Landscape.* Boston, MA, and London: Unwin Hyman.

Coones, P. 1979: Manufacture in pre-industrial England: a bibliography. *Journal of Historical Geography*, 5, 127–55.

——1985: One landscape or many? A geographical perspective. *Landscape History*, 7, 5–12.

Cosgrove, D. and Daniels, S. (eds) 1988: *The Iconography of Landscape: essays on the symbolic representation, design and use of past environments.* Cambridge: Cambridge University Press.

Darby, H. C. 1951: The changing English landscape. *Geographical Journal*, 117, 377–94.

——1953: On the relations of geography and history. *Transactions of the Institute of British Geographers*, 19, 1–11.

English, P. W. and Mayfield, R. C. (eds) 1972: *Man, Space, and Environment: concepts in contemporary human geography.* New York and London: Oxford University Press.

Entrikin J. N. 1984: Caul O. Sauer, philosopher in spite of himself. *Geographical Review*, 74, 385–408.

Hartshorne, R. 1939: *The Nature of Geography: a critical survey of current thought in the light of the past.* Lancaster, PA: Association of American Geographers.

Hewison, R. 1987: *The Heritage Industry: Britain in a climate of decline.* London: Methuen.

Higbee, E. C. 1952: The three earths of New England. *Geographical Review*, 42, 425–38.

Hoskins, W. G. 1955: *The Making of the English Landscape*. London: Hodder & Stoughton.

Isachenko, A. G. 1968: Fifty years of Soviet landscape science. *Soviet Geography*, 9, 402–7.

——1974: On the so-called anthropogenic landscapes. *Soviet Geography*, 15, 467–75.

——1975: Landscape as a subject of human impact. *Soviet Geography*, 16, 631–43.

——1977: L. S. Berg's landscape-geographic ideas, their origins and their present significance. *Soviet Geography*, 18, 13–18.

Leighly, J. 1978: Carl Ortwin Sauer 1889–1975. In T. W. Freeman and P. Pinchemel (eds), *Geographers: biobibliographical studies*, vol. 2, London: Mansell, 99–108.

Leser, H. 1976: *Landschaftökologie*. Stuttgart: Ulmer Verlag.

Lowenthal, D. and Binney, M. (eds) 1981: *Our Past Before Us: why do we save it?* London: Temple Smith.

Meinig, D. W. 1979: Reading the landscape: an appreciation of W. G. Hoskins and J. B. Jackson. In D. W. Meinig (ed.), *The Interpretation of Ordinary Landscapes: geographical essays*, New York and Oxford: Oxford University Press, 195–244.

Mikesell, M. W. 1968: Landscape. In *International Encyclopedia of the Social Sciences*, vol. 8, New York: Crowell Collier & Macmillan, 575–80.

——1976: The rise and decline of 'sequent occupance': a chapter in the history of American geography. In D. Lowenthal and M. J. Bowden (eds), *Geographies of the Mind: essays in historical geosophy in honor of John Kirtland Wright*, New York: Oxford University Press, 149–69.

Penning-Rowsell, E. C. and Lowenthal, D. (eds) 1986: *Landscape Meanings and Values*. London: Allen & Unwin.

Roberts, B. K. 1987: *The Making of the English Village: a study in historical geography*. Harlow: Longman.

Sauer, C. O. 1925: The morphology of landscape. *University of California Publications in Geography*, 2, 19–53.

Solot, M. 1986: Carl Sauer and cultural evolution. *Annals of the Association of American Geographers*, 76, 508–20.

Wagner, P. L. and Mikesell, M. W. (eds) 1962: *Readings in Cultural Geography*. Chicago, IL, and London: University of Chicago Press.

Whittlesey, D. 1929: Sequent occupance. *Annals of the Association of American Geographers*, 19, 162–5.

Williams, M. 1983: 'The apple of my eye': Carl Sauer and historical geography. *Journal of Historical Geography*, 9, 1–28.

——1989: Historical geography and the concept of landscape. *Journal of Historical Geography*, 15, 92–104.

Wooldridge, S. W. and East, W. G. 1951: *The Spirit and Purpose of Geography*. London: Hutchinson.

II.8
The Developing World
Stuart Corbridge

Mention of a Third World was probably first made in 1952 by the French scholar Alfred Sauvy. Sauvy's intention was to signal to the ex-colonial nations a third way between the free-market capitalism of the American empire and the socialist logics of the Soviet world. It was not until the late 1950s that the Third World came to define a group of Latin American, African and Asian countries struggling to escape a similar morphology of backwardness (as seen, for example, in low levels of gross domestic product (GDP) per capita, literacy and life expectancy).

The concept of a Third World had some validity at least until the late 1960s. Although we should be wary of assuming that Ghana, India and Brazil were ever so similar as to take shape as archetypal 'Third World' nations, these and other countries did share a certain quality of 'Third-Worldness'. The economic base of each country remained largely dependent upon the rural sector and upon the production of primary commodities for the world market. Further, each country sought the modernization of its economy through various forms of import-substitution industrialization. Politically, too, the state in many Third World countries was placed in an uneasy and ambivalent relationship to the processes of development. On the one hand, it was expected to be an impartial progenitor of growth; on the other hand, it was often compromised by local proprietary groups (landowners, for instance), and by the agents and institutions of a pervasive neocolonialism.

In the 1990s a tripartite division of the world system seems less convincing. The annual Development Report of the World Bank now provides information on six groups of countries – low income economies, middle income economies, upper middle income economies, oil-exporting economies, industrial market economies and non-market economies – in the process calling to mind a fragmentation (a changing geography) of the contemporary world order. In the 1970s and 1980s certain Third World countries attempted to join the ranks of the core economic powers (with varying degrees of success: compare South Korea and Brazil) while others, most obviously in sub-Saharan Africa, saw levels of food production per capita dip below the levels reached in the late 1950s. The economic successes of some East Asian newly industrializing countries (NICs) found a mordant echo in the famine- and debt-ridden landscapes of the so-called Dark Continent. An apparent and welcome development of some 'Third World' nations must be seen in the context of a continuing rise in the total number of people living in conditions of absolute poverty and suffering from some form of protein–calorie malnutrition.

This essay seeks, firstly and briefly, to describe the main trends and mechanisms which lie behind the changing map of global development since 1945. It next provides an outline of four grand theories which have attempted to describe and understand these changes. Finally, it offers a prospectus for development geography. It considers how geographers are obliged to change their models and theories of development in the light of recent changes in the global political economy.

Integration and Fragmentation: Trends and Mechanisms

The world economy which emerged after the Second World War was a world economy designed to fall into four parts. A first part, called the Second – or socialist – World, has been left to one side in most accounts of the development process. This is perverse because almost half the population of the developing world were, by 1980, living in countries committed to a version of socialism. In countries as diverse as China and Cuba, Angola and Nicaragua, contacts with the capitalist world economy were carefully regulated and rapid strides were made in the provision of basic needs (see Forbes and Thrift, 1987). Two problems besetting such countries (apart from foreign intervention) have been the following: (a) falling rates of agricultural productivity (possibly linked to local incentive structures), and (b) an often poor record with respect to civil liberties and human rights (although this is not an indictment which applies only to the socialist world). The rest of the world economy was supposed to comprise the United States, which would export manufactured goods to a rebuilt Europe and Japan, and an emerging Third World which would import manufactured goods from Europe and Japan while exporting primary commodities to the First World. More recently, this strict spatial ordering has come undone. West Germany and Japan have each come to challenge the economic might of the United States, and parts of the Third World have become major sites of manufacturing industry. The talk now is of a new international division of labour (Harris, 1986; Thrift, 1989).

The mechanisms which lie behind this tendency to a parallel integration and fragmentation of the world economy are well known. For convenience we can discuss an economic mechanism, a technological mechanism and a political mechanism, although in practice these motive forces overlap.

The economic mechanism is often referred to as the transnationalization or globalization of *capital*. The term capital is useful here because it points to a greater internationalization of economic activity than manufacturing alone; the internationalization of financial and other services is increasingly important. In simple terms we can point to three moments of the process of globalization. A first moment refers to the relocation to Europe and Japan of US industrial capital in the 1950s and thereafter. This moment was associated with the consolidation of US hegemony in the central regions of Fordism (see chapter II.9) and with the parallel emergence of international institutions and regimes to guard an open international economy (think of the North Atlantic Treaty Organization, the World Bank, the International Monetary Fund and the General Agreement on Tariffs and Trade). A second moment refers to the relocation to parts of the Third World of US, European and Japanese industrial and financial capital in the 1960s and thereafter. The East Asian NICs and parts of Latin America were especially favoured sites for the assembly operations of the modern transnational corporation (TNC) (plate II.8.1). A third moment refers to the emergence and diffusion of global corporations from within the Third World. Large South and East Asian corporations are today major players on a world stage. They signal the formation of a much more complex map of global industrial spaces than that which prevailed forty years ago (Lipietz, 1987; Jenkins, 1988).

The globalization of capital in turn

Plate II.8.1 In certain parts of the Third World, especially in south east Asia, factories employ large numbers of females on assembly lines for electrical components that will be supplied to the developed world.

entails, and is made possible by, a series of recent technological developments (Castells, 1989). Put simply, we live in a shrinking world. The emergence of jet passenger planes and telex facilities, of the microcomputer and the fax machine, means that our spatial and temporal horizons are drawing close as never before. David Harvey refers to this as a continuing process of time–space compression (Harvey, 1989). Others talk of the annihilation of space by time. However we refer to it, it is clear that the development of new technologies has greatly expanded the global reach of the modern corporation. Although places do still compete with one another to receive and service the TNCs, transportation costs now feature less prominently in the locational strategies of the giant corporation. In the world of the 1990s fictitious capitals (for example, credit monies) can be transferred around the world (often through fictitious spaces, as we might style some centres of offshore banking) through continuous 24 hour markets and exchanges. The meeting place for the modern world is more than ever the computer terminal and its video screen (Strange, 1986).

None of this means that states and development policies no longer matter. It is probably the case that states in our interdependent world have less autonomy than once was the case. Nevertheless, even in the developing world, where states are often weak because of a heritage of colonialism (and a legacy of unhelpful territorial boundaries), local politics can make a difference. The differential development of East Asia and South Asia, or of countries within East Asia, cannot fully be explained by reference to the strategic and locational needs of global capital. Low wages may be attractive to a TNC escaping the strong labour movements of Western Europe, but low wages prevail throughout the developing world. TNCs also look for stable educated workforces and for local markets – qualities which vary from one developing country to another and which may be influenced by local policy initiatives. Equally, local states, and local structures of ethnicity and class, will determine in part the sorts of development policies which a country is free to choose (Watts, 1984; Becker and Sklar, 1987). The fact that land reform ground to a halt in India in the 1950s – making a later Green Revolution inevitable – has everything to do with the power of India's wealthier farmers in that country's legislative assemblies (Rudolph and Rudolph, 1987).

Grand Theories

The effects of all these changes are etched firmly into the landscapes of our fragmented yet interdependent world. At one level they are visible in the development escalator which cuts across a sharp north–south divide. Between 1965 and 1987, South Korea and Taiwan each recorded annual average rates of growth of gross national product of 7 per cent or more, and annual average rates of growth of merchandise exports in excess of 18 per cent. These rates of growth are historically unprecedented and should not lightly be dismissed. Collectively, too, the so-called Third World is now supplying 20 per cent of the world's manufacturing capacity. This compares with less than 12 per cent in 1960.

But what are we to make of these changes? How are we to explain the fragmentation–integration of the modern world system? How do we account for the successes of some NICs, while recognizing the continuing dreadful levels of poverty and underdevelopment which mark out a much larger Third World? Can we even speak of 'success' and 'development'? There is space here only to sketch an outline of an answer to some of these questions. It is perhaps more important to recognize that such questions continue to excite great debates within the academic and policy-making communities, and that these debates continue to be informed by corresponding 'grand visions'. We might also note that these ideas and visions change. Just as the world economy has fragmented and become less geographically certain in recent times, so has academic and practical opinion become less convinced by any one grand vision. This is not to say that grand theory has no place today (there can, after all, be grand theories of the local); it is rather to suggest that the two grand theories which now dominate are substantially different from the two dominant grand theories of the 1950s and 1960s. A brief paragraph on each theory will make this clear (for more detail, see Toye, 1987; Corbridge, 1988). (Needless to say, these theories are not exhaustive. The greening of development theory is not discussed here, nor is due attention given to critical visions which continue to emerge within the non-Western world: see Hettne, 1990.)

In the 1950s and 1960s two schools of thought dominated the theoretical literatures on economic development: a school of modernization theorists (including Rostow, McClelland and Friedman) and a school of *dependencia* theorists (which in geography drew disproportionately from the zero-sum typologies of Gunder Frank and others writing about the *development of underdevelopment*).

Simplifying greatly, we can say that the modernization paradigm was committed to three sets of propositions: (a) that all countries lie on a continuum between an original state of tradition (typified by rural bias, low rates of saving, high death rates, low levels of GDP per capita etc.) and an end state of modernity (its opposite); (b) that traditional 'late-comers' could learn from modern 'pioneers' and could take from the latter the skills and capital necessary in the short run to promote a national big push or take-off into self-sustaining growth; and (c) that this orbital imagery of rapid development in turn entailed an interventionist state to channel scarce resources into import-substituting industrialization. In sum, the developing countries are urged to copy the developed world into an age of global modernity. The First World in turn makes available to the Third World foreign direct investment and foreign aid (see Chenery et al., 1986).

Set against modernization theories are the more pessimistic accounts of Frank and his school. In their judgement, the countries at the periphery of the capitalist world system are condemned to a continuing state of underdevelopment because the continued development of the core countries demands a continual transfer of wealth from the periphery. This process of exploitation is said to have begun in the fifteenth and sixteenth centuries in Latin America, and continues today through the conduits of international trade and the operations of the TNCs. By this logic, peripheral countries will only develop and industrialize on the basis of first severing their links to the developed world. (It should be noted that Frank's work of the 1960s is not representative of a wider *dependencia* school. Sunkel, Cardoso and others accept that the global political economy is asymmetrical – that Third World development is dependent upon the actions and resources of others – but they are also sensitive to the possibility of contesting global structures of exploitation by local political actions (see Kay, 1989).)

In hindsight, we can see that each of these schools of thought was responding to a particular history and geography. They both spoke to a world system divided into a clear core and a clear periphery. Either industrialization would come rapidly to all modernizing countries (Rostow), or it would fail to take root at all (Frank). More recent theories continue to be divided as to the effects of capitalism in the periphery, but they are also more attentive to the fact that the effects of globalization are worked out in different ways in different places. Some are also more sceptical of the claims of growth-centred industrial development policies.

On the political Right a curious heir to Rostow has emerged in the form of the new classical economics. The corridors of power in Washington (home to the International Monetary Fund and the World Bank) now beat to the drum of the market as a development panacea. As the *Economist* put it in September 1989, 'what the Third World needs most [*pace* Rostow] is less government'. A new counter-revolution in development theory and policy is urging developing countries: (a) to service the global market by maximizing a local comparative advantage (in, say, cocoa or textiles), and (b) to complement a growing external orientation of the economy by 'getting prices right' at home (e.g. by abolishing the marketing boards and urban food subsidies which some see as the main barrier to agricultural development in sub-Saharan Africa). In the case of the debt crisis the language of the counter-revolution is the language of structural adjustment. Indebted nations must adjust their overprotected stagnant economies to the chill wind of an allocatively efficient market (Corbridge, 1992).

This over-optimistic vision of market-led development in turn is opposed by the (new) Left. Mindful of the recent industrialization of some developing countries (*pace* Frank), a number of Left-inclined scholars now argue: (a) that most Third World countries are untouched by the blessings of globalization, and indeed suffer depradations and distortions because of unequal contacts in the market-place (Ross, 1983), and (b) that even in the NICs the blessings of capital are uneven and exaggerated. A distinction must be drawn between the more or less inclusionary, more or less autonomous development of the industrial economies and a local process of growth without development which excludes a majority of the population from those benefits which are on offer, and which depends largely upon the exploitation of cheap female labour. More positively, some Left scholars are combining with proponents of appropriate technology, sustainable development and indigenous agricultural revolutions to promote development strategies which seek to put the last first. Local needs must take precedence over global demands (see Chambers, 1983; Hecht and Cockburn, 1989).

Conclusion and Prospectus

This is not the place to begin an adjudication of these competing claims and visions, not least because such an adjudication would have to second-guess the results of a programme of research which is still evolving. A more appropriate set of conclusions is as follows: (a) the map of the world economy is fast changing; (b) these changes are related to the economic, technological and political effects of the processes of globalization; (c) the spatial extent of these processes can be exaggerated (*pace* the optimism of the New Right); and (d) the effects of globalization are mediated in particular places by local structures of class, ethnicity, territory and gender. The 'Third World' (as unitary space) may indeed be at an end; within the spaces of the Third World, however, local problems of poverty and dispossession continue to worsen and to haunt the imagination.

These conclusions in turn suggest a research agenda for present and future development studies. They suggest, first, that geography does matter: that a central task of new development studies is to provide an account of the increasing time–space compression which binds places together, while recognizing that the map of the world thus disclosed is always rewritten and reshaped according to local circumstances (of culture, gender, politics, ecology and demography). In more detail, we can see that a modern development geography will continue to form around a linked series of issues and debates of which the following list is by no means exhaustive.

A 'first' task is to understand the constitution and dynamics of a changing world system, and to pay close attention to its varied modes of accumulation and crisis formation in time and space. Already, there is work in this field: e.g. David Harvey's work (1982) on *The Limits to Capital*, Peter Dicken's work (1986) on industrial change in a turbulent world and the collection of essays on *A World in Crisis?* edited by Ron Johnston and Peter Taylor (1989). Of pressing importance here is the need for geographers to attend to the ecological crisis which threatens the world (as indeed they are: see Odell, 1989; Adams, 1990).

A second task is to examine what might be called the *spatiality* of the development process, or the joint production of places by forces which are at once local and supra-local. Again, there is progress to report: e.g. the work of Warwick Armstrong and Terry McGee (1985) on the production of particular urban geographies as theatres of accumulation or the work of Nanda Shrestha (1988) on the dynamics and consequences of migration in peripheral areas of South Asia.

A third task is to attend to the varied conditions of existence of national and international capital accumulation, i.e. to the extraordinarily rich and vital roles played by class and culture, gender and ethnicity, geopolitics and ecology, in making possible, or transforming, local maps of development. This third task promises to be a fecund area of geographical enquiry, and already it is presenting a set of issues which undergraduate geographers can begin to explore in the course of their dissertations. Here is a chance to examine, for example, a local political economy of soil erosion (see Blaikie, 1985), or the role of racist ideologies in the production of policed urban environments (Robinson, 1990). Here is a chance to study, as did Judith Carney, the effects of contract farming schemes on local systems of gender relations in West Africa (Carney, 1988), or to investigate the interdependences which exist in parts of Bangladesh between a disposition to have large numbers of children and local structures of environmental and social risk (Cain, 1986).

A fourth task is to describe better those patterns of political and cultural mobilization through which people in the periphery seek to understand and contest the structures of 'modernity' into which they are drawn. Once again, geographers are producing important pieces of research in this area. Jonathan Crush has written an excellent monograph on the battle for Swazi labour at the turn of the century (Crush, 1988); Michael Watts has looked into struggles over land, and struggles over meaning, in the Gambia (Watts, 1988); and Ramachandra Guha (not a professional geographer, but a scholar with a geographical imagination) has written forcefully about the *Chipko* movements of Himalayan India (where women in particular have hugged trees to protect their local environments against the activities of the logging companies (Guha,1989)).

A fifth task is to understand the role of the state in the Third World: to examine its (varied) constitution and territorial powers, and to investigate its capacities to oppress, and to mobilize and transfer resources. Again, there is work in progress, e.g. the work of Barbara Harriss on the merchant state in South India (Harriss, 1984) and the work of Carole Rakodi on the urban local state in Africa (Rakodi, 1986). At a wider spatial scale, John O'Loughlin has written to great effect on world-power competition and local political conflicts in the Third World – a disturbing account of the global geography of war and peace (O'Loughlin, 1989).

A sixth task is to investigate the territorial, economic and political possibilities and contradictions of socialist development strategies. What is there to be said for socialism in the 1990s? Can regional inequalities be greatly reduced in socialist developing societies, and if so at what cost? For one answer, see the work of David Slater (1982, 1989). What, too, can we say about food policies within different socialist societies, or about local patterns of employment, wealth and empowerment? Some interesting, if often rather different, answers can be found in the work of the geographer Doreen Massey (1987) and the humanist Germaine Greer (1988).

A final task is to connect an analysis of the provision of welfare in the periphery with theories of justice and spatial equity. Again, a start has been made in the work of Morag Bell (1988) and Peter Ward (1986), to name but two scholars.

By writing about such issues and debates, development geographers will continue to affirm the vitality of their subdiscipline, while exemplifying also a fine awareness of the connections between development theory and development practice. Here is a second message of this chapter. Nobody wants development geography to become part of an academic ghetto in which armchair practitioners write their papers blissfully cut off from the pressing problems of hunger and illiteracy, oppression and poor housing, which still face so many people. Nevertheless, a simple dichotomy between development theory and practice is not easily sustained. A decision on whether or not to give foreign aid to a country, and in which form and to whom, depends in part upon a series of prior 'academic' judgements. For some dependency scholars, aid is a form of imperialism; for some scholars on the radical right, aid is a prop for bureaucracy and an impediment to enterprise. The same point can be made of policies and practices with respect to the provision of employment, or land reform, or transport: how can we intervene, effectively, if we do not first describe and understand, debate and criticize? Debates within feminism may seem arcane to some students of development, but it is precisely debates such as these which have forced the major development agencies (including the World Bank) slowly to recognize that all development initiatives are gendered and that such initiatives can hold very different implications for women than for men.

To sum up: development geography should resist a seductive polarity between doing and knowing. It must, instead, re-tie the Gordian knot which binds theory to practice in the context of specific places. In and through its efforts, development geography can then continue to contribute to the vital intellectual, moral and practical debates which define the field of development studies. Therein lies the challenge.

References

Adams, W. 1990: *Green Development: environment and sustainability in the Third World.* London: Routledge.

Armstrong, W. and McGee, T. 1985: *Theatres of Accumulation: studies in Asian and Latin American urbanisation.* London: Methuen.

Becker, D. and Sklar, L. (eds) 1987: *Postimperialism: international capitalism and development in the twentieth century.* Boulder, CO: Rienner.

Bell, M. 1988: Welfare, culture and environment. In M. Pacione (ed.), *The Geography of the Third World: progress and prospect,* London: Routledge, 198–231.

Blaikie, P. 1985: *The Political Economy of Soil Erosion in Developing Countries.* London: Longman.

Cain, M. 1986: The consequences of reproductive failure: dependence, mobility and mortality among the elderly of rural South Asia. *Population Studies,* 40, 375–88.

Carney, J. 1988: Struggles over crop rights and labour within contract farming households in a Gambian irrigated rice project. *Journal of Peasant Studies,* 15, 334–49.

Castells, M. 1989: *The Informational City.* Oxford: Basil Blackwell.

Chambers, R. 1983: *Rural Development: putting the last first.* London: Longman.

Chenery, H., Robinson, S. and Syrquin, M. 1986: *Industrialization and Growth.* Oxford: Oxford University Press and the World Bank.

Corbridge, S. 1988: The 'Third World' in global context. In M. Pacione (ed.), *The Geography of the Third World: progress and prospect,* London: Routledge, 29–76.

——1992: Discipline and punish: the New Right and the policing of the international debt crisis. In S. Riley (ed.), *The Politics of Global Debt,* London: Macmillan.

Crush, J. 1988: *The Struggle for Swazi Labour, 1890–1920.* Kingston: McGill-Queens University Press.

Dicken, P. 1986: *Global Shift: industrial change in a turbulent world.* London: Harper & Row.

Economist 1989: A survey of the Third World. *Economist,* September, 23–9.

Forbes, D. and Thrift, N. (eds) 1987: *The Socialist Third World: urban development and regional planning.* Oxford: Basil Blackwell.

Greer, G. 1988: *The Madwoman's Underclothes.* London: Picador.

Guha, R. 1989: *The Unquiet Woods: ecological change and peasant resistance in the Himalayas.* Delhi: Oxford University Press.

Harris, N. 1986: *The End of the Third World: the newly industrialized countries and the end of an ideology.* Harmondsworth: Penguin.

Harriss, B. 1984: *State and Market.* Delhi: Concept.

Harvey, D. 1989: *The Condition of Postmodernity: an inquiry into the origins of cultural change.* Oxford: Basil Blackwell.

——1982: *The Limits to Capital.* Oxford: Basil Blackwell.

Hecht, S. and Cockburn, A. 1989: *The Fate of the Forest: developers, defenders and destroyers of the Amazon.* London: Verso.

Hettne, B. 1990: *Development Theory and the Three Worlds.* London: Longman.

Jenkins, R. 1988: *Transnational Corporations and Uneven Development.* London: Methuen.

Johnston, R. and Taylor, P. (eds) 1989: *A World in Crisis? Geographical Perspectives,* 2nd edn. Oxford: Basil Blackwell.

Kay, C. 1989: *Latin American Theories of Development and Underdevelopment.* London: Methuen.
Lipietz, A. 1987: *Mirages and Miracles: the crises of global Fordism.* London: Verso.
Massey, D. 1987: *Nicaragua.* Milton Keynes: Open University Press.
Odell, P. 1989: Draining the world of energy. In R. Johnston and P. Taylor (eds), *A World in Crisis? Geographical Perspectives,* 2nd edn. Oxford: Basil Blackwell, 79–100.
O'Loughlin, J. 1989: World-power competition and local conflicts in the Third World. In R. Johnston and P. Taylor (eds), *A World in Crisis? Geographical Perspectives,* 2nd edn. Oxford: Basil Blackwell, 289–332.
Rakodi, C. 1986: State and class in Africa. *Environment and Planning D: Society and Space,* 4, 419–46.
Robinson, J. 1990: 'A perfect system of control'. State power and 'native locations' in South Africa. *Environment and Planning D: Society and Space,* 8, 135–62.
Ross, R. 1983: Facing Leviathan: public policy and global capitalism. *Economic Geography,* 59, 144–60.
Rudolph, L. and Rudolph, S. H. 1987: *The Political Economy of Lakshmi.* Chicago, IL: University of Chicago Press.
Shrestha, N. 1988: A structural perspective on labour migration in underdeveloped countries. *Progress in Human Geography,* 12, 179–207.
Slater, D. 1982: State and territory in post-revolutionary Cuba. *International Journal of Urban and Regional Research,* 6, 1–33.
——1989: Peripheral capitalism and the regional problematic. In R. Peet and N. Thrift (eds), *New Models in Geography,* vol. 2, London: Unwin Hyman, 267–94.
Strange, S. 1986: *Casino Capitalism.* Oxford: Basil Blackwell.
Thrift, N. 1989: The geography of international economic disorder. In R. Johnston and P. Taylor (eds), *A World in Crisis? Geographical Perspectives,* 2nd edn. Oxford: Basil Blackwell, 16–78.
Toye, J. 1987: *Dilemmas of Development: reflections on the counterrevolution in development theory and policy.* Oxford: Basil Blackwell.
Ward, P. 1986: *Welfare Politics in Mexico: papering over the cracks.* London: Allen & Unwin.
Watts, M. 1984: State, oil and accumulation: from boom to crisis. *Environment and Planning D: Society and Space,* 2, 403–28.
——1988: Struggles over land, struggles over meaning: some thoughts on naming, peasant resistance and the politics of place. In R. Golledge and P. Gould (eds), *A Search for Common Ground,* Goleta, CA: Santa Barbara Geographical Press, 31–50.

II.9

Economic Geography in the 1990s: The Perplexing Geography of Uneven Redevelopment

Erik Swyngedouw

Towards a New 'Mosaic of Uneven Development'

Economic geography has undergone a dramatic sea change over the last two decades. Until the late 1960s and early 1970s, economic geography was mainly concerned with either location models based on spatial extensions of neoclassical economic interpretations (see Isard, 1960) or with descriptions of the spatial variation of economic activities. Variants of centre/periphery analyses (Myrdal, 1959; Perroux, 1961; Friedmann, 1966) tried to capture the dynamics of persistent geographically uneven development. Nevertheless, they traditionally looked at uneven development as fundamentally given and immutable, and failed to account for dramatic shifts in the 'mosaic of uneven development' (Walker and Storper, 1981) that constitutes the geography of the world economy. Some neo-Marxist interpretations of unequal exchange and uneven development (Emmanuel, 1972; Amin, 1976), the international spatial division of labour (Fröbel et al., 1980) or dependent development (Frank, 1972; Wallerstein 1974) equally assumed a particular historical–geographical structure of centre/periphery relations to be virtually constant over time (see chapter II.8).

Such rather static interpretations of the organization of space prove to be extremely difficult to maintain under the present conditions of international and national patterns of development. Indeed, spatial economic relations have been radically turned upside down over the last few decades. On a world scale, the Pacific Rim has considerably improved its power in the world system at the expense of Europe and, particularly, the United States. The latter has lost its hegemonic economic and political position of dominance over the (Western) world economy. The post-war 'Pax Americana', which structured most of the world's political-economic space for almost half a century, eroded away along with the declining importance of the dollar as the world currency and the falling competitive position of American industry. Changes in Eastern Europe promise a radical reshuffling of European political-economic space. In sum, the geopolitics of the world map are shaken in a way which fundamentally affects the pattern of uneven development on a world scale.

On the scale of nations and regions, similar dramatic shifts in the spatial structure of the economy have altered the map of spatial development. This is easily observable in the larger countries. The US snowbelt/sunbelt shift is a case in point (Sawers and Tabb, 1984). In Europe as well, equally radical changes in the pattern of uneven development can be observed. The remarkable growth of the Third Italy – the artisan-based flexible production districts in the Veneto, Emilia-Romagna and Tuscany regions – which took over the lead role in economic development from the 'golden triangle' of Milan–Genoa–Turin shook the geographer's trust in traditional interpretations of core and periphery (Scott, 1988). The Ruhr lost its economic predominance in the economic geography of Germany in favour of the high technology and advanced services production complexes of Bavaria (Brake, 1986). In the United Kingdom, the southeast became the core and prospering region of the country while the Midlands struggled for survival in the light of rampant capital flight and massive de-industrialization processes (Massey and Meegan, 1982).

The London Docklands development, the revitalization of inner-city areas or the regenerated motion picture industry district of Los Angeles illustrate that the reversal of spatial processes of growth and decline also takes place at the micro-geographical level (Storper and Christopherson, 1987).

In short, it seems as if the last decade has seen the map of economic decline and growth radically altered (Massey and Allen, 1988). Traditional spatial hierarchies are turned upside down and a new landscape of production emerges out of the restructuring processes that took place in the aftermath of the crisis of the 1970s. This process took place on a variety of geographical scales, which hints at potential theoretical connections that have to be made between the historical geography of spatial relationships on a world scale and on the national/regional level on the one hand and the wider political-economic structure on the other.

Economic-geographic Transformations

The onset of the great economic crisis after 1972, which had devastating effects on the economic fortunes of most regions while others forcefully resisted economic decline, began to preoccupy the research endeavours of economic geographers.

The restructuring during the 1970s and, in particular, during the 1980s (see Massey, 1984; Cooke, 1990) not only resulted in massive de-industrialization in a number of regions but was accompanied by the rise of entirely new industrial sectors (microelectronics, biotechnology, informatics), the growth of service industries and the mushrooming of new spaces of production and consumption in other areas. The wave of technological innovations that coincided with this restructuring drew attention to the geographical dynamics of technological and industrial-organizational change and innovation (Hall and Markusen, 1983; Castells, 1985, 1989).

The transformations in the spatial organization of most advanced capitalist countries reflect a more general tendency. These changes can be briefly summarized as follows (Moulaert and Swyngedouw, 1989).

1 The formation of spatial clusters of high technology complexes, or 'territorial innovation complexes', often in areas with no or little industrial tradition (Stöhr, 1986a, b): the M4 corridor in England (Breheny et al., 1983), Silicon Glen in Scotland (Haug, 1986) or Silicon Valley in California (Saxenian, 1983) are cases in point (plate II.9.1).
2 New methods of industrial organization: these can take the form of vertically quasi-integrated production complexes combined with new forms of organization of the production process such as just-in-time manufacturing (Swyngedouw, 1987a; Sayer, 1989). This organizational structure is based on the maximization of external relations through subcontracting and a minimization of transaction costs (Scott, 1988), i.e. of the costs associated with handling and organizing the flow of commodities. Benetton's

Plate II.9.1 Among the most important of the new, high technology industrial centres is Silicon Valley in California. This aerial view shows silicon chip factories near Palo Alto.

hollow corporation is the most extreme form of such organization. While key functions of the production process are performed in-house (research and development, product design and development, marketing), most actual production is contracted out to semi-independent small and medium-sized firms. This leads to new forms of inter-firm collaboration and the formation of strategic alliances (Cooke, 1988). These forms of industrial organization are associated with qualitative changes in the technical, social and spatial division of labour.

3. A considerable rate of growth in terms of output, employment or exports in a limited number of key industrial sectors – moreover, this growth performance is concentrated in only a few areas (Markusen et al., 1986).
4. The rise and consolidation of flexible artisan-based production districts: the clothing, mechanical engineering or specialized food production areas in the Third Italy or the spectacular rejuvenation of the Swiss watch industry (SWATCH!) along the French border have shown unprecedented growth rates and remarkable resistance against crisis tendencies (Piore and Sabel, 1984; Swyngedouw, 1987b; Scott, 1988; Storper and Scott, 1988).
5. The emergence of a new socio-organizational structure ranging from low-skilled unstable bottom-end activities, such as in catering and maintenance, personal services, retail, and low touch administrative activities, to highly qualified 'high-tech' services and manufacturing (business services, consulting, finance and insurance etc.) (Bluestone and Harrison, 1988). This coincides with a new social and spatial stratification structure characterized by a wide polarization and growing inequalities (Harrison, 1984; Swyngedouw and Anderson, 1987).

These observed changes coincide with major transformations in the economic, social and institutional organization of Western society, a process which started during the turbulent crisis years of the 1970s and has accelerated since the early 1980s (Harvey, 1989; Albrechts and Swyngedouw, 1989). In what follows, an attempt will be made to theorize about these transformations.

The Debate on Economic Crisis and Growth: From Fordist to Flexible Accumulation

If we look back to the geographical dynamics of economic growth over the last few decades, it is startling to see how the same political-economic system which gave the Western world (or, at least, large parts of it) unprecedented wealth and prosperity has plunged into a global crisis since the mid-1970s. The 'golden sixties' were succeeded by an era of global decline, characterized by a perplexing geographical reshuffling and spatial complexity. The key problem, then, is to try to elucidate the spatial dynamics which are part of this global process of restructuring.

Socio-economic development can be looked at as a historical succession of specific 'ensembles of productive forces and relations' (Scott and Storper, 1986). This refers to the interpretation of long-term economic growth as being structured around a set of systemic technologies and organizational forms, each of which is accompanied by a distinct spatial pattern. This can be exemplified by the textile age of the early nineteenth century, heavy industry around the turn of the century and standardized mass production of household durables (electronics, automobiles, housing) since the 1930s. The particular forms of firm organization, logistical and process organization, competition and so on constitute a 'regime of accumulation'. This is nothing more than a particular set of economic characteristics which dominate a given historical epoch.

But there is more. Any given stage of capitalism is characterized not only by a series of dominant economic characteristics but also by a particular 'mode of social regulation' or, in other words, by a particular dominant way of organizing capital–labour relationships, by a system of distribution of surplus over the various layers of society, by a particular form of state intervention and by a certain configuration of the world system. The combination, then, of a paradigmatic economic system and a particular set of institutional–regulatory mechanisms may result in a 'mode of development'. The latter concept refers to a specific and concrete way of organization and change in the political-economic system which allows for a more or less sustainable form of development (Lipietz, 1986; Leborgne and Lipietz, 1988; Boyer, 1989).

If we look at the most recent period, from 1930 until today, we can see the rise, demise and restructuring of such a particular mode of development. This mode of development is labelled 'Fordism'. Indeed, the period 1930–72 was dominated by a particular economic form: global vertically integrated firms producing standardized commodities on a mass scale. The automobile and electronics sectors exemplify this trend. At the same time, the institutional framework changed. In particular, the capital–labour relationship – the wage nexus – under rising pressure of the labour movement, took a new form. Wages were negotiated on a collective basis and rose in real terms, more or less in line with changes in productivity. The state itself started to intervene directly in the economic sphere through state investments (among others in infrastructure, defence and so on) on the one hand and demand regulation (through regulation of the credit system, a series of subsidies and incentives and so on) on the other. Moreover, the rise of the welfare state (characterized by a series of redistribution policies such as health insurance, pension schemes, unemployment benefits etc.) enabled the steady expansion of consumer demand for most social groups. The financial system was heavily regulated, both on a national and an international scale, among other ways through the Bretton Woods Agreement which guaranteed the gold convertibility of the US dollar and the building of a series of international financial institutions such as the International Monetary Fund and the World Bank.

This system generated a virtuous spiral of growth. Productivity increases enabled a continuous expansion of industrial output at falling costs per unit while real wage increases together with welfare state policies guaranteed a growing market to absorb the output (Aglietta, 1979; Coriat, 1979; Lipietz, 1986; Schoenberger, 1988).

This complex structure of capitalist production and regulation actually produced a specific and concrete form of spatial coherence in the regions where it was concentrated. First, a specific physical form, i.e. large manufacturing plants functionally and spatially separated from housing, recreation and commercial spaces, long-distance transportation networks to minimize the production and marketing time etc., was created and cemented into a set of zoning and other physical planning arrangements. The development of this mode of socio-economic production resulted in a quite specific built-up environment and spatial form (Walker, 1981; Florida and Feldman, 1988). Suburbanization, both of housing and industry, exemplifies this new structuring of space. The development of Detroit as 'Automobile City' (both in production and in consumption) in the United States and the industrial belt in the English Midlands still reflect such particular built-up forms. Moreover, this structure of the built environment was accompanied by the formation of a corresponding social fabric, characterized by, for example, the growth of negotiating powers of the labour movement, the dominance of left-wing or social-democratic political parties and increasing labour force participation rates (especially by women). Simultaneously, new consumption patterns emerged on the level both of the household (housing, consumer durables) and of the state (Keynesian demand management combined with redistributive welfare policies) (Lash and Urry, 1987).

In the context of such a construction of territorial coherence produced over a relatively long period of time but resulting in a relatively fixed spatial structure, capitalist firms operating in a competitive environment, have continuously to find new ways to avoid a falling profitability or to escape problems of overproduction (Harvey, 1985).

During an expansionist phase, capitalist firms can basically trade off space and technology in their strategies to overcome these problems (Harvey, 1982). They can take the route of technological change and thus increase their productivity in existing facilities or they may decide to relocate (part of) the production process spatially in search of either new markets or areas where the production cost (labour, resources) is lower or both. In reality, these two strategies, i.e. technological change versus spatial rearrangements, often take place simultaneously. For example, while colour television sets were introduced in the United States in the 1950s, the mass production of standard black and white sets was increasingly relocated to low wage areas in Asia or Latin America.

However, on the one hand, the scope of technological improvements, implemented within the context of a territorial coherence as described above, is limited by the historically produced physical, social, political and ideological rigidities of the local milieu. Put simply, the historically produced socio-spatial conditions cannot, without massive economic and social cost, be restructured overnight to adopt to the new standards and exigencies of profitable production. Moreover, the 'spatial fix' to the accumulation problem, on the other hand, produces new spatial rigidities. In fact, the very mechanism of using space to overcome growth problems produces a new fixed structure (e.g. transportation, communication, production and market networks), which, in turn, prevents the introduction of radically different (technological or organizational) changes.

Moreover, the spatial solution to gain a competitive edge is limited by the physical and social limits of space itself. The physical limit of space is self-explanatory. The socio-economic limits of space refer to the following factors, among others: (a) the limits of spatial market expansion (in a given system of distribution of income); (b) the limits of exploiting spatial variations of labour conditions, i.e. a firm in a market system loses the competitive edge gained by exploiting low wage areas at the moment that its competitors use the same strategy; (c) the limits of moving across space, which means that the time–cost compression of the barrier of distance is determined by the technical limits set by particular infrastructural forms of transportation and communication; (d) geopolitical limits, which are illustrated by the fact that, throughout the post-war period (until very recently), the communist world was literally closed off for penetration by Western capital.

In short, at the moment that the global world economy is integrated in a particular international spatial division of labour and given the form of institutional regulation and geopolitical organization, the possibilities of an extensive spatial fix become (temporarily) exhausted and a period of massive intensive technological–spatial transformation sets in.

It is exactly such a process that characterized the post-war development both within advanced capitalist countries and on an international scale. During this period, we actually observed a deepening spatial division of labour between core and peripheral areas on a national level and between advanced countries and less developed countries on an international scale. US firms were among the first to use direct foreign investment strategies in search of new markets and/or areas which permitted lower cost production, but were soon followed by European firms. This strategy actually became dominant in the 1960s and grew rapidly in importance. The spatial solution sought in low wage countries, however, was very limited in scope. The limitations of their internal market and the absolute failure to install a Fordist labour relation in the world's periphery reduced these locations to places for low cost production, while the commodities produced were exported to be sold on the high income markets of the advanced countries. This resulted in a major 'spatial' contradiction, i.e. the spatial division of labour threatened profitability in plants located in advanced countries resulting in a downward pressure on wages, restructuring, de-industrialization and unemployment. Demand for new goods became saturated while new markets in the periphery hardly grew because of the ruthless exploitation of the local labour force (Swyngedouw, 1989).

Indeed, by the end of the 1960s, the organizational, spatial and technical limitations of the Fordist mode of development and the countervailing practices of institutionalized regulation, along with the inherent tendencies towards over-accumulation and for the rate of profit to fall, threatened this accumulation regime in its existence. Investment levels started to drop, productivity growth slowed down and only a few years later a structural crisis became apparent. On the international scene, the intensified global competition resulted in the breakdown of the fixed exchange rates of the international financial system (the Bretton Woods Agreement). The subsequent volatility of the international financial markets with their rapidly changing exchange rates made offshore production extremely speculative, while the rebirth of protective trade practices and the successive oil shocks accelerated the process of crisis formation. At the national level, industrial restructuring processes accelerated, and unemployment and creeping inflation rates started to leap-frog up to a two-digit level. At the same time, traditional Keynesian policies of debt financing were jeopardized by the growing debt crisis of most nation-states (itself the result of the internationalization of Fordism and concurrent undermining of the industrial base and the competitive position of advanced nations), forcing them gradually toward austerity policies (Swyngedouw, 1990).

The effects of Fordist crisis formation in terms of regional restructuring are by now relatively well known: massive unemployment and social disintegration in old industrial regions, fiscal crisis of the cities and rampant de-industrialization. Traditional industries had to restructure or close down, deserting entire regions (Carney et al., 1981; Massey and Meegan, 1982). The decline of the American Frostbelt, the economic disintegration of the British north and Midlands, and the capital flight from the industrial areas in north France are all sorry reminders of this process.

The trade-off between space and technology, characteristic of the post-war development, was no longer possible and had to take dramatically new forms. Indeed, the weakened effect of the spatial solution within the Fordist paradigm necessitated a return to dramatic technological changes and innovations and to an intensive spatial 'fix', a fundamental restructuring of the geographical basis of development. It is no coincidence, therefore, that the subsequent period was characterized by rapid and revolutionary technological and organizational change, accompanied by drastic social re-regulation. Combined with new and more flexible forms of labour management and aligned changes in state regulatory mechanisms, the production process became more flexible in organizational and technological terms as well as in the variety of products produced. New high technology firms were created, and others restructured or reorganized their mode of production. In short, a new ensemble of forces and relations of production is being created.

The relative fixity of the built and social environment of former core areas, whose economic base was severely undermined in the course of Fordist crisis formation, hampers rapid reconstruction along the required more flexible lines. Therefore, capital seeks out new places to reconstruct space according to the new exigencies of the capitalist accumulation process.

In more concrete terms, new complexes of high technology production mushroomed in many countries, constructing a new technological–organizational pattern. Silicon Valley, the M4 corridor, the southeast of France, and Bavaria in Germany illustrate this new geography of production (Scott, 1988). Specialized industrial districts become new growth areas: e.g. in the 1970s the Swiss watch industry – spatially highly concentrated – was virtually dead; in the 1980s, SWATCH swept the market with its electronic fashion-designed watches. Those areas tend to be dominated by a few core firms, surrounded by a host of more or less dependent suppliers and subcontractors. Nevertheless, such territorial complexes not only make the regional economy highly sensitive to international boom and bust cycles in the production of final goods, but they also create a pyramid of small and medium-sized industries which are themselves highly vulnerable to international disruptions and changes in corporate policies of the dominating firm(s) (Feldman, 1989). Furthermore, the labour market in these and similar places becomes highly segmented as a consequence of the differences in labour force requirements of each layer in the production hierarchy. Moreover, this segmentation is reinforced and consolidated by changing state regulations concerning the organization of the labour process and individualized (on the firm and sometimes sector level) capital–labour negotiations. The breakdown of social resistance, in turn, enables the reconstruction of the labour force and labour market along these more flexible lines. The relatively 'virgin' regions, which hitherto were only marginally integrated in the capitalist production process, become the favoured areas for the location of these new technical–organizational complexes. These areas do not possess the rigidities prevalent in the older industrial spaces and, consequently, a new image and spatial landscape can be produced (Harvey, 1987): a geography which is much more attuned to the needs and demands of the new dominant forms of accumulation. A new spatial 'frontier' is constructed which permits a further deepening and expansion of the capitalist relations of production.

Conclusion: Recapturing the Space Economy

Capitalist development is fundamentally characterized by phases of (spatial) expansion and growth followed by a period of contraction and intense (spatial) restructuring, i.e. global crisis moments (1930s, 1970s). Each of these phases is characterized by a particular combination of technical–economic organization and institutional regulation which assures the steady expansion of the system. The core areas of such expansionist phases see their physical, social, economic and institutional landscape moulded in a specific way. In other words, space is constructed in a particular way which reflects the technical–economic characteristics and regulatory practices referred to above. However, the role of space therein transcends a mere container or recipient function. Space itself plays a key role in the expansion and restructuring of the accumulation process. This dynamic process can be illustrated by looking at the spatial dynamics of Fordism. For example, the expansion of Fordist regions brought high wages, high employment and a high degree of unionization in its core regions. Competitive pressures on private economic agents, however, demanded a continuing quest to maintain or regain a competitive edge. This triggered a process of selective decentralization and spatial division, a process which – under Fordism – resulted in the emergence and consolidation of the national and international spatial division of labour. The Fordist core–periphery structure was hereby deepened and became more complex, resulting in a multi-hierarchical structure of national and international core-periphery relationships. Space provided, as it were, a continuously moving frontier in which Fordist capitalism could penetrate further (Smith, 1984). However, given the particular technical–organizational and institutional structure of Fordism, capturing and penetrating the spatial frontier at any given time is conditioned by the technical–organizational and institutional characteristics of development prevalent at that moment. The spatial expansion of growth is limited, furthermore, by the combination of the limits of space on the one hand and the limits of the dominant accumulation regime on the other. In other words, once world space is integrated in a particular way, spatial expansion reaches its final limits. The mode of development collapses, the crisis spreads. It is at such moments that the whole global fabric of economic-geographic organization and institutional regulation is deconstructed and reconstructed in a profoundly new way. This is always accompanied by the deconstruction (or de-territorialization) of space in one place and the reconstruction (re-territorialization) of space in another place (Harvey, 1989); new spatial frontiers are constructed, and the accumulation process produces new spaces of production, new spaces of consumption and, consequently, a new spatial dynamic.

An Agenda for the 1990s

While economic geography in the 1970s and 1980s was essentially a geography of crisis, the late 1980s demonstrated the need for a geography of growth. In fact, economic growth and decline are two sides of the same coin. For too long, (critical) geographers have sought to elucidate the dynamics of spatial crisis formation (Massey, 1984; Martin and Rowthorn, 1986). It is time to redress the balance and to try, from the same critical perspective, to grapple with the dynamics of growth (see Massey and Allen, 1988; Storper and Walker, 1989). This challenge would eventually contribute to clarifying a long-standing debate in economic geography, i.e. the dynamics of uneven geographical development. In the end, it is the dialectic of growth and decline, of expansion and crisis, which produces that perplexing and continuously changing geography of uneven development.

This task is not an easy one. First, the dynamics of the heralded new spaces of production should be further unravelled through detailed historical-geographical research. Second, this includes a search for and an understanding of the mechanisms of industrial organization and restructuring (Scott and Storper, 1986; Gertler, 1988; Schoenberger, 1989) and their relationship with institutional and social regulatory forms. The role of particular forms of territorial organization, which is constructed in and through the combination of technical–organizational structures of the production process with socio-political relations, in fostering socio-economic growth needs further clarification.

Finally, the most pressing task ahead is the further development of a spatial theory of development. For too long, geographers have tried to spatialize economic theory rather than to integrate space and spatial relations from the very beginning in the relational formulations that constitute theory and explanation. The challenge, therefore, is not to write a geography of development, but to produce a real development geography.

References

Aglietta, M. 1979: *A Theory of Capitalist Regulation.* London: New Left Books.

Albrechts, L. and Swyngedouw, E. 1989: The challenges of regional policy under a flexible regime of accumulation. In L. Albrechts, F. Moulaert, P. Roberts, and E. Swyngedouw (eds), *Regional Policy at the Crossroads: European perspectives.* London: Jessica Kingsley.

Amin, S. 1976: *Unequal Development.* New York: Monthly Review Press.

Bluestone, B. and Harrison, B. 1988: *The Great U-Turn.* New York: Basic Books.

Boyer, R. 1989: *The Regulation Theory: a critical perspective.* New York: Columbia University Press.

Brake, K. 1986: Das 'Sud-Nord Gefälle' als Ausdruck epochaler Strukturvernderungen in Produktion und Territorium. *Raumplanung,* 34, 171–4.

Breheny, M., Cheshire, P. and Langridge, R. 1983: The anatomy of job creation? Industrial change in Britain's M4 corridor. *Built Environment,* 9, 1, 61–71.

Carney, J., Hudson, R. and Lewis, J. (eds) 1981: *Regions in Crisis.* London: Croom Helm.

Castells, M. (ed.) 1985: *High Technology, Space and Society.* Beverley Hills, CA: Sage.

——1989: *The Informational City.* Oxford: Basil Blackwell.

Cooke, P. 1988: Flexible integration, scope economies and strategic alliances: social and spatial mediations. *Environment and Planning D: Society and Space,* 6, 281–300.

——1990: *Back to the Future.* London: Unwin Hyman.

Coriat, B. 1979: *L'Atelier et le Chronomètre.* Paris: Christian Bourgois.

Emmanuel, A. 1972: *Unequal Exchange.* London: New Left Books.

Feldman, M. 1989: The flexibility thesis and vertical disintegration: some issues. Working Paper BV89-2, Graduate Curriculum in Community Planning and Area Development, University of Rhode Island, Kingston.

Florida, R. and Feldman, M. 1988: Housing in U.S. Fordism. *International Journal of Urban and Regional Research,* 12, 187–210.

Frank, A. G. 1972: *Le Développement du Sous-Développement.* Paris: Maspéro.

Friedmann, J. 1966: *Regional Development Policy: a case-study of Venezuela.* Cambridge, MA: MIT Press.

Fröbel, F., Heinrichs, J. and Kreye, O. 1980: *The New International Division of Labour,* Cambridge: Cambridge University Press.

Gertler, M. 1988: The limits of flexibility: comments on the post-Fordist vision of production and its geography. *Transactions, Institute of British Geographers, New Series,* 13, 419–32.

Hall, P. and Markusen, A. R. (eds) 1983: *Silicon Landscapes.* Boston, MA: Allen & Unwin.

Harrison, B. 1984: Regional restructuring and good business climates: the economic transformation of New England since World War II. In L. Sawers and W. K. Tabb (eds), *Sunbelt/Snowbelt,* Oxford: Oxford University Press.

Harvey, D. 1982: *Limits to Capital.* Oxford: Basil Blackwell.

——1985: The geopolitics of capitalism. In D. Gregory and J. Urry (eds), *Social Relations and Spatial Structures,* London: Macmillan.

——1987: Flexible accumulation through urbanization: reflections on 'post-modernism' in the American City. *Antipode,* 19, 260–86.

——1989: *The Condition of Post-modernity.* Oxford: Basil Blackwell.

Haug, P. 1986: U.S. high technology multinationals and Silicon Glen. *Regional Studies,* 20, 103–16.

Isard, W. 1960: *Methods of Regional Analysis.* Cambridge, MA: MIT Press.

Lash, S. and Urry, J. 1987: *The End of Organized Capitalism.* Cambridge: Polity Press.

Leborgne, D. and Lipietz, A. 1988: New technologies, new modes of regulation: some spatial implications. *Environment and Planning D: Society and Space*, 6, 263–80.

Lipietz, A. 1986: New tendencies in the international division of labour: regimes of accumulation and modes of regulation. In A. J. Scott and M. Storper (eds), *Production, Work, Territory*, Boston, MA: Allen & Unwin, 16–40.

Markusen, A. R., Hall, P. and Glasmeier, A. 1986: *High Tech America: The What, How, Where and Why of Sunrise Industries*. Boston, MA: Allen & Unwin.

Martin, R. and Rowthorn, B. (eds) 1986: *The De-Industrialization of Great Britain*. London: Macmillan.

Massey, D. 1984: *Spatial Divisions of Labour*. London: Macmillan.

——and Allen, J. (eds) 1988: *Uneven Re-development*. Milton Keynes: Open University Press.

——and Meegan, R. 1982: *The Anatomy of Job Loss*. London: Methuen.

Moulaert, F. and Syngedouw, E. 1989: A regulation approach to the geography of the flexible production system. *Environment and Planning D: Society and Space*, 7, 3, 327–45.

Myrdal, G. 1959: *Economic Theory and Underdeveloped Regions*. London: Duckworth.

Perroux, F. 1961: *L'Economie du XXième Siècle*. Paris: Presses Universitaires de France.

Piore, C. and Sabel, C. 1984: *The Second Industrial Divide*. New York: Basic Books.

Sawyers, L. and Tabb, W. K. (eds) 1984: *Sunbelt/Snowbelt*. Oxford: Oxford University Press.

Saxenian, A. 1983: The urban contradictions of Silicon Valley: regional growth and restructuring in the semiconductor industry. *International Journal of Urban and Regional Research*, 7, 237–62.

Sayer, A. 1989: Postfordism in question. *International Journal of Urban and Regional Research*, 13, 666–95.

Schoenberger, E. 1988: From Fordism to flexible accumulation: technology, competitive strategies and international location. *Environment and Planning D: Society and Space*, 6, 245–62.

——1989: Thinking about flexibility: a response to Gertler. *Transactions, Institute of British Geographers, New Series*, 14, 98–108.

Scott, A. J. 1988: *New Industrial Spaces*. London: Pion.

——and Storper, M. (eds) 1986: *Production, Work, Territory*. Boston, MA: Allen & Unwin.

Smith, N. 1984: *Uneven Development*. Oxford: Basil Blackwell.

Stöhr, W. 1986a: Territorial innovation complexes. In P. Aydalot (ed.), *Milieux Innovateurs en Europe*, Paris: Groupe de Recherche Européen sur les Milieux Innovateurs (GREMI), 29–54.

1986b: Regional innovation complexes. *Papers of the Regional Science Association*. 59, 29–44.

Storper, M. and Christopherson, S. 1987: Flexible specialization and regional industrial agglomeration: the case of the U.S. motion picture industry. *Annals of the Association of American Geographers*, 77, 104–17.

——and Scott, A. J. 1988: Work organization and local labour markets in an era of flexible production. Paper prepared for the International Labour Office, Research Program on Labour Flexibility, Geneva.

——and Walker, R. 1989: *The Capitalist Imperative*. Oxford: Basil Blackwell.

Swyngedouw, E. 1987a: Social innovation, organization of the production process and spatial development. *Revue d'Economie Urbaine et Régionale*, 3, 487–510.

——1987b: Planning? ... What planning! In J. J. M. Angenent and A. Bongenaar (eds), *Planning Without a Passport: The future of European spatial planning*, Netherlands Geographical Studies vol. 44. Amsterdam: Koninklijk Nederlands Aardrijkskandig Genootschap, 77–89.

——1989: The heart of the place: the resurrection of locality in an age of hyperspace. *Geografiska Annaler*, 71B, 31–42.

——1990: Perspective on a 'regulation approach' of spatial change and innovation: a case-study of the Hasselt/Genk metropolitan region. In P. Nijkamp (ed.), *Sustainability of Urban Systems: A cross-national evolutionary analysis of urban innovation*, Aldershot: Avebury, 247–86.

——and Anderson, S. 1987: Les dynamiques spatiales des industries de haute technologie en France. *Revue d'Economie Urbaine et Regionale*, 2, 321–49.

Walker, R. 1981: A theory of suburbanization; capital and the construction of urban space in the U.S. In M. Dear and A. J. Scott (eds), *Urbanization and Urban Planning in Capitalist Society*, Boston, MA: Methuen.

——and Storper, M. 1981: Capital and industrial location. *Progress in Human Geography*, 5, 473–509.

Wallerstein, I. 1974: *The Modern World System*. New York: Academic Press.

II.10
Social and Cultural Geography

Peter Jackson

Introduction

The combined field of social and cultural geography is currently going through a period of revitalization and renewal, making it one of the most buoyant areas of human geography, with new areas of research opening up all the time. Contemporary social geography is concerned with the social significance of space and place. It examines the humanistic elements that comprise a 'sense of place' and attempts to understand changing patterns and processes of social inequality, in terms of race and ethnicity, gender and sexuality, class and generation, especially where they result in spatial inequalities. But social geographers are no longer content with simply mapping the processes that other social scientists study. They are increasingly insisting on the importance of space in the constitution of social life, the spatial structuring of social relations (Gregory and Urry, 1985) and the reassertion of space in social theory (Soja, 1989).

Cultural geography is also undergoing a considerable renaissance, its earlier preoccupation with the study of landscape being increasingly supplemented with other approaches and concerns. Contemporary cultural geographers have begun to stress the idea of landscape-as-text (Cosgrove and Jackson, 1987), iconography and symbolism of landscape (Cosgrove, 1985; Cosgrove and Daniels, 1987), and a range of ideas from media studies and cultural politics (Burgess and Gold, 1985; Jackson, 1989).

Although social and cultural geographers are increasingly drawing on (and contributing to) contemporary theoretical debates in other disciplines, exploring new methods and approaches (Jackson and Smith, 1984), the roots of the subject extend far back into the history of geography itself. Contemporary social geography developed from French and German sources, including the *géographie humaine* of Paul Vidal de la Blache and Hans Bobek's interest in the relationship between culture groups (*Kulturgemeinschaften*) and patterns of social life (*Lebensformen*). During the 1950s, when geography was still defined in terms of 'areal differentiation', social geographers were concerned with 'the identification of different regions of the earth's surface according to associations of social phenomena related to the total environment' (Watson, 1953, p. 482). During the 1960s, more sociological definitions emerged, as social geographers became concerned with 'the study of patterns and processes in understanding *socially defined populations* in a spatial setting' (Pahl, 1965, p. 81). Humanistic and radical approaches also had their impact, with an increasing emphasis on human subjectivity and a growing realization of the structural basis of social inequalities. For the humanist, social geography was concerned with 'understanding the processes which arise from the use social groups make of space *as they see it*, and the processes involved in making and changing such patterns' (Jones, 1975, p. 7), while more radical geographers emphasized 'the analysis of social patterns and processes arising from the distribution of and access to *scarce resources*' (Eyles, 1974, p. 29). Today, then, social geography comprises an eclectic mixture of ideas, theories and empirical research (Knox, 1987), and the extent of ideological ferment within the subject can be taken as evidence of social geography's rude health rather than as a symptom of intellectual incoherence (Cater and Jones, 1989).

Social geography, even more than other branches of the discipline, is very much 'a European science' (Stoddart, 1986). In North America, the environmental basis of human societies has probably remained closer to the core of the discipline than in Britain or France, for example. One example of this tendency is the strength of *cultural geography* in North America, with somewhat different emphases from British or French traditions of social geography. Broadly speaking, the American tradition of cultural geography studies the human transformation of the physical environment and the evolution of cultural landscapes (see chapter II.7). In recent years the two fields have been converging as social geographers have begun to explore the cultural construction of everyday life (including concepts of gender and sexuality, ethnicity and race) while cultural geographers are beginning to show more sensitivity to current theoretical debates, adding an interest in contemporary urban environments to their long-standing interest in rural landscapes of the past.

Origins and Development

Modern British social geography received its impetus from Emrys Jones's classic study of Belfast (Jones, 1960) in which he outlined the historical and cultural basis of residential segregation by religious group. His work was pursued by Fred Boal and others, who examined the behavioural patterns of Protestants and Catholics in Belfast, including territorial markers such as graffiti and the geography of violent crime (Boal, 1969). This early emphasis on religious groups was soon to be superseded by studies of ethnic segregation in British and American cities, drawing theoretical inspiration from the Chicago school of urban ecology (plate II.10.1).

The ideas of the Chicago sociologist Robert Park (1864–1944) were particularly influential at this point. He drew an important distinction between social and physical distance, arguing that by reducing social relations to relations of space it would be possible to apply to human relations the fundamental logic of the physical sciences. By reckoning human relations in terms of distance, he attempted to translate complex proceses

Plate II.10.1 Ethnic segregation is a prominent feature of the human geography of many large cities, as shown in Chinatown, New York, USA.

of social segregation and ethnic assimilation into spatial patterns of residence or intermarriage. By modelling these patterns in terms of index measures of spatial mixing and applying them to small-area census data, social geographers gained a statistical handle on formerly intractable aspects of social interaction. At the height of the 'quantitative revolution', social geographers made great strides in measuring spatial patterns of ethnic segregation and deliberating on their social significance (Peach, 1975; Peach et al., 1981).

Empirical analyses of residential segregation have now been challenged from two directions. Some have taken a more ethnographic approach, based on qualitative first-hand fieldwork, following the pioneering work of David Ley in Philadelphia's black inner city (Ley, 1974; Eyles and Smith, 1988), while others have focused on the social construction of 'race' and the territorial basis of specific forms of racism (Jackson, 1987). The latter work draws on earlier studies of ghetto formation by Marxist geographers such as David Harvey (1973) and on Doreen Massey's recent work on spatial divisions of labour (Massey, 1984).

New Directions

Cultural studies began in Britain with the work of Richard Hoggart, a tireless champion of working-class culture, and Raymond Williams, who consistently emphasized the connections between culture and society, developing his own brand of cultural materialism. Their work was developed by members of the Centre for Contemporary Cultural Studies at the University of Birmingham, originally under the directorship of Richard Hoggart and later that of Stuart Hall. Hall defined culture as the way people 'handle' the raw material of their social and material existence. He refers to the 'codes' with which meaning is constructed, conveyed and understood, and to the 'maps of meaning' by which social life is made intelligible (Jackson, 1989).

These ideas, together with other branches of social, cultural and literary theory, are giving rise to a 'new' cultural geography, distinguishable by its insistently critical political edge (Gregory and Ley, 1988). For example, Stuart Hall and his colleagues have rejected an elitist and unitary conception of culture and focused instead on the plurality of cultures that flourish in contemporary capitalist societies. They are interested in popular culture as well as in the culture of the elite, in gay and lesbian subcultures as well as in the culture of the heterosexual majority, in the cultures of black people, women and other oppressed groups.

Developing the ideas of the Italian Marxist Antonio Gramsci (1891–1937), they recognize that cultures can be ranked hierarchically along a scale of power: that dominant groups exercise power not just through physical violence and coercion but, more subtly, through culture and ideology. Gramsci's conception of hegemony explores the way that particular readings of society come to dominate others, serving the interests of the elite. Dominant ideologies operate by representing particular ideas and values as part of everyday 'common sense', beyond the realm of rational debate. But hegemony is never complete; it is always contested, both ideologically (in ritualized or symbolic form) and in more directly instrumental, political ways. Thus, working-class youth subcultures can be understood as 'rituals of resistance' (Hall and Henderson, 1976), issuing a symbolic challenge to their economic and political subordination. Geographers have contributed to this literature by tracing the material circumstances in which particular cultural forms emerge, and by recognizing that a plurality of cultures will generate a plurality of landscapes.

The cultural studies literature encourages geographers to explore alternative approaches to culture besides those that focus exclusively on landscape. For culture is not just socially constructed and geographically expressed; it is *spatially constituted*. Recent examples which demonstrate this process include studies of the geographically variable basis of patriarchal gender relations in different labour markets (McDowell and Massey, 1984); the territorial constitution of gay politics in San Francisco (Castells, 1983); and political struggles over symbolic space in nineteenth-century Paris (Harvey, 1985).

Feminist Challenges

As these remarks suggest, the recent reinvigoration of social and cultural geography can be traced back to specific political challenges, dating from the 1960s: the movements for civil rights, women's equality, and the rights of gay men and lesbians. In each case, a political movement has given rise to a range of intellectual challenges, transforming the way we study, teach and research, as well as the way we live our lives. For these social movements demand a personal commitment to change as well as requiring fundamental changes at the institutional level. Not least among these social movements, feminism reminds us of the extent to which *the personal is political*: a lesson that cultural and social geographers have been slow to appreciate.

We concentrate here, then, on the feminist challenge to social and cultural geography which began by demonstrating the simple fact that women are significantly under-represented within the profession of geography, particularly at higher levels (Zelinsky, 1973; McDowell, 1979). Comprising roughly half the number of geography undergraduates in Britain and the United States, for example, the proportion of women falls off steadily as one ascends the hierarchy from lecturer to professor, with women concentrated most heavily in short-term posts as research assistants, with few prospects of career advancement, limited autonomy and no security of tenure. Only recently has there been any institutional response to these stark inequalities with the formulation of an equal opportunities policy within the Institute of British Geographers (Jackson et al., 1988).

With the benefit of hindsight, it is possible to trace a number of 'stages' within the development of feminist geography (for bibliographies, see Lee and Loyd, 1982; Zelinsky et al., 1982; Women and Geography Study Group, 1984; Bowlby et al., 1989). The fact that women were 'hidden from geography' as well as from history gave rise to a series of studies of the 'geography of women' (e.g. Tivers, 1978), including studies of women's unequal access to a range of public goods and services. Women's subordination was traced to the undervaluation of domestic work and to simplistic distinctions between 'productive' and 'reproductive' labour. Attempts were also made to understand the spatial underpinnings of the gender division of labour through the characteristic separation of 'home' and 'work' in capitalist urbanization (McDowell, 1983).

A transition has gradually occurred from the study of *women's roles* to the study of *gender relations*, establishing the fundamental connection between gender and power (Connell, 1987) and focusing on the structured inequalities between men and women. More recently there have been major debates about the nature of patriarchy (Foord and Gregson, 1986; Walby, 1986) and about the intersections of class and gender in different labour markets (McDowell and Massey,

1984; Lancaster Regionalism Group, 1985; Bowlby et al., 1986; Bagguley, 1990). Most recently, there has been a recognition of the need to theorize about the intersection of 'race' and gender, acknowledging the significance of different forms of oppression among different groups of women (McDowell, 1991).

From the beginning, and particularly in Britain, there has been an explicit commitment to the development of *feminist theory*, as well as to the conduct of theoretically informed *empirical research*. With many authors writing from the perspective of socialist feminism, there has also been a consistent commitment to *political and social change*, through collective struggles, rather than simply to the advancement of knowledge through purely individual scholarship.

Many of these developments can be traced at the institutional level, particularly in Britain and the United States but also, now, more widely. After some initial resistance, the Women and Geography Study Group was established within the Institute of British Geographers in 1982. In the United States, a similar group exists within the Association of American Geographers, dedicated to the development of geographic perspectives on women. Finally, in 1988, the International Geographic Union set up a Study Group on Gender to co-ordinate research, publish working papers and assist feminist geographers to create an international network. These developments, though welcome, do not disguise the level of resistance that feminist geographers continue to face within the discipline. Feminist ideas are still marginalized by those who regard themselves as part of the geographic 'mainstream', dominated by white middle-class men.

Taken seriously, feminism forces a transformation in every aspect of our geographical work. One cannot simply 'add women and stir' as one observer recently remarked; instead, one must recognize that social relations are always and everywhere gendered, though to different degrees and with different consequences at different times and places. From an initial concern with the economic subordination of women in the 'developed' world (Women and Geography Study Group, 1984), feminist geographers have begun to diversity their interests to include cultural and historical issues (Rose and Ogborn, 1988), exploring the relationships between geography and gender in the 'developing' world (Momsen and Townsend, 1987; Brydon and Chant, 1989), with specific studies of low-income housing (Moser and Peake, 1987), urban politics (Little et al., 1988) and family farms (Whatmore, 1990). Future work can be expected to explore the cultural construction of gender identities (masculinities and femininities, heterosexualities and homosexualities) and to examine the contradictory effects of postmodernism and the 'new times' (Smith, 1989; Bondi, 1990) for gender relations in the 1990s and beyond.

Current Research

Contemporary research in social geography is focused on a range of empirical questions concerning housing and neighbourhood change; health and health care; crime, the police and the community; ethnicity, 'race' and racism; educational and social welfare. Theoretical questions include the geographically differentiated impact of economic and social restructuring at the local level (Cooke, 1989); the reciprocal relationship between social relations and spatial structures (Gregory and Urry, 1985); the interplay of gender and sexuality, ethnicity and race, class and generation in particular places, at particular times (Castells, 1983); and the development of spatially-sensitive approaches to cultural studies, including the media and popular culture (Burgess and Gold, 1985).

In cultural geography, too, there is a growing dialogue between humanistic and materialist approaches; a concern with contemporary urban landscapes as well as with the evolution of rural landscapes; an awareness of the plurality of cultures besides the dominant culture of the elite; and a recognition that the cultural is political. These developments bring social and cultural geography closer together. Recent arguments about post-modernism, for example, raise cultural questions about symbolic representation as well as social questions about the economic significance of increasingly 'flexible' modes of capitalist accumulation (Soja, 1989). The research agenda is increasingly broad, concerned with every aspect of culture and society where 'geography matters', not just because of the pattern of local or regional variation but also because social relations and cultural politics are often spatially structured.

References

Bagguley, P. 1990: *Restructuring: Place, Class and Gender*. London: Sage.

Boal, F. W. 1969: Territoriality on the Shankill–Falls divide, Belfast. *Irish Geography*, 6, 30–50.

Bondi, L. 1990: Feminism, postmodernism and geography: space for women? *Antipode*, 22, 156–67.

Bowlby, S., Foord, J. and McDowell, L. 1986: The place of gender relations in locality studies. *Area*, 18, 327–31.

——Lewis, J., McDowell, L. and Foord, J. 1989: The geography of gender. In R. Peet and N. Thrift (eds), *New Models in Geography*, vol. 2, London: Unwin Hyman, 157–75.

Brydon, L. and Chant, S. 1989: *Women in the Third World: gender issues in rural and urban areas*. Aldershot: Edward Elgar.

Burgess, J. A. and Gold, J. R. (eds) 1985: *Geography, the Media and Popular Culture*. London: Croom Helm.

Castells, M. 1983: *The City and the Grassroots*. London: Edward Arnold.

Cater, J. and Jones, T. 1989: *Social Geography*. London: Edward Arnold.

Connell, R. W. 1987: *Gender and Power*. Cambridge: Polity Press.

Cooke, P. (ed.) 1989: *Localities*. London: Unwin Hyman.

Cosgrove, D. E. 1985: *Social Formation and Symbolic Landscape*. London: Croom Helm.

——and Daniels, S. J. (eds) 1987: *The Iconography of Landscape*. Cambridge: Cambridge University Press.

——and Jackson, P. 1987: New directions in cultural geography. *Area*, 19, 95–101.

Eyles, J. D. 1974: Social theory and social geography. *Progress in Geography*, 6, 27–87.

——and Smith, D. M. (eds) 1988: *Qualitative Methods in Human Geography*. Cambridge: Polity Press.

Foord, J. and Gregson, N. 1986: Patriarchy: towards a reconceptualisation. *Antipode*, 18, 186–211.

Gregory, D. and Ley, D. 1988: Culture's geographies. *Environment and Planning D: Society and Space*, 6, 115–16.

——and Urry, J. (eds) 1985: *Social Relations and Spatial Structures*. London: Macmillan.

Hall, S. and Henderson, J. (eds) 1976: *Resistance Through Rituals*. London: Hutchinson.

Harvey, D. 1973: *Social Justice and the City*. London: Edward Arnold.

——1985: *Consciousness and the Urban Experience*. Oxford: Basil Blackwell.

Jackson, P. (ed.) 1987: *Race and Racism*. London: Allen & Unwin.

——1989: *Maps of Meaning*. London: Unwin Hyman.

——and Smith, S. J. 1984: *Exploring Social Geography*. London: Allen & Unwin.

——Johnston, R. J. and Smith, S. J. 1988: An equal opportunities policy for the IBG? *Area*, 20, 279–80.

Jones, E. 1960: *A Social Geography of Belfast*. Oxford: Oxford University Press.

(ed.) 1975: *Readings in Social Geography*. Oxford: Oxford University Press.

Knox, P. 1987: *Urban Social Geography*. London: Longman.

Lancaster Regionalism Group 1985: *Localities, Class and Gender*. London: Pion.

Lee, D. and Loyd, B. 1982: *Women and Geography: Bibliography*. Cincinnati, OH: Socially and Ecologically Responsible Geographer (SERGE), University of Cincinnati.

Ley, D. 1974: *The Black Inner City as Frontier Outpost*. Washington, DC: Association of American Geographers.

Little, J., Peake, L. and Richardson, P. (eds) 1988: *Women in Cities*. London: Macmillan.

Massey, D. 1984: *Spatial Divisions of Labour*. London: Macmillan.

McDowell, L. 1979: Women in British geography. *Area*, 11, 151–4.

——1983: Towards an understanding of the gender

division of urban space. *Environment and Planning D: Society and Space*, 1, 59–72.
——1991: The baby and the bathwater: diversity, deconstruction and feminist theory in geography. *Geoform*, 22, 123–34.
——and Massey, D. 1984: A woman's place. . . . In D. Massey and J. Allen (eds), *Geography Matters!*, Cambridge: Cambridge University Press, 128–47.
Momsen, J. and Townsend, J. (eds) 1987: *Geography and Gender in the Third World*. London: Hutchinson.
Moser, C. O. N. and Peake, L. (eds) 1987: *Women, Human Settlements and Housing*. London: Tavistock.
Pahl, R. 1965: Trends in social geography. In R. J. Chorley and P. Haggett (eds), *Frontiers in Geographical Teaching*, London: Methuen, 81–100.
Peach, C. (ed.) 1975: *Urban Social Segregation*. London: Longman.
——Robinson, V. and Smith, S. J. (eds) 1981: *Ethnic Segregation in Cities*. London: Croom Helm.
Rose, G. and Ogborn, M. 1988: Feminism and historical geography. *Journal of Historical Geography*, 14, 405–9.
Smith, S. J. 1989: Society, space and citizenship: a human geography for the 'new times'. *Transactions of the Institute of British Geography, New Series*, 14, 144–56.
Soja, E. W. 1989: *Postmodern Geographies*. London: Verso.
Stoddart, D. R. 1986: *On Geography and its History*. Oxford: Basil Blackwell.
Tivers, J. 1978: How the other half lives: the geographical study of women. *Area*, 10, 302–6.
Walby, S. 1986: *Patriarchy at Work*. Cambridge: Polity Press.
Watson, J. W. 1953: The sociological aspects of geography. In G. Taylor (ed.), *Geography in the Twentieth Century*, London: Methuen, 463–99.
Whatmore, S. J. 1990: *Farming Women*. London: Macmillan.
Women and Geography Study Group 1984: *Geography and Gender*. London: Hutchinson.
Zelinsky, W. 1973: Women in geography: a brief factual account. *Professional Geographer*, 25, 151–65.
——Monk, J. and Hanson, S. 1982: Women and geography: a review and a prospectus. *Progress in Human Geography*, 6, 317–66.

II.11

Urban Geography in the 1990s

Ruth Fincher

Someone important once said that if an urban geographer were parachuted into the centre of a large city, with no idea of which city it was, that urban geographer would find his or her way around quite successfully. This was not a claim that urban geographers have some special biological capacity to orient themselves. But it was a view that the descriptive models of contemporary cities produced in urban geography identify similarities in urban spatial structure that apply to most cities. Thus the city in which the lost urban geographer had found him or herself would probably have these recognizable characteristics too.

Until the early 1970s, the most influential generalizations about urban spatial structure, or the distribution of different zones of land use across metropolitan areas, came from the concentric zone model of the 1920s Chicago sociologists (reviewed by Badcock, 1984, ch. 1). The model maps the central business districts in the centre of the metropolitan area, encircled successively by zones of warehouses and factories, working people's housing, middle-income housing and commuter suburbs. In the 1970s and 1980s the generalizations about contemporary cities produced in urban geography have focused more on the processes by which urban forms of different sorts come about. The concentric zone model is still firmly imprinted in the minds of urban geographers as a description (more or less accurate in particular cases) of the capitalist metropolis of the early to middle twentieth century. But the debates of the moment are about *explaining* the similarities and differences we observe in our cities, applying a range of social theories in doing so.

Cities in capitalist countries are more widely studied in English-language urban geography than cities in non-capitalist countries. Four trends in them seem to be of major concern at present. Each focuses on the way that changes in urban built form are occurring, how these are explained by the exercise of power by certain groups and how they are affecting the life circumstances of certain (usually other) groups. These trends are (a) the movement of people outwards towards city borders and into residual suburbs, and the apparent counter-trend of the gentrification of inner city areas as people take up inner city living; (b) the changing locations of workplaces within cities as manufacturing firms and offices of different sorts appear and disappear; (c) the important role of the state in provision or withdrawal of public services for urban populations in different parts of the city; and (d) the growth of a local politics of cities in which territorially based groups protest against modifications to their localities. I shall consider these four matters in turn, noting as I do that they are interlinked and interdependent. Suggestions about future research directions will be included in the discussion of each area of research.

Suburbanization and Gentrification

In North America and Australia, where cities have vast areas of low density residential suburbs, the explanation and implications of suburbanization have received considerable attention. Important Marxist analyses of American residential decentralization have been contributed by Walker (1981) and Harvey (1977). The timing of greatest suburban expansion in cities of the United States is explained with reference to national government fiscal strategies at the end of the 1930s Depression. It was a time when the American economy could be bolstered by increased consumption of goods. The suburban built form and way of living, in which small household units each contain their own equipment in kitchens, laundries, workshops etc. rather than sharing these with other families, is clearly a style of daily life that maximizes consumption of manufactured goods. Government insurance for the mortgages of would-be home-owners and tax arrangements to encourage construction firms to become involved in road building and provision of utilities coincided with the appearance of technologies enabling rapid and inexpensive construction of single family homes. A mass movement of middle-income households to the suburbs occurred, and national consumption rose with it. In Australia, institutional encouragement for the decentralization of house building and subsidy of home ownership so that people could purchase the houses has been noted as well to underpin the mass outmigration of Australian households since the 1940s (Beed, 1981).

Associated with the growth in residential suburbs has been growth of the ideology of home ownership: in Australia home ownership is termed 'The Great Australian Dream' (see Kemeny, 1981), and it is commonly accepted that everyone aspires to it; in the United States the ownership of a new suburban tract house has been explained as part of a cultural 'love of the new' (see Muller, 1982). Debates exist as to whether the primary explanation for suburbanization in the middle to late twentieth century is consumer and cultural preferences, or rather the structuring of government policies to make the suburban decision advantageous for a range of interested groups (would-be home-owners, construction companies, washing machine manufacturers etc.).

In the wake of this interest in the explanation of suburban growth, feminist analysts have provided great insight into the implications of the suburban lifestyle for different gender groups, particularly women who are not wealthy. These implications apply for home-owners in suburbs, and for renters of public and private housing. Women who live in outer suburbs, without access to cars, suffer greatly because of the usually inadequate transport services on which they rely for shopping, getting to health facilities, workplaces and so on. Travelling to distant locations with children on public transport is even more stressful. Many post-Second World War housing policies seem to have assumed that women will spend their time in domestic chores in houses in suburbs (McDowell, 1983; Allport, 1986); even if women do work in the home, they make many trips outside it in order to do that work, and so require the means to have ready access to other places. The point is that suburban living, by definition, means a greater reliance on travel (by car or public transport) to conduct one's daily life. Whilst travel to work in a central location is often not too difficult, travel anywhere else at other than peak times can be most tiresome.

Gentrification is generally understood as the movement into dilapidated inner city areas of middle-income home-owners, who 'trendify' (the Australian

Plate II.11.1 Inner city areas in many cities are, as in Sydney, Australia, undergoing the process termed 'trendification'.

term) old houses (plate II.11.1), and may displace lower-income long-term residents in this process of 'upgrading'. Its extent and exact nature is still disputed; Smith and Williams (1986) air most of the angles on the issue. Useful feminist insights into the gentrification process have included the suggestion, made by Rose (1989) from her research in Montreal, that the assumption is quite invalid that all who move into inner city housing are wealthy upwardly mobile displacers. Rather, she argues, many women single heads of household are moving to the inner city to ensure better access to transport and other facilities and to be close to their employment which is often in the public sector offices of the central areas. An examination of just who are the gentrifiers, and what are their circumstances, seems overdue.

Urban Workplaces

Gentrification and suburbanization clearly are related to the location of urban workplaces. Two aspects of urban workplaces, of recent note, are (a) that manufacturing workplaces are declining in number but those that remain are decentralizing within metropolitan areas and (b) that office or clerical workplaces are appearing in greater numbers, and in some places are arranging themselves in a spatial hierarchy in which management functions are centralized and clerical processing functions are decentralized.

There is little doubt that there are now fewer manufacturing jobs in cities which depended on them earlier in the twentieth century, cities generally thought of as 'traditional' manufacturing strongholds. Most national statistics on the history of employment in different economic sectors of Western capitalist countries will show a decline since the start of the 1970s in the absolute numbers of manufacturing jobs held. There are regional shifts in those manufacturing jobs that remain: for example the shift in the United States from frostbelt to sunbelt (see Bluestone and Harrison, 1982). There are also redistributions of manufacturing jobs within metropolitan areas – certainly away from the inner city manufacturing zones noted in the old Chicago School models, and usually towards the outer suburbs. Debate exists about whether jobs went first to the suburbs or to the people wanting new homes there. Debate also exists amongst those interested in the manufacturing sector in cities about the significance of small flexible firms in parts of metropolitan areas that were not previously manufacturing centres. Los Angeles has been observed from this angle, and Boston also – especially with regard to the agglomerations of 'high tech' firms that occur on their boundaries (Storper and Scott, 1989).

Geographers are asking why these firms and workplaces are moving out. Are they seeking new labour forces – perhaps the pool of relatively immobile female labour residing in the suburbs? Gordon (1978) has argued persuasively that American manufacturing firms since the nineteenth century have moved in response to rounds of class conflict between workers and employers in particular spatial settings. On the other hand, the large decentralized greenfield sites of major manufacturers obviously house the technology of major auto manufacturers and the like better than do multistoried inner city factory buildings.

Offices are increasingly the workplaces of urbanites. They are shifting around also. Nelson (1986) argues that San Francisco 'back offices' are locating in the suburbs to obtain access to a docile female labour force that is unavailable (at the wages offered) in the central city. In different cities, researchers have observed a spatial and functional separation of offices, in which head offices are in central business districts and the 'processing' arms or regional administrative offices are in the suburbs. Sometimes, it has been argued that head offices are moving to the suburbs as well, enabled to do so by improvements in communications technology and freeway systems. It remains unclear, however, which sorts of firms, in which industrial sectors are locating where in different cities. These are empirical questions, requiring resolution before we can produce general explanations of the geography of office workplaces.

Studies of workplace location do tend to focus on firms' decisions about the best site, rather than on why people choose to work in those workplaces and locations. This is clearly because there is often little 'choice' for would-be employees. But questions may usefully be asked, for example, about why female labour force participants are available in the suburbs – what is it about the sexual division of labour in our cities that renders them less mobile than men? Also, if there really is (as is beginning to be identified) a trend to people engaging more in paid work from their homes, what are the circumstances under which certain groups of people would choose to do this?

Access to Services in Cities

The manner and shortcomings of the government's provision of public services to urban residents and workers has long been of interest to urban geographers (Pinch, 1985). One's quality of life may be greatly enhanced or lessened depending on the location of one's home or workplace and whether that location is supplied with good transport, library, educational, health and leisure facilities or not. The property value of a residential location is affected by the quality of public service provisions, then, as is the 'real income' one earns (Harvey, 1973; Badcock, 1984).

Geographers have long compared different municipalities and regions for their levels of service provision, devising useful measures of 'efficiency' and 'equity' in doing so. Pinch (1985) reviews this work. With contemporary cutbacks in government services provision in all Western capitalist countries, particularly in social services' provision of items like child care and care for the aged, a range of important new questions are being posed by urban geographers. These centre on the distributional impacts of the so-called 'welfare state restructuring', and depend on precise documentation of the ways in which reduction in government spending on public services is being achieved.

Examining the process of de-institutionalization provides one lens through which the impacts of reductions in government spending can be observed. This is the process in which residents of institutions (prisoners, aged or disabled persons requiring full-time care) leave those institutions to live, as independently as possible, in the community. Evidence reveals that this works splendidly for some individuals. But the process results as well in burdening certain 'communities' and certain people within them (see Dear and Wolch, 1987), as they struggle to support the ex-residents of institutions without adequate government services to help them.

Another example through which welfare state transformation might be studied is that of 'privatization', the phenomenon of services that were previously publicly provided being supplied by commercial or non-profit, non-government operators. The complex ways in which these services are supplied varies greatly between areas and has important distributional impacts for people who use the services and people who work to provide them. Volunteers often work to provide services – what does this mean if the supply of volunteers suddenly dries up, for example through increased female labour force participation? Wolch (1989) notes the emergence of a 'shadow state' of charitable organizations, desperately trying to keep up the standards of services in parts of California following reductions in federal spending.

Territorial Politics in Cities

Local political activism in cities has been of great interest to urban geographers, and of course to other urban analysts. This political activity most often concerns modifications to the urban environment to which local residents object – urban renewal schemes, for example, or the siting of nuclear power plants or overhead powerlines. In the 1970s, geographers documented cases of 'locational conflict', in which local residents objected to the siting of noxious facilities (or facilities perceived to be objectionable) and therefore came into conflict with those trying to establish such facilities in particular places. As the de-institutionalization process began, for example, much locational conflict occurred over the establishment in residential suburbs of community residential units and halfway houses for ex-residents of institutions; the NIMBY ('not in my backyard') response to this is now well known. Cox and Johnston (1982) review and give examples of this literature, noting its theoretical limitations.

Research in the 1980s has had stronger conceptual roots. The sociologist Castells (1983) has written an influential study of 'urban social movements', which are cases of territorial politics in which a progressive transformation of decision-making about the city is sought, favouring collective local control rather than anonymous, centralized and often conservative modes of decision-making. Cox (1981) has done much to develop a new line of Marxist thinking about the 'communal living spaces' of cities. He claims that the neighbourhoods in which we live are becoming commodities, and that characteristics of residential locations such as quietness, and even views of Pacific sunsets, are purchased and sold by property capitalists. In this context, the question of whether urban social movements or political activism over the communal living space are conservative or progressive is indeed hard to determine. As Cox (1981, p. 436) points out, it is easy to understand the stances of 'no growth' movements in Amercian territorial politics, as property capitalists move to create more profits from particular urban environments. These local activist groups exhibit 'resistance to annexation; resistance in the voting booth to the sale of bonds for expansion of sewer or water capacity; defensive incorporation; resistance to public housing; "gold-plated" subdivision regulations; and, of course, the notorious exclusionary zoning'. Whether these groups are actively 'protecting' their localities in order to preserve the communities they use in their daily lives, or in order to enhance their own property values (as their opponents always accuse them of doing), the exclusion of others from the localities often has the effect of rendering them more desired commodities than ever.

Interesting class alliances have formed in certain urban social movements, the Sydney Green Bans movement being one of the most interesting for the way it welded together groups of high-income women, working-class inner city residents, a set of powerful construction unions and the National Trust to stop high-rise commercial developments displacing older housing and parks in many parts of the inner city (Jakubowicz, 1984). Harvey (1985) has emphasized the usefulness of Marxist theory for the analysis of urban politics, identifying, as important questions to ask about local class alliances, who are their participants, how are their interests shaped and expressed politically, and why do they remain unstable.

Other important questions remain in the analysis of territorial politics, however. One startling silence in the literature concerns the important role of women in urban social movements, as participants in these neighbourhood conflicts and, more recently, as organizers of certain of them (Fincher and McQuillen, 1989). The class characteristics of these movements, in different places, is of interest: Are middle-class gentrifiers of inner cities objecting to changes to the inner cities they have reclaimed, and is this different from past movements? What sorts of class alliance occur in different places and over different issues? With the rise of national-level environmental politics in Western capitalist countries, the status and role of urban political activism, which has often taken up the cause of local environmental change, will be something to watch.

References

Allport, C. 1986: Women and suburban housing: postwar planning in Sydney, 1943–61. In J. McLoughlin and M. Huxley (eds), *Urban Planning in Australia: critical readings*, Melbourne: Longman Cheshire.

Badcock, B. 1984: *Unfairly Structured Cities.* Oxford: Basil Blackwell.

Beed, C. 1981: *Melbourne's Development and Planning.* Melbourne: Clewara Press.

Bluestone, B. and Harrison, B. 1982: *The Deindustrialization of America.* New York: Basic Books.

Castells, M. 1983: *The City and the Grassroots.* Berkeley and Los Angeles, CA: University of California Press.

Cox, K. 1981: Capitalism and conflict around the communal living space. In M. Dear and A. Scott (eds), *Urbanization and Urban Planning in Capitalist Society*, London: Methuen.

— —and Johnston, R. 1982: *Conflict, Politics and the Urban Scene.* London: Longman.

Dear, M. and Wolch, J. 1987: *Landscapes of Despair.* Princeton, NJ: Princeton University Press.

Fincher, R. and McQuillen, J. 1989: Women in urban social movements. *Urban Geography*, 10, 604–13.

Gordon, D. 1978: Capitalist development and the history of American cities. In W. Tabb and L. Sawers (eds), *Marxism and the Metropolis*, New York: Oxford University Press, 25–63.

Harvey, D. 1973: *Social Justice and the City.* Baltimore, MD: Johns Hopkins University Press.

— —1977: Government policies, financial institutions and neighbourhood change in United States cities. In M. Harloe (ed.), *Captive Cities*, London: Wiley, 123–40.

— —1985: The place of politics in the geography of uneven capitalist development. In his *The Urbanization of Capital*, Baltimore, MD: Johns Hopkins University Press, 125–64.

Jakubowicz, A. 1984: The Green Ban movement: urban struggle and class politics. In J. Halligan and C. Paris (eds), *Australian Urban Politics*, Melbourne: Longman Cheshire.

Kemeny, J. 1981: *The Myth of Home Ownership.* London: Routledge & Kegan Paul.

McDowell, L. 1983: Towards an understanding of the gender division of urban space. *Environment and Planning D: Society and Space*, 1, 59–72.

Muller, P. 1982: The role of suburbs in the contemporary metropolitan system. In C. Christian and R. Harper (eds), *Modern Metropolitan Systems*, Columbus, OH: Charles E. Merrill.

Nelson, K. 1986: Labor demand, labor supply and the suburbanization of low-wage office work. In A. Scott and M. Storper (eds), *Production, Work, Territory*, Boston, MA: Allen & Unwin, 149–71.

Pinch, S. 1985: *Cities and Services.* London: Routledge & Kegan Paul.

Rose, D. 1989: A feminist perspective of employment restructuring and gentrification: the case of Montreal. In J. Wolch and M. Dear (eds), *The Power of Geography*, Boston, MA: Unwin Hyman.

Smith, N. and Williams, P. (eds) 1986: *Gentrification of the City.* Boston, MA: Allen & Unwin.

Storper, M. and Scott, A. 1989: The geographical foundations and social regulation of flexible production complexes. In J. Wolch and M. Dear (eds), *The Power of Geography*, Boston, MA: Unwin Hyman.

Walker, R. 1981: A theory of suburbanization: capitalism and the construction of space in the United States. In M. Dear and A. Scott (eds), *Urbanization and Urban Planning in Capitalist Society*, London: Methuen.

Wolch, J. 1989: The shadow state: transformations in the voluntary sector. In J. Wolch and M. Dear (eds), *The Power of Geography*, Boston, MA: Unwin Hyman.

II.12
Political Geography

Peter J. Taylor

As we come to the end of the twentieth century we are experiencing an extra-ordinarily fluid political world. Political patterns and processes that were widely accepted just a decade ago have been jettisoned. Change has replaced stability as the hallmark of our rapidly approaching *fin de siècle*. For political geography – the study of the ways in which space and time are constituted within political processes, how territories are implicated in the pursuit of power – this fluidity represents both a challenge and an opportunity. Old ideas have to be reassessed and sometimes even discarded; new ideas have to be developed and tested. One thing is beyond dispute – this is an exciting time for political geography.

The basic document of political geography is that most familiar of all maps, the world political map of international boundaries. This spatial expression of the inter-state system shows the sovereign territories of all the states of the world. In political geography these territories form the pivotal geographical scale of study. Above the state level international relations are considered in the *geopolitics* of the inter-state system which covers both geostrategic and geoeconomic conflicts between states. Below the state level the *political geography of localities* considers the conflicts and strategies of local communities. At the pivotal scale itself the *geography of the state* has been concerned with the way in which governments and state bureaucracies conduct policies of conflict and consensus within their territories. It is with this three-scale structure of political geography that we can confront contemporary political change to make sense of our fluid world.

Geopolitics

Undoubtedly the most dramatic recent political change has been the demise of the Cold War. From both sides of the former 'iron curtain' politicians are proclaiming 'the end of the Cold War'. In political geography terms this means that a geopolitical world order – the Cold War – is in the process of disintegrating. Geopolitical world orders are relatively stable international political patterns that provide the basic assumptions underlying global-scale politics. Such world orders are clearly expressed in patterns of conflicts and alliances. In the Cold War geopolitical world order, for instance, a bi-polar division of the world has NATO and the Warsaw Pact at the centre of global politics and all other states have had to adapt their policies to this fact to become pro-United States, pro-USSR, neutral or non-aligned.

Although geopolitical world orders are stable they are not eternal. Changes between one world order and the next typically occur in relatively brief periods which are termed geopolitical transitions. For instance in 1938 the fate of the world seemed to hang on negotiations between Britain and Germany at Munich; a decade later it was the United States and the USSR confronting each other at Berlin that seemed set to plunge the world into war once again. In between these two events the political world had been 'turned upside down'. The United States had replaced Britain as the world's 'leading' country and the USSR had replaced Germany as the world 'challenger'. A fundamental political transition had occurred bringing into being the Cold War geopolitical world order. This new world order divided the world along ideological lines to produce a very entrenched political pattern. After all, the term Cold War suggests a 'frozen' structure and adversaries from both sides agreed that this world order represented a final contest of alternative ways of life. That is what has made the current demise of the structure so surprising. Like other world orders before it, the Cold War has had its time span and we are entering a new geopolitical transition.

We know we are experiencing a geopolitical transition when what had seemed 'impossible' actually occurs (plate II.12.1). In 1989, for instance, a non-communist government was installed in Poland with USSR approval. Such an event was unthinkable just a few years ago when several members of the new government were actually in gaol. Subsequently communist regimes in the remainder of eastern Europe either transformed themselves into pluralist political systems (e.g. Hungary) or were overthrown in the attempt to resist (e.g. Romania). These revolutions of 1989 were sealed by the breaching of the Berlin Wall, symbol of a divided Europe, on 9 November 1989. In 1990 Germany was united and in 1991 communism finally collapsed in the USSR itself. It is not surprising, therefore, that there is a widespread sense of entering a 'new world'. But we do not know what that world will be. Geopolitical transitions are packed with surprises. Since old assumptions are cast aside we have little basis upon which to make predictions.

Plate II.12.1 The geopolitical transformation of Europe has been one of the great events of recent years. Symbols of the old order, exemplified by this statue of Lenin in Romania, have been swept away.

Nobody in 1938, for instance, could possibly have foretold the Cold War. We are at the equivalent point in the current transition: the total disintegration of the 'old' is here; the construction of the 'new' is still some way off. Perhaps we shall return to a less ideological organization of global politics. One possible future world order that some political geographers think may arise is another bi-polar world that pits a Pacific Rim power bloc centred on a Japan–United States–China axis against a new united European power bloc that eliminates the old iron curtain to include the USSR. Such a world order would cut across the ideological logic of the Cold War with 'capitalist' and 'communist' states in both camps. But we do not know that this will come about. That is what makes contemporary geopolitical studies so interesting. We can only monitor trends and make imaginative guesses while world politics reserves its most audacious surprises – more 'impossibles' – for the near future: watch this (geopolitical) space.

The Geography of the State

Below the global level geographical studies of the state have had to come to terms with equally far-reaching political changes. These can be viewed both generally as processes affecting all or most states, and specifically in terms of their actual expression within a particular state. Until recently it was generally assumed that most states in the world had assimilated their populations to a degree that had eliminated separatist national challenges. Today nationalism is back at the centre of our political geography agenda as the spatial integrity of states is being challenged on all continents from Québec to Tibet. Even the three members of the inter-state system with the longest history of power politics – Spain, France and Britain – have experienced national separatist revivals (e.g. Basques, Corsica and Scotland). The most dramatic of such revivals has been in the USSR where the 'new' nationalisms may have profound implications for the new geopolitics.

These nationalisms point towards a break-up of existing states while conversely there are concurrent political moves towards combining states into larger units. The European Community is the most well-known example but on all continents new economic and political arrangements are being negotiated at this 'regional' scale. In addition there are new social movements, notably the 'Greens' whose attitudes to the state itself are quite ambiguous. For the first time in centuries we may be entering a period when the state will be seriously challenged as the focus of politics and the 'natural' recipient of our political loyalties.

These state-scale political processes are expressed differently in different countries because of their various roles in the inter-state system and their alternative institutional arrangements. In Britain a century of relative decline has culminated in a new politics generally termed 'Thatcherism'. This refers to the breakdown of the post-1945 political consensus in Britain which centred upon state intervention to produce a welfare state and mixed economy. In the 1980s this was overturned by cutting the welfare state and promoting a free-market economy. The result has been a much more polarized society whose geographical expression has been the 'north–south divide'. The political implications of this for the spatial integrity of the British state can only be speculated upon at this time. Certainly in terms of the electoral geography of the country the old 'uniform swing' at General Elections, when relative changes in votes were approximately equal across all or most regions, has long gone. The

Conservatives have consolidated their hold on the south where Labour are almost impotent but the latter party has grown in strength in the north. The most fundamental change is in Scotland where just thirty years ago the Conservatives had a majority of Members of Parliament. Today they hold only nine out of the seventy-one Scottish seats and have no Scottish representation at all in the European Parliament. A new political geography of Britain is being created but it is not yet clear what it actually means for the future.

The Political Geography of Localities

Within the states of the world further political changes are occurring in the regions and local communities. Once again we can identify both general processes that operate across many states and specific expressions in a particular state. There are two main questions that political geographers have focused upon in locality studies across states. The first relates to conflicts between local communities and the central state. In federal states such as the United States sovereignty is split between the centre and the provinces (e.g. US 'states') so that local communities may be protected from central dictat. Nevertheless the balance of power between the two levels of sovereignty remains a contested politics (e.g. US 'states' rights' debates). The distribution of power is much simpler in unitary states such as Britain where the central government has total control of the state. In such situations other levels of government only exist with the permission of the centre. Local government has no protection from the whims of central government.

The second general political process relates to conflicts among the local communities themselves. For instance in many countries it has been shown that suburban populations benefit from central city services without contributing to their costs. Such 'free-riding' is particularly important in the United States. In recent years studies of competition between communities have concentrated on 'growth coalitions' where local political parties and interest groups sink their differences to promote their community. This new local politics may be led by either business or labour as traditional animosities are transcended, producing another break with past norms. The purpose of such local government is to attract investment to a particular locality in preference to its rivals, neighbouring localities. A new political geography of community competition is emerging in different countries throughout the world.

In the particular case of Britain the breakdown of the post-1945 consensus has been most dramatically expressed in centre–local government relations. Local government was at the centre of politics in the 1980s as never before. The outcome of the conflict in this unitary state was a victory for the centre, but not until after a series of major disputes. In the process a whole range of functions has been removed from local government control in the fields of transport, planning, education and housing, and one whole tier of local government, the metropolitan counties including Greater London, have simply been abolished. In the late 1980s local governments have had to reassess their role, and economic promotion strategies, often in conjuction with central policies, have become important. Growth coalitions are appearing in both Conservative-dominated local governments in the south and Labour-dominated local governments in the north. Perhaps we are moving towards a new political geography of localities where neighbouring communities are the chief 'enemy', not central government:

Glasgow is 'smiles better' . . . than Edinburgh.

We are living in a period of massive political changes and this is clearly expressed in the three geographical scales of political geography. No wonder this previously neglected subdiscipline of geography is undergoing an unprecedented revival at the present time. But we must be careful not to treat the three scales as autonomous political patterns and processes. Rather, they link together as one overall set of changes. 'Thatcherism' as a political phenomenon, for instance, has been portrayed above as part of the geography of the state but it is obviously an important component in recent trends in geopolitics – the 'Iron Lady' – and in the curbing of local government. In fact whereas other political studies typically treat politics as a set of disconnected processes, the promise of political geography for the 1990s is to produce a rigorous and vigorous, more holistic, body of knowledge to make sense of our rapidly changing political world.

Further Reading in Political Geography

Articles on Political Geography can be found in all the mainstream geography journals. In addition *Political Geography Quarterly* (founded in 1982 and just *Political Geography* from 1992), is an international journal specializing in the subdiscipline. Here are some of the main books published in this area in the last decade which illustrate the revival.

Agnew, J. 1987: *Place and Politics: the geographical mediation of state and society*. Boston, MA: Allen & Unwin.

Anderson, J. (ed.) 1986: *The Rise of the Modern State*. Brighton: Wheatsheaf.

Archer, J. C. and Shelley, F. M. 1986: *American Electoral Mosaics*. Washington, DC: Association of American Geographers.

——and Taylor, P. J. 1981: *Section and Party: a political geography of American presidential elections*. New York: Wiley.

Blake, G. (ed.) 1987: *Maritime Boundaries and Ocean Resources*. London: Croom Helm.

Blaut, J. M. 1987: *The National Question*. London: Zed.

Chaliand, G. and Rageau, J.-P. 1985: *Strategic Atlas: world geopolitics*. London: Penguin.

Chase-Dunn, C. 1989: *Global Formation*. Oxford: Basil Blackwell.

Cox, K. R. and Johnston, R. J. (eds) 1982: *Conflict, Politics and the Urban Scene*. London: Longman.

Dalby, S. 1990: *The Coming of the Second Cold War*. London: Pinter.

Freedman, L. 1985: *Atlas of Global Strategy*. London: Macmillan.

Goldstein, J. S. 1988: *Long Cycles: prosperity and war in the modern age*. New Haven, CT: Yale University Press.

Hoggart, K. and Kofman, E. (eds) 1986: *Politics, Geography and Social Stratification*. London: Croom Helm.

Johnston, R. J. 1985: *The Geography of English Politics*. London: Croom Helm.

——Knight, D. and Kofman, E. (eds) 1988: *Nationalism, Self-Determination and Political Geography*. London: Croom Helm.

——Pattie, C. J. and Allsopp, J. G. 1988: *A Nation Dividing? The Electoral Map of Great Britain 1979–1987*. London: Longman.

——Shelley, F. M. and Taylor, P. J. (eds) 1990: *Developments in Electoral Geography*. London: Routledge.

——and Taylor, P. J. (eds) 1989: *World in Crisis? Geographical Perspectives*, 2nd edn. Oxford: Basil Blackwell.

Kirby, A. 1982: *The Politics of Location*. London: Methuen.

Modelski, G. 1987: *Long Cycles in World Politics*. London: Macmillan.

Mohan, J. (ed.) 1989: *The Political Geography of Contemporary Britain*. London: Macmillan.

O'Sullivan, P. 1986: *Geopolitics*. New York: St Martins Press.

Pacione, M. (ed.) 1985: *Progress in Political Geography*. London: Croom Helm.

Paddison, R. 1983: *The Fragmented State: the political geography of power*. Oxford: Basil Blackwell.

Parker, G. 1985: *Western Geopolitical Thought in the Twentieth Century*. London: Croom Helm.

Pepper, D. and Jenkins, A. (eds) 1985: *The Geography of Peace and War*. Oxford: Basil Blackwell.

Short, J. 1989: *Introduction to Political Geography*. London: Routledge.

Sloan, G. R. 1988: *Geopolitics in United States Strategic Policy, 1890–87*. Brighton: Wheatsheaf.

Slowe, P. 1990: *Geography and Political Power*. London: Routledge.

Taylor, P. J. 1989: *Political Geography: world-economy, nation-state and locality*. London: Longman.

——1990: *Britain and the Cold War: 1945 as geopolitical transition*. London: Pinter.

——and House, J.W. (eds) 1984: *Political Geography: recent advances and future directions*. London: Croom Helm.

Thrift, N. and Williams, P. (eds) 1987: *Class and*

Space: the making of urban society. London: Routledge.
Wallerstein, I. 1984: *The Politics of the World-Economy*. Cambridge: Cambridge University Press.
Williams, C. H. and Kofman, E. (eds) 1989: *Community, Conflict, Partition and Nationalism*. London: Routledge.
Wolch, J. and Dear, M. (eds) 1989: *The Power of Geography: how territory shapes social life*. Boston, MA: Unwin Hyman.

Part III
How to Study Geography

Doing geography offers the student the opportunity to master a large and growing range of techniques. In recent years computing, geographical information systems (GISs) and remote sensing have added to the skills available to geographers. Many students will be required to undertake some original research as part of their degree. This provides the chance to put into practice some of the new technologies, but also the chance to learn the more traditional crafts of interpreting landscapes, searching through archives and interviewing people. This section presents a review and introductory guide to many of the skills geographers can acquire. Alan Jenkins and John Gold offer advice on how to get the best out of a course by learning how to learn. Barbara Kennedy then offers some salutary advice to students embarking on research of their own. The remaining essays can be read as guides on how to get started on geographical research using a particular skill and what pitfalls await the budding researcher.

III.1

Effective Learning: A Traveller's Guide

Alan Jenkins and John R. Gold

Geographers know that things change as they travel from one place to another. Familiar landscapes are replaced by unfamiliar landscapes, known landmarks disappear, and the inhabitants have different cultures. A student entering higher education is also destined to go on a journey from familiar to unfamiliar territory. If you are about to embark on that journey, we set out here a traveller's guide to help you to organize your learning effectively and to gain the most from your studies.

We begin by explaining something about the differences between school and college learning, before providing practical guidance about lectures, small-group teaching, reading, writing and examinations. We stress throughout the view that effective learning requires you to become an active participant in the learning process. Quite simply, learning is not something that happens to you, but something that you do. The more actively you participate in the task, the more likely you are to learn and the better you will enjoy what you do.

The Ground Rules

At first sight, you may not appreciate the full differences between school and college, since there will be some similarities with the school that you attended. Despite course titles, geography is still normally taught as a subject in its own right and divided into subdisciplines with familiar names like human geography or geomorphology. You will again be supplied with a weekly timetable, divided into periods into which teaching is scheduled. There will still be teachers, classrooms, libraries and laboratories. Yet, in other ways, appearances are deceptive.

Take, for example, the timetable. At school, that timetable structured your day, with formal teaching sessions occupying much of the week. At college, by contrast, the timetable normally leaves you considerable amounts of time to study independently, often with little obligation even to attend for the relatively small number of hours for which teaching is scheduled. *You* now have to take greater responsibility for organizing your own learning.

Teaching is also arranged differently. At school, you may well have been part of a class of ten to thirty pupils, all specializing in geography, who worked together as a unit for several years. Teaching would have been handled by a small number of staff, sometimes just a single teacher, who covered the full breadth of the geography curriculum. School teachers will have got to know you well, have understood your strengths and weaknesses, and have had the time to encourage and cajole.

Perhaps you thought that higher education was the same, since college prospectuses invariably feature photographs of students learning in small groups and receiving instruction on a one-to-one basis with a teacher in the laboratory or at a computer terminal. The reality, however, can be very different, especially if financial cuts have eroded the amount of individual instruction offered. You may be assigned to an individual tutor to look after your general welfare, but the rule in most institutions is that large lecture groups are the primary forum for teaching and learning, with smaller seminar groups used only in a supporting role. In the first year in particular, these lectures may be attended by seventy to 200 students, who are enrolled in a variety of courses and come together at no time other than for a small number of shared lectures. Moreover, a lecture series may well feature a number of teaching staff, with their varying styles and modes of delivery. The impression of being surrounded by a sea of unknown faces and taught by a never-ending procession of lecturers can be unnerving. Again, it is often hard to escape the feeling that you are on your own.

That feeling is reinforced when you discover that teachers are less accessible than at school. Some seldom seem to be in at all or, if they are in, their doors are firmly closed. Others may actively encourage students to discuss problems and create a climate in which you can feel that you can approach them, both in and out of the classroom. Yet even these teachers may only be available at specific times in the week, perhaps by appointments made days in advance.

Why should there be these differences? To make better sense of things, let us return to the notion that entering higher education is akin to moving to a new country. Three broad notions are helpful in understanding its culture.

First, there are different cultural assumptions about learning. The ethos of many institutions of higher education is that the students should 'learn to learn independently'. For some, this is interpreted as meaning little more than that students with the qualifications to enter higher education are bright enough to fend for themselves, and able to use the framework of lectures and the guidance of prepared booklists to *read* for their degrees. Others would argue that the goal is actively to facilitate the student's own learning, so that the student leaves college enthused by what geography tells us about the world and fascinated by learning itself. Either way, however, there is a heavy emphasis on the role of the student as an 'autonomous learner'.

Second, the culture requires teachers to do a variety of other things besides teaching. As part of their jobs, they are frequently expected to conduct research on topics of scholarly and applied interest; to act as course, departmental or institutional administrators; to involve themselves in prestigious professional activities beyond the bounds of the department; and, increasingly, to attract funds for research or consultancy into their departments. They divide their time between the various aspects of their job and sometimes these tasks require them to be away from their rooms or from the department altogether.

Third, the culture offers few direct rewards for teaching quality. Unlike their school counterparts, higher education teachers are rarely required to have professional teaching qualifications; indeed, the institution may not even value such qualifications. Most academic staff came into higher education from a background in postgraduate research or industry. They see, and are encouraged to see, the central part of their jobs as creating, extending and applying knowledge. Their career prospects are more likely to rest on the strength of their research or consultancy achievements than on their teaching. They are human and respond to what the culture of their institution rewards.

These aspects of higher education's culture are rarely acknowledged and deserve fuller debate than they normally receive. The current authors, for example, are highly critical of elements that diminish the importance of quality teaching in higher education, but fully endorse the cultural view that higher education's mission is to help students to become independent learners, a mission which, as teachers, it is our responsibility to facilitate (see Gold et al., 1991). Yet, for present purposes, one point is clear. If you can quickly understand that these values exist and learn the ground rules of this new culture, then your own transition to becoming an effective learner will be that much quicker. This thought underpins much of the material in the ensuing sections.

Learning from Lectures

Now I lay me back to sleep,
the speaker's dull, the subject's deep.
If he should stop before I wake,
give me a nudge for goodness sake.

Anon

Lectures are the heart of geography teaching in higher education, with the lecture programme normally providing the framework for a student's working week. Among other things, teachers use them to map out what to study and to supply a basic grounding in aspects of geographical knowledge and how to approach independent reading. Lectures can vary markedly in format, ranging from uninterrupted monologues, through illustrated talks based around audiovisual aids or handouts, to presentations that the lecturer divides into segments to enable students periodically to ask questions or carry out lecture-related tasks. Whichever forms you encounter, the fact remains that lectures suffer from the severe limitation of being essentially one-way transmissions of information, typically lasting fifty to fifty-five minutes, which require the audience to do little more than just be there. No matter how good the lecturers or how much they try to retain your interest, you are going to find it hard to sit there passively and maintain concentration for that time.

To improve matters, we recommend that you think carefully about how you can improve your own learning by taking a more active participating role. You can do this by the following:

1 Arrive in good time to ensure that you have a comfortable seat, can write easily and can see the boards or video screens if used.
2 Briefly review the notes that you made in earlier lectures. We learn better if knowledge can be related to what we already know.
3 If a handout or, better still, lecture synopsis is supplied, read or scan it immediately. This triggers previous knowledge and arouses expectations about what will follow.
4 Decide what notes to take. There are arguments for not taking notes at all, for example that the act of note-making distracts from listening attentively, but we would argue that they are valuable. Notes help you to remember the lecture, provide a basic structure of ideas and information that you can apply to your reading, and help you to maintain attention. Moreover, you may find that lectures contain material that cannot be obtained elsewhere, perhaps because it represents research that the lecturer has not yet published. The value of notes, however, depends on thinking through carefully what forms meet *your* needs. Experiment with different ways of taking notes, varying the length and layout. A useful idea here is to compare your notes with those of a friend and analyse any areas in which you both might improve your note-taking. You should quickly be able to find a method that suits you.
5 Not all information given in lectures is of equal significance. Listen attentively for those portions that explain the lecture's basic theme, for the key ideas that exemplify that theme, for

the evidence offered in support of them and for recommended reading.

6 Leave space in your notes to add material *after* the lecture. Then, as soon as possible, review the notes that you have taken. Try to draw out links with previous lectures or seminars. Write down points that you did not understand or want clarified. Look for extra information that supports or questions the lecturer's arguments. Remember it is easiest to do something positive with lecture material when the lecture is fresh in the memory, rather than by simply filing notes away and next taking them out on the night before an examination.
7 Periodically review your notes and any handouts associated with the lectures. Quite apart from better retention of information, this will help you to understand the structure of the course and follow the direction in which it is going.

These then are some suggestions for improving your learning in lectures. Above all, you should recognize that, though you can have little power over a lecture's direction and organization, you can take steps to ensure that you get the most out of it. In the next section, we discuss situations in which you can take even more responsibility for the effectiveness of your own learning.

Small-group Teaching

Watermouth makes students nervous; you never know what to expect. . . . There are classes where the teacher, not wanting to direct the movement of the mind unduly, will remain silent throughout the class awaiting spontaneous explosions of intelligence from the students; there are classes indeed where the silence never gets broken.

Malcolm Bradbury

Virtually all college geography courses provide timetable slots for small-group discussions. These sessions can involve anything from three to thirty students and have various aims, which include exploring in greater depth material covered in lectures, drawing links between the different elements of a course, developing speaking and presentation skills and providing an opportunity for more relaxed interaction between students and teachers.

At its best, learning through discussion in small groups can be highly enriching, but, when used badly, it can also be immensely frustrating. A case in point occurs when the teacher, perhaps in response to initial student reticence to speak, so dominates discussion that the session turns into an unprepared and badly delivered lecture. Another example is when a student is asked to initiate discussion by reading a paper – a situation which can easily degenerate into a conversation between the teacher and that student, with the rest excluded. Too often other students leave these sessions without any clear sense that they have learned anything.

The responsibility to ensure effective learning is shared. The teacher is responsible for structuring the session, for providing any appropriate materials, for indicating what is required and for suggesting how you can contribute, but there is much that you can do, individually and collectively with your colleagues, to ensure the effectiveness of your own learning.

1 Come to the session prepared, even if you are not personally required to present anything formally. Such sessions can seem an easy option, but they can be made more valuable by doing background reading or even just by looking back through your lecture notes. When doing so, try to jot down any problems you are having with course material, questions that you want discussed, or points you want to raise, and take the list along with you

to the session. There will invariably be opportunities to have them discussed.

2 Arrive punctually and attend regularly.
3 Help to ensure that the seating is arranged so that everyone can hear and see one another. Research shows that seating arrangements do influence the interaction that results.
4 Be willing to break the ice. As at a party, someone needs to be willing to speak first, make a joke or even ask the seemingly naive question. Many students will silently thank you for this: it's the simple questions that many of us don't really understand but are hesitant about asking.
5 Be involved. Involvement can mean asking questions, contributing ideas, performing a role to the best of your ability. It can also mean helping others to contribute and being willing to listen to their point of view.
6 Carefully prepare any formal presentation. Do not intone from a prepared script; most people find it very difficult to follow material that is read at them. Outline the structure of your talk clearly and consider using a handout to communicate detailed material. Speak succinctly and with enthusiasm, maintaining eye contact with the group. Try to involve others by suggesting issues to discuss or areas that you found interesting.

After the session, and while the memory is still fresh, try to find time to review the material covered in the class and link it back to previous sessions and forward to the next. Even the best sessions gain from later reflection.

Reading Effectively

Some books are to be tasted, others to be swallowed and some few to be chewed and digested.

Francis Bacon

We commented earlier about the notion that you are at college to *read* for a degree. By definition, reading is a vital activity and one which involves a major commitment of time, but how do you decide what to read and how to read it? (See also chapters IV.2 and IV.3.)

We begin by stating a simple rule: you can never read everything relevant to your studies. Your time is short and libraries are large. Effective reading in higher education means reading those sources that provide you with sufficient information and insight to tackle a subject to the best of your abilities. In finding the formula by which you can do so, you should consider some broad guidelines.

1 Think through what you have read in the last six months, including reading for pleasure. Which of it could you now explain clearly to a friend? Analyse what it is that you have retained and where you feel you did your best reading. What clues does it give you for effective reading in the future?
2 Look for guidance about what to read, but approach course reading lists with caution. Teachers in higher education are notorious for giving lists intended to impress rather than inform. Listen out in lectures for key sources and keep a list of essential sources that you currently need to find and consult.
3 See if you can obtain guidance about how to use the library. Some college libraries have highly idiosyncratic reference systems and specialize in placing bound volumes of periodicals in unfathomable places. Take a guided tour of the library if one is offered and study any written guides that may be available: the investment of a small amount of time at the outset of your studies will subsequently repay you handsomely. When in doubt, ask a librarian. Many are themselves specialists in particular fields and can

often give you advice that goes beyond merely the location of books and journals.

4 Having found the sources that you require, read them with strategy. As ingrained as the habit of reading a book from cover to cover may be, it is not a particularly useful habit here. When approaching a book or article, think through what you already know about the subject and why a particular source has been recommended to you. Thereafter study the list of contents and skim the text to discern its structure. Read the abstract if one is supplied and glance at the introduction and conclusion – it isn't cheating and can tell you how much time to give a particular piece of reading.
5 As with lectures, consider taking notes of key ideas and arguments. Don't assume that you have to make copious notes, but do consider writing your own critique of what is being said.
6 Occasionally consider using a structured, SQ3R, approach to reading a text. This consists of *scanning* to gain a sense of the overall argument; *questioning* whether it needs to be read in full, in part or at all; *reading* in the light of those questions; *recalling* what you remember of the text; and *reviewing* what you have learned in relation to what you already knew, in the light of how it relates to the course and what it still leaves unanswered. This device may not be to everyone's taste, but at least give it a try. If nothing else, it again underscores the message of this chapter – that you can take greater responsibility for your own learning.

Written Assignments

Writing can help learners . . . we are not likely to sit down with empty minds or no plans, but the plans do change and new thoughts come to us. Certainly, we are all aware of the feeling of having finally mastered and possessed forever after, an idea when we wrote it out.

Peggy Nightingale

Writing is central to learning geography effectively. Being confronted by a blank sheet of paper is a powerful way of finding out what we know and what we have still to learn. It is also the principal way that you communicate what you have learned to your teachers. Indeed, although some courses will assess you through, say, oral presentations or statistical exercises, most of your assessment work will be written. It pays to be an effective writer.

The problem is that, as with reading, it is often expected that you can already write effectively, in the sense of organizing and structuring an argument and correctly handling syntax and grammar. If the activity does not come naturally, or you feel that you have deficiencies that could be improved, there are sources and people to whom you can turn. There are manuals on improving your writing and many colleges offer writing courses, perhaps including related skills such as word-processing. You can also learn by closely scrutinizing 'quality' journalism, which contains many conventions and stylistic devices that you can bring to your own writing. Above all, writing is a craft which can be learned and improved by dint of hard work.

In tackling a written assignment, the following may be helpful.

1 Instructions for written assignments are not always clear. As soon as you receive the assignment, read carefully through what is required and try to identify the key ideas, what research you require to do and what timetable you need to follow. If unsure about any of these points, try to discuss them with your tutor as soon as possible.
2 Plan out the structure along with first

thoughts about the content. The traditional school essay scheme – say what you are going to say, say it, and then say that you've said it – is still relevant, but be willing to experiment with other types of structure and ways of writing.

3 Use your initial plan to guide your reading, but remain receptive to new ideas as they arise. Certainly, most people find that their understanding of an issue changes significantly as they come to write about it.
4 Find out whether there are any rules or conventions that you are required to follow. For example, does the assignment have to be word-processed? Should it have an abstract? Is a formal bibliography needed?
5 Attempt a first draft and show it to a fellow student. Most articles that you read in geographical journals are refereed by other academics and subsequently rewritten to incorporate their suggestions. You should also be willing to take advice. In particular, ask your colleague whether the structure is sufficiently clear and whether the style is appropriate for the task in hand.
6 Ensure that you have sufficient time to amend or rewrite your draft, taking into account any critical comments that you have received.
7 When you receive the assessment back, don't only look at the mark but also read through the tutor's comments. Recognize that tutors may not have the time to comment at length on what you write. If you want further comment, go to discuss the comments with them. If you are unhappy with the quality of what you submitted, it may be possible to resubmit the work, not necessarily for a revised mark but to extend your grasp of the topic.

To conclude this section, we would ask you to recall the statement by Peggy Nightingale with which we started: 'we are all aware of the feeling of having finally mastered and possessed forever after, an idea when we wrote it out'. To express that differently, the act of writing geography assignments is itself a powerful and effective way to learn geography.

Examinations

The naïve teacher points to the beauty of the subject and the ingenuity of the research; the shrewd student asks if he is responsible for that on the final exam.

Paul Goodman

However effective you think your learning may have been, the formal assessment of that learning normally rests on your teacher's judgements of your performance in end-of-course or end-of-semester examinations. And, although higher educational practice is now changing to include a wider range of assessment criteria and techniques, assessment will still probably take the form of unseen written examinations.

In approaching this type of examination, we are acutely aware that you will have little control over how you will be assessed or the content of assessment, but there are positive steps that you can take to improve your chances.

1 Find out as soon as possible how you will be assessed. Some teachers will clarify this at the initial meeting of a class, perhaps providing you with a specimen examination paper from a previous year. If they don't, consult previous years' examination papers (there are usually copies in the library). Provided these are to the same syllabus and format, they can supply vital information about what is expected. You can then make decisions about how to approach lectures, seminars and your own independent study. Having said this, we also stress that knowledge of assessment procedures

is different from total preoccupation with them. Your final performance may well be substantially better if you first give yourself an opportunity to read around a subject and let your own ideas and interests develop. Exclusive concern with issues associated with assessment can ultimately prove counterproductive.

2 Ask teachers to give you some indication of what could constitute good quality in an examination answer: what would gain a First, or A grade, mark as opposed to just scraping a pass mark. Academics have conventions about what constitutes a first class answer and it is extremely valuable to know what they regard as excellence in this respect.
3 Listen carefully for ‘cues’. Many teachers would blanch at setting examinations that are totally unseen, in the sense of asking questions about absolutely anything, no matter how trivial, covered by a course. Hence, they frequently indicate, subtly and otherwise, what they regard as important and on which students can expect examination questions. Take notice and act accordingly; they have little reason to wish to mislead you.
4 Well before any actual examinations, sketch out some answers to questions from past papers. It is a useful indication of how well you have learned the material and what you still need to do.
5 Have confidence in your ability at examinations. Anyone who has got this far must already be competent in this matter and the well-tried rules of thumb learned at school still apply – read the question carefully, identify key words, plan your answer, maintain a consistent structure, carefully apportion your time, make sure your handwriting is clear and so on. However, be aware that simple regurgitation of notes may have been acceptable at school but will now be a recipe for achieving, at best, indifferent marks. Achievement of good grades in higher education depends on being able not just to cite the opinions of others but also to produce your own considered and well-argued interpretations of the evidence. It is this maturity of opinion that frequently distinguishes the good performance from the mediocre.

Becoming a Native

The aim is to be so actively involved in educational processes that one fully belongs to – is possessed and transformed by – that engagement.

Joe Powell

We started this chapter by likening your experience as a student entering college to that of a traveller moving to a new country. In conclusion, let's return to that analogy. We hope that you will enjoy your years studying geography as part of this new ambience and culture, but how deeply will you be affected by the experience? Perhaps it will be just like a holiday romance, pleasurable at the time but now only dimly remembered, even with the aid of faded photographs. Yet, if you really come to grips with this new culture, its values and practices, the end product will be something that you will carry with you wherever you go and whatever you do. You will quickly forget the content of many of the geography courses that you take, but you will continue to value both knowledge itself and the active pursuit of new knowledge. The Australian geographer Joe Powell, whom we quoted above, argued that the purpose of a geographical education should be to involve students so actively in educational processes that they fully belong to, are possessed and transformed by ‘that engagement’. To achieve that engagement, you have to take responsibility for your learning.

Further Reading

As well as many useful articles in the *Journal of Geography* and the *Journal of Geography in Higher Education* the following guides, amongst many others, set out useful information on how to study effectively.

Dunleavy, P. 1986: *Studying for a Degree in the Humanities and the Social Sciences*. London: Macmillan.

Gold, J. R., Jenkins, A., Lee, R., Monk, J., Riley, J., Shepherd, I. and Unwin, D. 1991: *Teaching Geography in Higher Education*. Oxford: Basil Blackwell.

Marshall, L. A. and Rowland, F. 1983: *A Guide to Learning Independently*. Milton Keynes: Open University Press.

Meredeen, S. 1988: *Study for Survival and Success*. London: Paul Chapman.

Phipps, R. 1983: *The Successful Student's Handbook*. Seattle, WA: University of Washington Press.

III.2

First Catch Your Hare . . . Research Designs for Individual Projects

Barbara A. Kennedy

Many geographers nowadays first embark on individual project work while they are still at school, and many of those reading for first degrees in universities and polytechics are required to produce a dissertation based on original research of some kind. Clearly, this is an area where 'practice makes perfect' and, equally clearly, there will be a huge variety in the scope and nature of topics attempted, but in this section we try and set out some of the critical steps which need to be considered in designing an individual research project.

Choice of Topic

There are two basic rules: first and most crucial, *you* must be interested in the project at the outset; second, you should be able to see how the particular aspect you wish to investigate relates to some broader area of geographical enquiry, so that *other* geographers will potentially be interested in your findings.

Far too many undergraduate dissertations emerge as 'stand-alone' accounts of some area of local concern – such as new building development in a particular centre – with no framework of general discussion or references to comparable studies. This omission makes it extremely difficult for the reader to evaluate the significance of the case study, however well it is presented, and leads to the general reaction 'so what?'

In an ideal scientific work, the broad question should come first and the particular case study should be chosen so that it will provide maximum information. So, if you wish to study the effects of point pollution emissions on soil, vegetation and water, you would *first* identify the key relationships you wished to investigate and only secondly search for a site or sites where you could hope to analyse the question with a minimum of interference by extraneous factors (e.g. variable surface and solid geology, land use etc.). In practice, particularly for many undergraduate studies, the cost of getting to and living in the 'ideal' site(s) whilst conducting the investigation would prove prohibitive, so that the choice of topic tends to be determined by the range of possibilities in an easily accessible target area (usually that close to the student's home or in a region where an expedition is being mounted). In the real undergraduate world, then, logistics usually outweigh 'pure' considerations. However, it is still highly important to ensure that the topic is going to be of some general significance.

Having decided upon an area of interest, of whatever form, the second step is to focus upon a set of questions which you will hope to answer. It is at this point that it becomes absolutely critical to have a broad view of the *general* field of enquiry before you start detailed planning.

If you are going on an expedition to a semi-arid area where irrigation is practised, and you are interested in aspects of this form of water management, then you should begin by looking *both* at general texts about dry land farming *and* at recent detailed studies of irrigation projects, to see exactly what other geographers think are key topics or problems or what kinds of information they think need to be collected. In the process you will also, of course, get some idea of the practical difficulties likely to be encountered, as well as the variety of ways in which you can hope to acquire information. Unless you do this preliminary reading carefully your study is likely to fall into one or both of two traps: reinventing the wheel; or barking up the wrong tree. The former is a self-evident problem: there is little point bounding back from the boondocks with the 'study of the century' only to find (or to have your supervisor or examiners find) that what you have sweated to do, others did earlier (and probably more thoroughly). The latter difficulty can cover a multitude of situations, but it will arise from the well-known operation of Murphy's law (anything that can go wrong, will) and the likelihood of finding that you have to change direction in mid-investigation: only if you have a good *general* grasp of the broad topic will you be able to modify your research plans in a sensible and fruitful direction, when apparent disaster strikes.

General Outline of Approach

The choice of topic and of 'testing ground' then, have to proceed hand-in-hand; from these choices, the general and particular questions to try and answer will emerge. Figure III.2.1 endeavours to set out two alternative ways in which undergraduate studies can be categorized. Figure III.2.1(a) shows the preferable or 'wine glass' format, where a broad initial survey of the topic leads on to a narrowly focused core of original work and analysis and then *back* to a broader review of findings in the light of the initial ideas and the general questions with which the study began. Like a good wine glass, this is a very stable structure and the proportions of 'bowl', 'stem' and 'base' can be modified according to need and taste: if you are working in a relatively new field but have a great deal of your own material, then the 'stem' can be proportionally extended; if, on the other hand, you hit enormous problems with collecting material, you can expand the general discussion and conclusions. Figure III.2.1(b) in contrast, is what all too many dissertations resemble: it is what I would term 'My home town: its role in song and story' or the 'All you wanted to know about new houses in Lower Slobovia but were afraid to ask' approach. It may be superficially attractive, but is actually lethal and you are strongly advised to shun it.

Why is this second approach not a good idea? First, it is often not set in any true general context. At the practical level, this means that if the house-builders of Lower Slobovia will not talk to you, nor its government allow you access to maps, then you are stymied. Second, because there is no general context, there are likely to be no easily discerned general questions other than 'where are the new houses and when were they built?' This in turn leads almost inevitably to a *narrative* and, contrary to the pious hopes of generations of undergraduates, pure narrative is both exceedingly difficult to write and almost impossibly difficult to write well: if you do not believe this, then look at some of the eighteenth- and nineteenth-century travel-cum-science narratives (those by De Luc (1810–11), Pallas (1802–3) or Saussure (1779–96) are good examples) and see how gripped you are *even* in cases where landscapes and their inhabitants are being described for the first time. Pallas and Saussure succeed to a remarkable, though not universal, degree, but De Luc's account – even though it is imbued with a high theoretical purpose – is one of the most

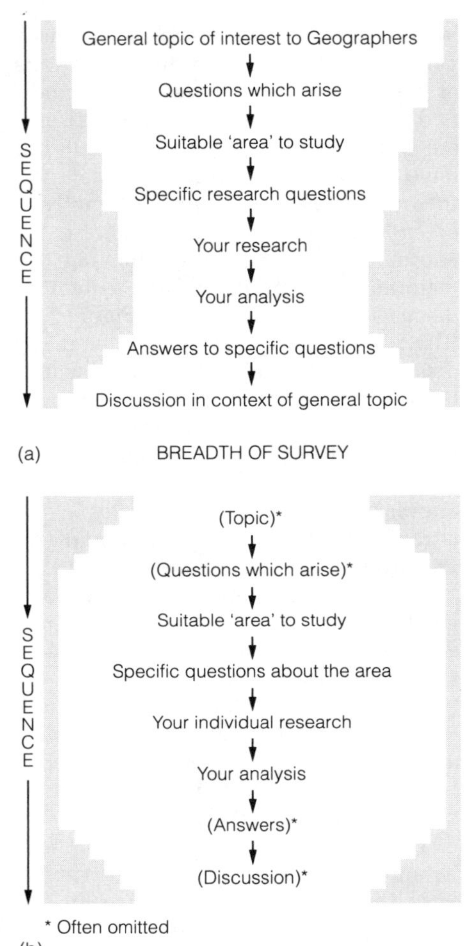

Figure III.2.1 Two general research plans: (a) the 'wine glass' which produces focused study in a broad framework; (b) the 'splodge' with little or no general context but a great deal of specific detail.

dismally boring pieces of work you are likely to encounter. Figure III.2.1(b) is deliberately drawn to resemble nothing more than an amorphous splodge and that is, all too often, what ill-defined dissertations are like. Avoid this approach.

Designing Your Data Collection Programme

All research is undertaken within constraints of time and money. The most important things, then, to establish before you are totally committed to spending three months in Lower Slobovia are the following: What data will you need to collect in order to try and answer the questions you are interested in? How long will that take? How long will the material take to analyse? Your research design needs to take a clear account of the nature of any 'bottleneck' in the sequence, although this will vary from topic to topic. Let us look at some examples.

A very common limitation, especially in historical projects or those involving access to official statistics or company records, is the sheer availability of information. Do the Parish Registers or County Rate Books exist? Will local or national government allow you to inspect statistical returns? Will the management of companies you target for investigation let you set foot on the premises? Given the realities of the sources actually available and accessible, can you hope to reach any worthwhile conclusions?

Second, there is the constraint of the time and/or expense of actually acquiring the material. If your total budget is £100 and twenty days and the 'cost' of visiting each source is going to be £20 and four days, you can only hope to obtain five units of information: will that prove adequate, given the question you are hoping to answer?

Third, it is often very much easier to *obtain* raw information than to process it in a fashion which yields appropriate material. A single, unstructured, question-and-answer session may take only an hour, but leaves you with eight or ten pages of transcribed material from a tape. A single soil sample may take fifteen minutes to collect but two days to analyse for trace metals. A single run of a computer model may take two minutes, but assessment of the significance of the parameters may take two weeks.

The only effective way to try to foresee where the crucial bottleneck will occur, and therefore design the most cost-effective research programme, is to carry out a trial run or *pilot project.*

This is probably the single most important practical measure that you can take to ensure the success of your research. You should take a small amount of material from a source which is comparable to – but preferably not identical with – the one you hope to use for the main study and analyse it. This is, of course, much easier to do in some fields than others: if you are to study river channel morphology on a tributary of the Amazon, you can nevertheless try out measurement, recording and analytical procedures on the Nether Wallop brook. What you do when your project involves asssessing market trader patterns in Burkina Faso is, clearly, more problematical, but it is still likely to prove useful to visit a local market near your home and try to assess the situation; you should also, in this case, make sure that you read all available published sources on the markets of Burkina Faso and those adjacent areas *and*, if possible, talk to someone who has actually worked there. Even in the latter case, the experience of trying to disentangle what is going on in one section of the real world is likely to provide useful insights into the nature of the situation in another.

Probably one of the most valuable aspects of any pilot survey is the practice it provides in regulating the mechanics of data collection. It is by no means always apparent how to order your information: what is going to be the 'recording unit'? Which material is critical? Which needs to be quantitative? How will you later assess qualitative statements? Whether you are making physical measurements, merely counting, just observing, or extracting details from documents, the experience of the pilot project should help you to construct some kind of 'booking form' or standardized format for entering your material. This can save hours of agony at a later date. Equally significant is to try to *extract* data from your pilot collection and see just how easy or difficult it will be to analyse. Do you need more precise locations, in time or space, than you initially supposed? What will you do to cope with missing observations? What are the likely upper and lower limits of numerical values in your data sets?

Far too many undergraduate projects 'start from cold' and, in consequence, find that most of the early material gathered is desperately inadequate. Even the most humble pilot scheme can prove immensely time and cost effective and should allow you to design a realistic and appropriate data collection programme. It may well be, of course, that the pilot project shows up really fundamental flaws in your initial scheme. If this happens, *abandon it.* The whole point of a trial run is to indicate whether your ideas will work in practice and, if they clearly will not, you should cut your losses and try a different tack. It is a fallacy that research proceeds in a straight line, from conception to results: in practice, one is working round a series of interlocking 'loops' and difficulties encountered anywhere may demand complete rethinking of prior stages in the research design.

Collecting Your Material

The only useful general guidance one can give here is: before you plunge into the details of questionnaires, measurements or participant observation, *make a general survey of the situation.* A simple example should indicate the need for and merits of this procedure. Suppose you wish to study the hydraulic geometry and sediment load of three small streams. You have good maps, but drawn from air photos flown in 1965 and last field-checked in 1980. A day or half-day actually walking around the three catchments will show you whether a new reservoir, or a new road, has been put in since 1980. Merely starting your field programme at one end of one valley may lead to a very nasty surprise on the last day of your project. This can be disconcerting, even if you are working close to home, but could be calamitous if you are in a remote and/or foreign location.

Although this advice is particularly important for 'field-based' research, it applies equally to archival work. Look through the whole mass of documents available; do not just start on the first one in the pile.

A general survey will not always pinpoint difficulties: it should give you a reasonably clear picture of the overall context of your research, however, which is invaluable.

Analysing Your Material

The first rule is: never throw material away or amalgamate categories early on. Anyone working with census data will rapidly grasp the problems which arise from the aggregation of units; and it is easy to put yourself in the same position by cavalier treatment of your material. This is not to say that every piece of information you have collected will ultimately prove valuable; indeed, knowing when you must – regretfully – jettison extraneous or unsound information is a very important part of conducting research, but it comes at a late, not an early, stage.

As indicated above, you should have a fairly clear idea of how you wish to analyse material before you collect it and your pilot project should have helped you to clarify the best methods to use. Nevertheless, whether you wish to use statistical, cartographic, graphical or computer-modelling techniques to handle your material, you must *check* that the data you have actually acquired will, in fact, be suitable for the method you wish to use. If your documents do not allow you to locate half the woods or fields of a seventeenth-century parish, then you are clearly going to have difficulty presenting a sequence of dated maps. It is problems like these which both demand that you have retained all your initial data and yet may necessitate, in some cases, your having to jettison some of them or having to amalgamate them into broader categories than seemed initially desirable.

Finally, the golden rule of all analysis is that its merit is entirely dependent on – and must be closely related to – the quality of the raw material. Over-elaborate analysis of lousy data is just as much a mistake and a waste of time as lousy or non-existent analysis of good material.

Interpreting Your Results

Whether for reasons of time or undue modesty, far too many undergraduates skimp or even neglect altogether the vital stage of evaluation and interpretation, yet this should be the most significant section for both the author and the reader. At the most basic, you should state clearly what your findings are and how these relate to the initial questions that you set out to try and answer. However, it is extremely important to stop and look carefully at the material you have acquired. Very often the processes of adjustment between your original idea and the cussedness of reality (described above) will have resulted in the fact that you have actually answered rather different questions from the ones you started out with. It is obviously very important indeed to check through and clarify what your project *has* achieved, rather than to blunder along, trying to demonstrate simply what you initially *hoped* to find.

What is most important is the assessment of your evidence. There are three major pitfalls to try to avoid: over-interpretation; under-interpretation; bias.

Over-interpretation, or trying to wring the last morsel of generally highly dubious significance from the thinnest pieces of information, is perhaps more common in PhD theses than undergraduate dissertations, but it is at best irritating and at worst misleading. If your analysis of 300 years of Parish Records has turned up only one unambiguous reference to the birth of twins, or death from smallpox, then you are ill-advised to parlay this into a disquisition on the relative frequency or significance of such events in the period under review. Similarly, if you have a matrix of 20 × 20 simple correlations and are using the 95 per cent significance level, you should bear in mind that you can expect, on average, 5 per cent of the 'significant' figures to be statistical artefacts.

Under-interpretation, clearly, errs in the opposite direction and is often the result of inadequate analysis or display of information, especially where this involves large data sets which have been statistically manipulated. A very common fault which arises from studies using questionnaires is a failure to try and correlate categories of answers: figures are given for those answering 'Yes', 'No' and 'Don't know' to Question One, but no attempt is then made to relate these to the responses to other questions (this is a situation where a simple contingency table (Siegel, 1956, pp. 196–202) may prove illuminating).

Equally, a large correlation matrix may be searched for significant figures with no attention given to those relationships which are statistically insignificant; yet these may be of equal interest in the light of previous work.

Both under- and over-interpretation are best avoided by keeping a clear grip on the geographical realities of the situation and, whatever your analytical techniques, returning frequently to the *absolute values* of variables. Just as 'one swallow doesn't make a summer', so a statistically significant increase of mean temperature of 0.01 °C may not necessarily imply that global warming is upon us.

Finally, there is the question of bias. All investigations, obviously, begin from some particular viewpoint and all of us are conditioned in what we see in the real world by expectation and experience which are constrained by the time and place in which we live. However, the object of a research project is not to reach 'The Truth' but merely to provide a 'consistent and responsibly-supported explanation' (Nagel, 1961) of the phenomena investigated. In the limit, this means that the interpretation must match the evidence and, where the latter is ambiguous or incomplete, competing possible interpretations should be provided. This very basic device is a long-established one used by a whole range of scientists and is generally known as the *method of multiple working hypotheses* (Chamberlin, 1890; see Kennedy, 1985). You would be well advised to read what Chamberlin has to say and to try to follow his advice.

Presentation

At the end of the day, the way in which you present your material will have an enormous impact on the reader.

Is there a clear flow of argument, from section to section? Are maps, figures, plates or graphs provided and referred to at the appropriate places in the text and are they clear? Are references accurate and are there adequate cross-references between sections of text? Have you proof-read the work properly?

The simplest way to ensure that you have met all these requirements is to get someone who does not know anything about the topic to sit down and read the text critically. All too often, by the writing-up stage, the author is so close to the problem that he/she can fail to realize that terms are undefined or that arguments lack some vital segment: an alert and critical non-specialist can generally recognize these important faults.

Conclusion

These guidelines have, of necessity, been couched in very general terms. However, even given the enormous range of topics tackled by undergraduate projects, they should provide you both with an indication of the main pitfalls you are likely to encounter and, it is hoped, a clear idea of the most time- and cost-effective ways in which your original ideas can be translated into a worthwhile and satisfying research project.

References

Chamberlin, T. C. 1890: The method of multiple working hypotheses. *Science*, 15, 92–6. (Reprinted in C. C. Albritton (ed.) 1975: *Philosophy of Geohistory*. Stroudsburg, PA: Dowden, Hutchinson & Ross.

De Luc, J. A. 1810–11: *Geological Travels*, 3 vols. London: F. C. and J. Rivington.

Kennedy, B. A. 1985: Indeterminacy. In A. S. Goudie (ed.), *Encyclopaedic Dictionary of Physical Geography*, Oxford: Basil Blackwell, 241–2.

Nagel, E. 1961: *The Structure of Science*. London: Routledge & Kegan Paul.

Pallas, P. S. 1802–3: *Travels through the Southern Provinces of the Russian Empire, in the Years 1793 and 1794*, translated, 2 vols. London: T. N. Longman and others.

Saussure, H. -B. 1779–96: *Voyages dans les Alpes*, 4 vols. Neuchatel: L. Fauche-Borel.

Siegel, S. 1956: *Nonparametric Statistics*. New York: McGraw-Hill.

III.3 Thinking Statistically

Stan Gregory

In any concern with geographical enquiries, it is essential to understand how and why results and arguments in print have been obtained, i.e. the processes of scientific reasoning that have been followed, so that the validity and value of the published material can be assessed and evaluated. Equally, it is often necessary to carry out one's own enquiries so as to arrive at conclusions, sometimes firm and sometimes only tentative, on the basis of sound and acceptable working and analytical techniques. When such studies involve, even in part, data that have been measured or counted, and especially if those data are only samples of a larger potential data set, then 'thinking statistically' must become part of the geographer's approach.

Types of Geographical Phenomena and Data

Types of geographical phenomena and data are many and varied. Which techniques should be used for their analysis will partly depend on the aims and objectives of the study, but in addition the nature of the geographical phenomena under review and also of the data that reflect such phenomena can be critical. They may impose certain limitations on the choice available, or make specific demands in terms of analysis and ultimate interpretation.

Phenomena distributed in space can be scheduled into three categories. Some, including those that vary in all directions, such as slopes, soils, temperature, rainfall, vegetation, and the processes that act upon them, are distributed continuously. Other continuous distributions may be linear in nature, such as rivers, coastal features, transport routes and flows in general, whilst yet others vary continuously but in discrete blocks. The latter include rural and urban land use, administrative units, geological outcrops and large-scale soil and vegetation units. A second category would be features that are essentially discontinuous but are often viewed as continuous. For example, population, employment, industrial production or social relationships are phenomena that occur at points within areas but which are often recorded as applying to those areas as a whole. Third, there are those features that are studied as discrete or discontinuous, such as particular landscape or townscape phenomena, or people as individuals, families or households.

Comparable distinctions can be made in relation to geographical phenomena varying in time. Thus climatic conditions, river discharge or population, although they may be recorded at discrete moments in time, are in fact varying continuously. Again, other phenomena, such as agricultural outputs or financial allocations, tend to occur in discrete but virtually regular stages, whilst yet others are intermittent and irregular, e.g. policy decisions and their implementation.

These distinctions are relevant not only to the issue of research design but also to the problems posed by the very mixed and varied nature of the data that need to be considered and analysed within the one geographical study. Four types of data are usually envisaged.

1 *Nominal* This implies data that fall into categories or groups, such as landforms, land-use types, occupational categories, diseases or nationalities. These cannot be measured, but their frequencies in given areas can be counted.
2 *Ordinal* In these cases the relative magnitude of the phenomena can be known, so that they can be ordered from largest to smallest, even though actual values are not available. This is often because of difficulty in obtaining exact measurements, either due to instrumentation problems or because of limited access to more detailed data.
3 *Interval* With interval data the actual magnitudes of the phenomena are available, such as slope angle, rainfall, production figures or population.
4 *Ratio* These include proportion or percentage measurements, such as voting percentages, relative humidities or chemical concentrations, as well as a variety of indices, such as pH values. These ratio data may have finite upper or lower limits, which obviously affect the way they can be analysed statistically.

Another distinction between different data sets that affects the potential for analysis is between continuous and discrete variates. Thus for costs, prices, distances, ratios, indices, temperature and rainfall, for example, all values of measured data are possible, including fractional or decimal parts. In contrast, discrete variates involve the counting of occurrences, as with people, factories, votes, trees or floods. Interpretation of results for the latter requires common sense, such as when describing an average number of children per family as 2.4, or the average number of floods per year as 3.6. Finally, a third distinction between types of data is between individual and grouped data. The former means that each item is recorded or measured separately, thus permitting analysis and conclusions to be explicitly related to individual items. Grouped data implies that information is in the form of the number of occurrences within specified magnitude limits. This may be because only certain points on the magnitude scale can be relied upon or are available, or because for operational or legal reasons data are made available only in this grouped format.

Sampling, Description, Inference and Testing

When considering a geographical problem, it may sometimes be possible to make a total enumeration of the relevant conditions. Thus for a specific locational area of interest all occurrences of some clearly defined feature may be recorded or measured, especially if it is a spatially or temporally discrete feature. More commonly, however, such total enumeration is not possible, perhaps because the locational area or the number of occurrences are too large, or the time available is too short. More critical, however, is the case when the phenomenon under study varies continuously in space or time, when it is in effect impossible to enumerate totally.

Under these circumstances it is necessary to study, record or measure only part of the total data set and thus to work with simple data. It is, of course, then possible simply to analyse and describe the characteristics of these sample data – to be concerned with *descriptive* statistics. Thus the central tendency or mean conditions can be specified, together with such other characteristics as the variability of the individual occurrences around that mean value and the nature of the overall frequency distribution of the observations. Presented numerically or graphically, these will provide a description of the data that have been gathered, whether this be for totally enumerated data or for only a sample of these data.

Useful as this may be in a limited sense, it often fails to ensure maximum scientific benefit from the data collection that has been carried out. Ideally, the items forming the sample, and the sample itself, should be as representative as possible of the total data set from which the sample has been selected. If this representativeness can be ensured, then – within certain limits – the characteristics of the sample can be used to infer the characteristics of the total data set, i.e. of the population from which the sample has been drawn. This move to *inferential* statistics represents the main benefit and advantage of using a properly structured and selected sample.

There are numerous ways of setting up a sampling procedure that will provide a representative and unbiased sample of the population. All of them involve, to a greater or lesser extent, the idea of a *random* sample, implying that every item in the population (or at least in the *target* population that can in fact be sampled) has an equal probability or chance of being selected as a sample item each time that such selection is made. Provided that the size of the sample is large enough, this should ensure that a satisfactory cross-section of the target population is selected. In many specific geographical problems, however, this is difficult to effect. Climatic data are for sites that already exist and records are for a continuous but finite period; the latter is essentially true for all time series data, whilst the availability of data on such things as glacial deposits or unpublished records, for example, may well be intermittent and fortuitous rather than random. In such cases it is important to ensure that there is at least no deliberate bias, that as large a random element as possible is involved in the data themselves, and to remember the limitations when drawing conclusions.

When inferring characteristics of the population from the observed characteristics of the sample, recourse is made to the known mathematical properties of various frequency distributions. The normal frequency distribution, or the Gaussian curve, is most widely used, it being assumed that very many phenomena under study display frequencies of occurrence around the mean value that approximate to this curve. By utilizing its properties it is then possible to infer the parameters of the population data from the sample characteristics within statistically specified margins of error.

This concept of 'error' also permits comparisons to be made, and conclusions to be arrived at, in terms of what is called 'statistical significance'. It must be appreciated that if two separate random samples are taken from the same population the averages of these two samples are almost bound to be different from one another. Yet each is representative of the same population. It is therefore clearly necessary to have some method by which to assess whether the observed difference between sample results is sufficiently large for it to reflect selection from two different populations or whether it could simply be the result of chance differences for samples from the same population.

This process of testing the statistical significance of an observed difference is fundamental to making decisions about the results derived from sample studies. A large number of statistical tests exist, each applicable to specific types of data. For example, the chi-squared test and Fisher's exact probability test are both applicable to nominal data; the Kolmogorov–Smirnov and Mann–Whitney U tests are used with ordinal data; whilst for interval and ratio data both Student's *t* test and Snedecor's *F* test can be used. The latter two tests are referred to as 'parametric' tests, for they make assumptions about the parameters or characteristics of the population data, especially that they approximate to the normal frequency distribution. In contrast, tests used for nominal or ordinal data are called 'non-parametric' as these assumptions about normality are not involved.

Critical values for all of these tests (and many others) have been calculated and are published in standard statistical tables. Such critical values vary with what are termed the relevant 'degrees of freedom' (usually related to the size of the sample taken) and the 'level of significance' required. The latter is a function of the frequency distribution underlying the test. When testing for differences, one first postulates that there is no real difference between the sample results, whatever the apparent difference may be. This is referred to as the 'null hypothesis'. The test then indicates the probability that this null hypothesis is wrong. The commonly used probabilities are 0.05, 0.01 and 0.001 implying, for example, that there is no more than a 5 per cent chance (0.05 probability) of being wrong when accepting that there is a real difference between the two samples and the populations from which they were drawn. It is argued, therefore, that the odds must be heavily in favour of the difference being real before it is accepted – testing to establish that sample results are indeed the same requires a slightly different approach.

Other Areas of Consideration

Thus it can be seen that 'thinking statistically', in geography as in any other field, implies a *probabilistic* approach to evaluation and decision making. The level of probability accepted is a matter of judgement by the researcher, but that decision necessarily controls the validity and value of any conclusions, and the reliability of any application of those conclusions in an applied sense. The assessment of probability can be derived from a variety of mathematical probability functions apart from the normal frequency distribution. For example, there are those applied in hydrological and climatological studies of extreme events, whether of floods, high rainfalls, droughts or extreme cold conditions. A very wide range of other analytical methods is also open to the geographer who is willing to think statistically. Thus when data fall into a categorical framework (as with nominal data) the occurrences can be cross-tabulated to interrelate two sets of conditions in what are termed contingency tables. A whole literature exists presenting methods of analysis by which such data sets can be made to yield useful results.

Beyond this, however, is the extensive field of multivariate techniques. In most geographical problems, factors and causes do not operate singly, but rather in intricate interaction with one another. It is the unravelling of such interactions, and the evaluation of the relative importance of each contributory element, that often provides the focus and the very *raison d'être* of the geographical enquiry. Whether the aim is simply to understand such relationships within the context of the recorded items (i.e. in a descriptive sense) or by using sample data to infer relationships within a larger population (i.e. an inferential approach), statistical methods exist to facilitate understanding. Perhaps the commonest technique is to expand to the multivariate stage the idea of correlation and regression that at its simplest is used to relate the variations in one variable with those in another (possibly causative) variable. The ability to incorporate several partial causes into the one expression is invaluable, especially when both the 'effect' (or dependent variable) and all the 'causes' (the independent variables) can be measured in interval or ratio form. If the independent variables can only be expressed in nominal or ordinal form, however, other methods such as analysis of variance present themselves as possibilities. Once again, it is essential that methods be selected that are appropriate to the nature of the data sets involved, and that at the sampling design stage data be gathered in such a way as to permit those statistical methods to be employed effectively.

Two other major geographical issues must also be remembered. Geographers are concerned with describing and analysing the distribution of phenomena in a spatial context. Most standard statistical techniques, such as those indicated briefly earlier, have not been designed with spatial issues in mind, but nevertheless they are used on spatial data. The major limitation and problem is that standard statistical techniques assume that the items being recorded or measured are independent of one another, and that the occurrence of a high or low value for one item does not tend to lead to a high or low value for the next item. Yet for so many geographical phenomena just such a tendency does exist in a spatial context. High rainfall at one point tends to mean high rainfall at adjacent points, although high rainfall in one year does not equally tend to mean high (or low) rainfall in the preceding or succeeding years. Thus whereas independence may appear in the time series, in the spatial pattern there is very strong spatial autocorrelation. If population data be considered instead, there is clearly very strong temporal autocorrelation for the population in one year necessarily controls to a large extent population in the next year. Spatially, there may also be considerable autocorrelation as one moves from one unit to another in an urban area, or equally from one unit to another within a rural environment. Where town and country are adjacent, however, such spatial autocorrelation breaks down. It is therefore once again very important that the geographer relates the methods used to the nature and characteristics of the phenomena and data involved, and that if a perfect match is not possible caution is exercised in the interpretation of results.

Finally, a further objective of geographical enquiry is often to attempt to summarize the complexity of reality by grouping or classifying locations or conditions. The reduction of large data sets to a more limited and more easily understood format, concentrating on major characteristics, is made possible by such techniques as principal components analysis (or factor analytical techniques in general). Further than this, there are many procedures for grouping the results of such analyses, or the original data themselves, into quasi-homogeneous clusters which may well provide a formal system of classification or even of regionalization.

Thus many of the aims of geographical study and enquiry can be facilitated by 'thinking statistically'. In most geographical work, at least some of the data used are of a quantitative nature, whilst the very complexity and variability of most phenomena that are of interest to geographers ensures that sample data rather than total enumeration are used. The potential benefits that can be derived from this situation by the use of statistical techniques of analysis and by adopting a statistical mode of thought have been outlined very briefly. To be able to apply this approach requires study of the methods and ideas in detail, perhaps at an advanced level. To assist in the beginning of this journey into 'thinking statistically' a selection of essentially *introductory* books, all but one written for the geographer, is appended below.

Introductory Books

Ebdon, D. 1985: *Statistics in Geography*, 2nd edn. Oxford: Basil Blackwell.

Gregory, S. 1978: *Statistical Methods and the Geographer*, 4th edn. London: Longman.

Hammond, R. and McCullagh, P. S. 1978: *Quantitative Techniques in Geography: an introduction*, 2nd edn. Oxford: Clarendon Press.

Johnston, R. J. 1978: *Multivariate Statistical Analysis in Geography*. London: Longman.

Norcliffe, G. B. 1977: *Inferential Statistics for Geographers*. London: Hutchinson.

Siegel, S. 1956: *Nonparametric Statistics for the Behavioral Sciences*. New York: McGraw-Hill.

Taylor, P. J. 1977: *Quantitative Methods in Geography: an introduction to spatial analysis*. Boston, MA: Houghton Mifflin.

Unwin, D. 1981: *Introductory Spatial Analysis*. London: Methuen.

Williams, R. B. G. 1986: *Intermediate Statistics for Geographers and Earth Scientists*. London: Macmillan.

III.4
Computing for Geographers

Derek Thompson

The Nature of Computers and Computing

I wrote part of this chapter on a recent visit to Edinburgh University, Scotland. If I had been able to bring a Macintosh portable computer with me from College Park, Maryland, USA, I could have quickly typed and edited the draft without ever using a pen, using my computer as a word processor. Then I could have inserted graphs, maps and tables, using the same processing tool, turning my computer into a desktop publishing appliance. Alas, the popular portable was not delivered to me by the store in time before I left the United States!

But I *was* able to use a personal computer at Edinburgh as a terminal to connect to an international electronic communications service, INTERNET, via a cluster of small shared minicomputers and workstations, and in this way send a message to my wife who read it on our personal computer connected by modem to my university's electronic mail system. Also, while in Edinburgh I was making enquiries at computer dealers and via magazines purchased at the John Menzies store on Princes Street as to how I might connect my computer in my office in College Park, Maryland, to that of my brother in County Durham so that we may both evaluate computer programs for geographic education. I ran into some difficulties here because he has a BBC ACORN Archimedes and I have a Macintosh. However, we may be able to work things out by having both machines act as IBM brand personal computers.

The world of computing is changing rapidly (plate III.4.1); and computing is helping to change the world – facilitating international communications, fostering scholarly exchanges and allowing many actions to be performed faster than they could just a few years ago. A recent issue of the *Washington Post* newspaper stated that two computer companies had created a computer chip smaller than a plastic credit card, containing 4 million transistors and capable of performing 200 million simple operations (e.g.

Plate III.4.1 The first computer, built by Babbage, now in the Science Museum in London.

adding) per second. Professors giving lectures with electronic blackboards and video cameras can be seen and heard by students tens of miles away from the lecture hall. For less than £500 or $1000 you can now buy yourself an electronic notebook for note-taking in classes or in the field, or for completing an essay or thesis while you fly off to the Canary Islands or the Caribbean region for a winter holiday.

The computer today is what *you* make of it – it is an appliance, a machine, a tool, a toy. It is simple or very complex, it is small or large, it is light or heavy, it is personal or shared, it is general or special purpose. But whether you work with numbers, words, pictures, graphs, maps, sounds or entire multimedia documents, you have a very rich resource available, founded upon something very fundamental. The collection of electronic devices and computer software that allows you to take advantage of the rapid speed with which basic arithmetic and logical operations are performed, i.e. the *computer system*, comes in very many varieties today compared with just ten years ago, but consists of just a few fundamental components founded upon the use of positive or negative electrical charges to process data using a two-state, or binary, number system.

Data (numbers, words etc.) and instructions to the computer (the lines of code called programs or routines) utilize a set of BInary digiTS (bits), the smallest unit of information. Thus a 00000001 stands for the number one; 00100000 stands for 32. Alphabetic characters and special symbols are represented by particular combinations of these digits, e.g. 01000001 is the letter A in the standard code known as ASCII (American Standard Code for Information Exchange), and 01111111 stands for line delete. Other types of data are represented by other arrangements of the bits.

Sets of bits are put together to form a second-level unit for processing – the word or *byte*. Originally computers generally processed 8 bits at a time; newer machines process 16 or 32 bits, and are thereby able to do basic operations much faster. Large quantities of information are usually referred to as kilobytes or megabytes. One kilobyte, 1024 or 2 to the tenth power bits, is today considered a small quantity. The magnetic or optical tapes and discs used today for storage can easily handle hundreds of megabytes (million bytes) – and they need to, because ten colour pictures need about 30 megabytes, and a large encyclopaedia can take up 500 megabytes. Remote sensing databases for the globe are measured in gigabytes (thousands of millions of bytes).

The computer industry hardware manufacturers are now delivering reliable products to allow us to store and process large quantities of varied information very quickly. The software creators are making it easier for users to work with the great variety of data forms that we encounter in the non-computer world. The good software to go along with the electronic devices is what is making the computer system a very powerful resource for learning and other purposes.

Computers at Work; Computers at Home

Computers are tools for improving personal productivity as well as resources you will encounter in school, college or other learning situations. Many colleges provide desktop computers in student accommodation as well as in on-campus laboratories. Some universities require students to buy personal computers. Much work is still done by shared computers, either minicomputers or supercomputers.

While most personal computers are still used primarily for word-processing, other main categories of functions are increasing as the computer systems today can process more data faster for less money. Some figures provided in a September 1988 issue of the *Financial Times* showed that word-processing was the most widely used application of personal computers in the 'home office', followed by database and file management and the use of spreadsheets. Certainly today high quality documents can be prepared at home on computer systems costing about £2500 or $4000. The current software provides spelling checkers, varied fonts, integration of graphs and tables into textual documents, and page layout tools.

Students with access to greater financial resources can equip their study at home or at college with the hardware and software to make maps or scientific graphs, to do statistical analysis, to create, store and manage large amounts of numeric and other information in databases or to access other computers via telecommunications.

For personal use at home the student has control over the computing environment. It is likely to be relatively simple. In contrast, the institutional computing environment can be very complex, more challenging and possibly quite intimidating; Geography departments, increasingly well endowed with their own computing resources, may offer a great variety of machines for student use. For example, Edinburgh University has a set of four Digital Equipment Corporation VAX computers operating in a cluster, with software for thematic mapping, geographic information systems and database management, and several graphics digitizers and plotters. My own department has a dozen IBM and five Macintosh computers, primarily for undergraduates, and special purpose workstations with the UNIX operating system for graduate students and faculty for advanced work in remote sensing, spatial analysis and geographic information systems.

Increasingly, pedagogically oriented software is used by academic staff as part of classroom activities or for homework assignments. Tools for word-processing, numerical analysis, data processing, graphics and image processing are commonly used as well as computer-assisted instructional software such as drill and practice, demonstrations, tutorials, simulation or instructional games. Such tools, beginning in the late 1960s on computers of that time, today called mainframes (e.g. the Association of American Geographer's land-use simulation), are now represented by, for example, the Birkbeck College's tutor on geographic information systems, GIST (see chapter III.7), and a commercial product for simulating urban growth, SIMCITY, available in the United States and Britain, which was originally marketed as a home use game but which has found use in colleges and schools.

Moving into the future

While there has been substantial activity at various levels of geographic education in Britain, New Zealand and the United States, and elsewhere, learning-oriented software is still in its infancy. An American study in 1988 showed that computers were used in high school particularly for programming, word-processing, and drill and practice, not for more creative subject-oriented learning. In the absence of any recent data, a guess on my part puts geography use as still dominated by map making and statistical analysis, as well as word-processing, among the college population.

However, with the advent of hypermedia tools, i.e. the merger of sound and graphics with an ability to manage electronically large collections of integrated information, like HYPERCARD for the Macintosh computer, it is now much easier for teachers to prepare classroom presentations or individual student use courseware for geography learning, and a new round of simulation software is expected to be forthcoming.

The concept of the geography student's workstation is near to reality – a desktop computer allowing students to work easily with a variety of data, to test hypotheses empirically and to undertake 'what if' kinds of alternative scenario computations, made possible by geographic information systems software, Macintosh style desktop interfaces and hypermedia–multimedia browsing.

The subservient relationship of student to teacher, marked by a hierarchical service delivery pattern from master computer to terminal, is gradually being replaced in instructional computing, as in other organizational computing, by a node in a network arrangement. Individual computers, whether at home or in the laboratory, are tied together in networks, facilitating the communication of results of learning or enquiries among individuals.

The Output from Computers – Products and/or More Knowledge

To many computer users the end result of accomplishing a particular task is a product; a table of numbers, a thematic map, a picture or an entire document which may include graphs, drawings, statistical tables and narrative text. To other people, the desired end result is a set of data that may be used for another purpose. To yet others the end result is a less tangible one, an improvement in knowledge, or a better understanding of a process, or learning a concept more speedily. For our purposes I shall concentrate on particular tangible items, returning later only briefly to the pedagogic issues.

Physical output may vary according to medium, which is a principal factor where cost is concerned, and other features such as colour, size and precision. A second major factor is the type of object being created by the computer – words, drawings, pictures etc. Different devices handle different kinds of data in different ways, and with varying efficiency and cost. Twenty years ago most output consisted of text and crude pictures and maps printed in black on white paper. Today there are colour printers, output in image form to photographic film or synthesized speech.

There are many alternatives for producing 'hard copy' output, on paper, photographic film or plastic material. Screen displays, for example (from which hard copy can be produced via a 'screen dump'), vary by size of monitor, use of colour, picture quality, precision and medium. Also, digitally encoded data may be converted to analogue form, as for many output monitors or for recording on videotapes or videodiscs.

An understanding of output alternatives, particularly if comparative evaluations have to be made, must be based on a knowledge of how pictures are created electronically. There are two main methods – object oriented (vector) and pixel (raster). For the former a screen display consists of turning lights on (or marking paper) for tiny points on a fine mesh or grid, as necessary to produce an irregular object like a line, or giving instructions to a pen to join dots with particular co-ordinate locations. Such object-oriented graphics are best thought of as a series of commands that would be used if you were telling someone how to draw a map by hand (e.g. go to a co-ordinate position 14,27, put the pen on the paper, move to position 22,25 etc.). The vector devices most commonly encountered in geography are pen plotters or photoscribers.

In pixel (standing for picture element) displays the entire graph or map is made up by colouring or shading individual cells in a rectangular array of rows and columns ('electronic graph paper'). Each element in this raster (sets of rows) may have a singular or multiple set of digital values, from 1 to 32 bits, or even more. For monochrome displays the cell will have a binary (on, off) state – the picture in computer memory is a series of ones and zeros. With 8 bits of information, 256 (2 to the eighth power) grey scale equivalents can be used; with 24 bits, over 16 million (2 to the sixteenth power) colour combinations are possible. Consequently the range of colour depends on storage capacity which in turn is a primary influence on the cost of the display monitor.

Alphabetic and numeric characters are commonly handled by this same method of 'bit-mapping'. Text-like graphics are treated as a series of black squares in a matrix of pixels. A common matrix has 640 rows and 480 columns, giving about 50 to 75 dots (square or rectangular pixels) per linear inch. Such a resolution may produce crude objects compared with devices like laser printers or colour thermal wax or dye transfer printers that work at about 300 dots per inch, or photo film recorders with 2000 or more dots per inch.

However, the underlying graphics operations for drawing can also be used for text display, being referred to as stroke (equivalent to vector) characters as opposed to block (bit-mapped) characters. If characters are treated like objects they too can be re-sized, rotated and moved just like line objects in a map, giving rise, for example, to different font styles and usually referred to as WYSIWYG (what you see is what you get) displays.

Input and Storage – Varied Data and Large Volumes

The geographer with a traditional bent for maps and pictures will find the input of data to computers as varied, interesting and challenging as that for creating output. There is much more to putting information into electronic digital form than typing words or numbers for the co-ordinates of county boundaries. The forms of digital encoding accepted by different machines or programs can vary considerably; most data, particularly maps and pictures, are currently available only in analogue form as paper maps or 35 mm slides; and the equipment needed for input of digital data with or without conversion from analogue varies a great deal in terms of price, quality, mode, resolution, accuracy and reliability.

It is useful to think of two forms of input – the data in the form of words, maps etc., and the instructions needed for the operation of a program. Independent pieces of data of the first category can be 'captured' by bar-code readers, electronic digitizing tables, scanners, keyboards, cameras, automatic dataloggers, microphones, remote digital recorders aboard satellites or other computers connected via networking.

Existing maps, architectural and engineering drawings, and photographs, often several linear feet in dimension, are usually digitalized by semi-automatic or fully automatic electronic devices that are of one of two modes, vector or pixel, as for output devices. Frame-based cameras and line scanners, still regarded as not very cost effective for large coloured documents, will produce data as a set of binary values for rows or arrays of pixels. The vector capture mode, usually simply called digitizing, uses devices which have electric or electromagnetic ways of identifying position via the equivalent of graph paper with a fine mesh of horizontal and vertical lines. More manual effort is required for digitizers than scanners because of the labour needed to move a hand-held mouse over the objects on the document. Scanners and digitizers can, of course, be used for other forms of data, such as letters and numbers.

Today there are many optical (and some aural) instruments in use for data capture, although most require a conversion from the waveform type of encoding associated with sound or vision to the digitial data form required by most computers. While spacecraft remote sensing usually records earth-based radiations via digital equipment, as in the Landsat multispectral and thematic mapper series of the United States or the French government's SPOT imagery, many analogue devices are still used – aerial photography cameras, camcorders, 35 mm cameras and the newer still video cameras which write to 2 inch floppy discs rather than photographic film.

For such media not only is there a need to convert from analogue to digital, but there is usually a major difference in resolution because standard video image pictures of the television broadcast industry standards (of which there are three principal forms of encoding: NTSC in the United States and other countries, SECAM in France, the USSR and other countries, and PAL in Britain and other countries) have much lower resolution than that of photographic film. Computer monitors today have better resolution than television sets, but the latter are widely accepted and used for video data. The integration of picture data into multimedia databases, or the combined use of computer and video technologies as represented by interactive videodisc equipment, has still to deal with the low resolution of the television world, as well as different basic electronics.

Specialized data are occasionally collected automatically by dataloggers. Usually incorporating some electronic microprocessors, such devices are used in the field or the laboratory for physical geography data capture, such as weather observations, sediment particle size or river flow quantities. Small computers can now be used for human event logging in the field, as for questionnaire surveys, or special devices such as the Ferranti Market Research Terminal, a hand-held tablet, can be used. Some loggers are so bulky and costly that they are only rarely used in geography.

Not only is there great variety in data formats, logical and encoded, but there is today an increasing variety of forms of data storage. The geographer's computing needs are very demanding of processing power and input–output operations, but they are also often storage bound. Large quantities of data are inherently a feature of remote sensing capture and motion video, even after compression (the use of special coding techniques to reduce the number of digits needed to represent a particular object). The increasing interest in the databases of geographic information systems has meant that users want to produce larger databases of almost any kind of data.

The electronics industry, though, is up to the challenge. Today we have a large range of media and models, magnetic and optical, and increasing recording densities. The magnetic tape (characterized by sequential reading and writing as for audio tapes) and disc (direct access to a position on the platter) generally hold less data than newer optical media although they have the advantages of the ability to erase and rewrite and fast access, and they are less costly.

Optical discs (tapes are not yet easily obtainable) offer a large recording capacity per platter, greater longevity than magnetic media and reasonably quick access. There is no direct physical contact between the platter and sensor – laser light is the only thing that touches the recording surface. Analogue optical discs have been successfully used in business and educational organizations for several years. The 1988 BBC Domesday project produced two such discs of mixed data for Britain, the storage capacity of one disc being equal to 108,000 still pictures or sixty minutes of motion video. Schools and colleges in the United States sometimes use a special videodisc on earth science topics.

The main advantage of magnetic media over optical is the ability of a user to prepare the material and rewrite it. Soon users will be able to do this for themselves using optical media, instead of having commercial companies write (i.e. master) the disc. Rewritable optical discs or optical discs or tapes in combination with computer chips will soon be available to hold the very data intensive motion video.

The user–computer interface, the second main category of input to the computer system, is also more exciting than it was just five years ago. Then interaction with the computer was almost always via typing commands at a keyboard. Today there are WIMPS – window, icon, mouse, pull-down menu interfaces, popularized commercially by the Apple Macintosh computer. Such so-called graphical user interfaces, becoming the standard on many different computing platforms, provide a more intuitive way of giving instructions, that is less dependent on human memory. Physical input is usually via a mouse for pointing to the menu boxes or icons, but touch-screens, trackballs, joysticks or even foot-pedals are also available and have some special advantages for particular purposes.

Definitely we have a graphics-oriented rather than character-oriented computing world. However, as a recent article in the *Academic Computing* magazine suggests, there may be unanticipated influences on human behaviour – for example, styles of writing appear to be influenced by the form of the input or tools in use when the script is prepared.

Applications Software for Geographers

Computer programs, sets of instructions for simple or complex tasks, are necessary for input to and output from the various devices that make up the computer system, but I shall pay attention here only to the software for managing and using the data of various forms. In between input and output, processing can be done by general or special purpose routines, individual programs, modular packages or individual programs in libraries. These programs usually today operate at a 'high level', i.e. they expect input in a form, usually everyday language, that is conducive to easy use; the fundamental computer arithmetic and logical operations, input and output and processing control, are undertaken behind the scenes.

Software may be classified first of all according to the main functions performed and, secondarily, according to the types of data being used. For example a graphics program may make diagrams or may make maps, but may not be able to do statistical analysis also. For our purposes we can think of the major categories of special purpose software as text processing, document preparation, statistical analysis, scientific graphics, thematic mapping, remote sensing, spatial analysis and modelling, and number data computations (spreadsheets), all of which the computer industry refers to as 'application programs'.

There are also general purpose application programs, usually allowing use of a variety of data and having more than one main kind of processing capability, or functionality. For example, there is 'integrated software' for personal computers, i.e. one package which has tools for text processing, numeric computations, graphing and communications to other computers. Generally speaking, special single purpose programs will have more features than the equivalent part of the integrated software. The user must choose between the range of functionality on the one hand, and the power on the other, or pay a higher price to get both.

Some general purpose programs increasingly used by geographers are the geographic information systems (see chapter III.7). Available for handling vector or raster geographic data, these systems generally contain many tools – for input by digitizing maps, for structuring data so that objects and spatial relations of objects can be recognized and used, for managing large quantities of data, for doing mathematical computations, for making graphs and maps or tables of numbers as output, and for undertaking a few kinds of spatial analysis or modelling.

Database management systems allow, and greatly facilitate the use of, a variety of data types, particularly numeric but also words and in some cases pictures, generally set up in different tables; the software tools facilitate drawing on several tables of different, but related, information at one time. For example, a student of mine did a study of travel in Baltimore using data in six large tables – for households, for persons, for trips made, for vehicles, for highway network and for origin and destination zones. With a database management system he was able to retrieve data directly at one time from all tables for a compound piece of information such as the number of vehicles owned per household of size two adults and two children making together more than forty trips per week between a residence in a suburban zone on the north side of the city to central Baltimore by inter-state highway.

Information retrieval is also very much at the heart of a newer category of programs known in the computer industry as 'hypermedia' tools. Best thought of, on the one hand, as providing tools to create databases of mixed form data, such as graphs, numbers and text, often with a metaphor of the (3 by 5 inches or larger) index card, they do, on the other hand, provide an alternative mode to working with information in an unstructured way. That is, the user can make connections between the separate pieces of information by doing the equivalent of relating one index card to another, or the user may request more detail on a subject by choosing one word in a paragraph and then may choose one more word in the paragraph revealed in response to the first word chosen. This 'hyper' concept provides a basis for browsing through a complex web of information.

However, while there are many exciting new resources for working with multimedia data in single packages, the world of computing still does not have perfect harmony or standardization. In particular, for graphics and image types of data the user must still appreciate that there are many different ways of encoding data. Individual thematic mapping programs, for example, have their own proprietary or 'native' format for the map features and the graphic display for those objects. There is no single dominating format for most of the graphics-oriented work undertaken by geographers. Rather than list the twenty or so different types, suffice it to say that the user should be aware of the difficulties that may be encountered in moving graphic or image data from one program and/or computer platform to another, and that there are special purpose programs called data translators that can assist in this task.

Not only can geographers draw on a very rich range of software today, for home use or in college laboratories, but they can also acquire commercial or public sets of data in word, number or graphic form. Some companies are marketing electronic atlases consisting of data for countries of the world and software to make maps for those data. The United States Census Bureau will make 1990 census of population data available on compact disc media, and other businesses sell boundary data (co-ordinates) or street network data for particular areas. Some picture databases are available in videodisc form for physical and cultural landscapes. We can now have an orientation towards information as opposed to or along with application programs.

The Future

Indeed, one of the trends most promising for geographic learning is the convergence of computing and video technologies so that it is possible to create and, of course, use mixed media databases incorporating still video camera pictures, movies stored on videodiscs, remotely sensed satellite or other datalogger observations and census data already in digital form. The multimedia computer will soon be available for college and university laboratory use.

The computing environments for personal productivity and, more especially, for learning are evolving rapidly. We, student and teacher, will be able to access rich multimedia information resources as nodes in an egalitarian network. We will be able to go on journeys of discovery through hypermedia navigation paths. Alongside the traditional 'scientific' computing characterized by formal logic, algebraic equations, numeric computations and application programs, we have a 'humanistic' computing environment, less structured, less formal, less product oriented, less comput(at)ing. Whether or not there will be substantial pedagogic benefit remains to be seen, but it is fun to explore broader horizons with the new types of virtual travel!

Supplementary Material

Your search for more details about computing can most usefully begin with the 1989 book by David Maguire of Leicester University: *Computers in Geography*, published by Longmans in Britain (ISBN 0-582-30171-8) and John Wiley in the United States (ISBN 0-470-21194-6). This work has chapters on computers in geography, geographical data collection, geographical data management, statistical analysis, computer cartography, remote sensing and image analysis, simulation, word-processing, communication, geographical information systems, computer hardware and computer software.

There are other books for specialized topics, or guides to computing in general. While the Maguire book is a good source for other books, I will mention just a few items here: M. Batty, *Microcomputer Graphics: Art, Design and Creative Modelling* (Chapman & Hall, London, 1987), M. J. Kirkby, Burt, T. P., Naden, P. S. and Butcher, D. P., *Computer Simulation in Physical Geography* (Wiley, Chichester, 1987), T. M. Lillesand and R. W. Kiefer, *Remote Sensing and Image Interpretation* (Wiley, New York, 1987) and I. D. H. Shepherd, Z. A. Cooper and D. R. F. Walker, *Computer Assisted Learning in Geography: Current Trends and Future Prospects* (Council for Educational Technology with the Geographical Association, Sheffield, 1980). If you are interested in doing some programming or learning more of programming principles a useful book is J. A. Dawson and D. Unwin, *Computing for Geographers* (David & Charles, London, 1976).

Several journals, academic and professional association newsletters, and trade magazines can be consulted for software reviews and information on new products or computing industry events and trends. Among this type of source are *The Professional Geographer* (software reviews), *The American Cartographer* (software reviews), *Environment and Planning* and *Computers, Environment and Urban Planning* (articles and reviews), *Computer Graphics Today* and *Computer Graphics World* (trade magazines) and two professional journals, *Journal of Geography in Higher Education* and *The International Journal of Geographic Information Systems.*

There are many general and brand-specific monthly or weekly magazines available in good bookstores for personal computers if you need to find more details to help you make a purchase decision or keep up with events for a particular computer you already own. General magazines include *Byte, Infoworld, PC Week, Personal Computer World*; and there are several magazines for each of the IBM, Macintosh, Archimedes, Amiga, Amstrad and other computers.

III.5
Remote Sensing
Nigel Gardner

Introduction

Like so many of the techniques used by modern geographers, remote sensing is relatively new. A decade ago, leading US remote sensing specialist Jack Estes, writing with colleagues from his Santa Barbara laboratory, argued that 'remote sensing is a reality whose time has come . . .; we predict it could change our perceptions, our methods of data analysis, our models, and our paradigms' (Estes et al., 1980). Whether or not the grand agenda for remote sensing encapsulated in Estes' comments has been accomplished is debatable; what is clear is that the battery of new techniques and approaches afforded by remotely sensing has emerged as important components of modern geographical method.

Remote sensing is the term used to refer to the variety of techniques for acquiring and recording environmental data from points which are distant from the phenomena of interest. The human eye and brain thus constitute a remote sensing system, albeit one which is only capable of operation in the visible spectrum and whose capacity for permanent data recording and reliable data retrieval is limited! The release of a radioscope with attached thermometers which measure ambient air temperature as the sonde rises through the lower atmosphere is *not* remote sensing, as in this instance no element of distance separates the sensor (the thermometer) and the object of study (the air). But the assessment of ground temperature using a thermal scanner mounted on an aircraft is a remote sensing technique. Similarly, the assessment of sea floor topography through sonic profiling, the monitoring of seasonal Sahelian vegetation change by repeated satellite linescanning, or the assessment of flood damage to cropland by false colour infrared photography from aircraft, *are* all examples of remote sensing; in each instance the object of study is remote from the sensor, and the technology includes some capacity for data retention.

The rapid development of remote sensing has been predicated upon developments in four technologies over the last three decades. The first of these is information technology (IT). Falling costs have paralleled massively enhanced performance to make advanced computing technology accessible to many researchers and even students, who, just a few years ago, would have viewed such facilities as something for which only the leading research groups might aspire. These developments in computing have made it possible for researchers to deal efficiently and promptly with the huge streams of digital data afforded by modern remote sensing systems.

The second core technology has been communications technology, although with rapid technical and regulatory development in the world of telecommunications, IT and communications technology are rapidly converging. Many commentators argue that it is no longer realistic to see these technologies as independent. New communications technologies allow us to transmit and exchange data around the globe at unprecedented speed, whether by international ground telecommunications networks, by microwave transmissions or by satellite link.

Thirdly, developments in aerospace technology have afforded a whole new range of platforms for the acquisition of remote sensing equipment (plate III.5.1). Geographers now routinely exploit remote sensing data derived from aircraft overpasses, from manned satellite missions and from a large range of unmanned satellites launched bya variety of nations. Developments in launch, guidance, system monitoring, tracking and data relay capabilities have all contributed to making high-altitude and space remote sensing systems an increasingly attractive source of data for geographers.

Finally, developments in imaging and sensing technologies have opened up new areas of the electromagnetic spectrum for exploitation in remote sensing studies. Wholly new categories of sensor have been developed, such as sophisticated push broom scanners, lidar and advanced radiometers, allowing us to collect data from different spectral regions.

Geography is by no means the only discipline to have benefited so widely from developments in these core technologies and in remote sensing. Global geoscience can truly claim to have been revolutionized by these developments, as can the atmospheric sciences and oceanography. And, although not greatly affected by the growth of remote sensing *sensu stricto*, other areas of science such as astronomy, engineering, metallurgy

Plate III.5.1 Since the 1960s geographers have had access to a wealth of remote sensing imagery, and most geography courses include a consideration of the importance of such imagery and teach some of the basic methods of image analysis.

and medicine are keenly exploiting new opportunities proffered by the development of technologies which are cognate to remote sensing.

Electromagnetic Radiation – The Physical Basis to Remote Sensing

There are three main media which have been used to convey information from environmental phenomena to targets. These are electromagnetic energy, sound waves and force fields. The latter two have a number of highly specific applications which are relevant to the environmental and earth sciences. Thus ocean depth and sea floor characteristics have long been assessed using echo sounders which emit sound waves; and geomagnetic surveys, which are frequently used in mineral prospecting, are an example of the use of force fields in information transfer for remote sensing. Notwithstanding these notable exceptions, the great majority of applications of remote sensing of interest to geographers rely upon electromagnetic radiation to effect the transfer of data from target to sensor. Accordingly, the remainder of this chapter deals exclusively with electromagnetic remote sensing.

Electromagnetic radiation occurs naturally at a variety of wavelengths in and around the earth and its atmosphere. Remote sensing is by no means limited to that narrow portion of the spectrum to which the human eye is sensitive, although remote sensing in the visible spectrum remains a valuable tool for geographers. Just beyond the visible spectrum lies the near infrared; remote sensing studies in this portion of the electromagnetic spectrum have proved especially useful in a wide range of environmental applications. At yet longer wavelengths, the thermal infrared also has demonstrable utility in environmental remote sensing. At even longer wavelengths, the level of naturally occurring radiation in the earth–atmosphere system is very low, and *passive* remote sensing, in which the sensor relies entirely upon natural radiation reflection of emission, becomes very difficult. Accordingly, *active* systems such as imaging radar have been developed, in which the sensor relies upon measuring the backscatter by the target of radiation transmitted in intermittent pulses from the sensing device. The use of a flash gun in ordinary photography is of course another example of active remote sensing. Figure III.5.1 summarizes the principal spectral regions used by geographers for remote sensing, along with the dominant sensing technologies appropriate to each of these portions of the electromagnetic spectrum.

Sensors and Satellites

The conventional photographic camera remained the dominant form of sensor used until the advent of earth orbital satellites for remote sensing in the early 1960s. At that stage, the practical considerations associated with film retrieval, coupled with an increasing imperative to explore other spectral regions beyond the visible, spurred the development of new sensing technologies. Photography is still used, particularly for remote sensing from aircraft, in the visible and near infrared; the latter requires the use of special false colour infrared film. But now almost all space imagery and an increasing percentage of aircraft imagery is acquired using linescanners, radiometers or other sensors that provide digital format data which is easily transmitted back to ground receiving stations.

Over the last two decades remote sensing has matured considerably, and a relatively small number of major satellite programmes have come to provide the bulk of space imagery used by

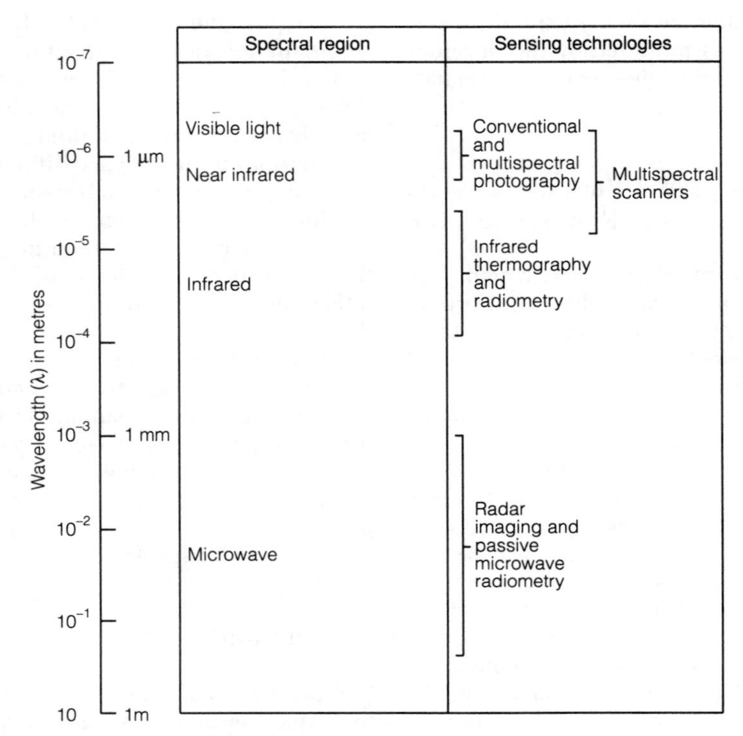

Figure III.5.1 The electromagnetic spectrum showing the principal spectral regions exploited by geographers using environmental remote sensing.

geographers. These major programmes are summarized in table III.5.1. It is worth highlighting in particular the US National Oceanic and Atmospheric Administration (NOAA) polar orbiting satellite series; since the launch of TIROS-1 in 1960, NOAA have provided frequent repetitive global coverage at the coarse resolutions demanded for routine weather forecasting; it is images from this series, and from the geostationary METEOSAT satellites, that we have now come to expect to see as a routine accompaniment to our television weather forecasts.

Useful though data from these satellites are for meteorological applications, the manned space programmes of the 1960s, such as the various Apollo missions, indicated the potential utility of much more detailed imagery. Thus in

Table III.5.1 Significant space programmes in the development of environmental remote sensing

Satellite	Main sensors	Orbital cycle	Lifetime
GMS	VISSR	Geostationary (hourly)	1978–now
Landsat	MSS	16 or 18 days	1972–now
		18 days	
	TM		1983–now
SPOT	MSS	26 days	1986–now
NOAA	AVHRR	Twice daily	1978–now
NIMBUS	CZCS	6 days	1979–86
HCMM	HCMR	16 days	1978–80
SEASAT	SAR	Limited life	1978
ERS	AMI/RA	3 days	1991–now

In addition to the unmanned satellites listed above, some or all of the missions in the following manned space programmes have carried a significant remote sensing component: in particular, Gemini (4–7 and 10–12); Apollo (in particular, 7 and 9); Skylab; Soyez (in particular post-1976 flights – Soyez 22 and following); the Space Shuttle.

1965 the US National Aeronautics and Space Administration (NASA) established the Earth Resources Survey Program. The first satellite launched under that programme was launched in July 1972; it was known originally as ERTS-1 (Earth Resources Technology Satellite One), and was later renamed Landsat-1. The first imagery from this satellite was a revelation; like some of the earlier photographs taken by Apollo crews, it drew the attention of a worldwide audience to the beauty and apparent fragility of the earth as viewed from space. But the first Landsat images highlighted a richness of variation in texture and colour of the earth's surface which was to provide a new and unparalleled data source for geographers. Four more Landsat satellites followed over the next dozen years, and the most recent continues to provide coverage of most of the globe every sixteen days. On all five Landsat satellites, a four-channel multispectral scanner system (MSS) offers continuous digital data collection in the visible and near infrared spectral regions. The instantaneous field of view of the MSS is a ground resolution cell which measures 79 metres along each side. On Landsat-4 and Landsat-5 the MSS (which has been included on each satellite in this series largely for data continuity) is supplemented by a much more advanced scanner called the thematic mapper (TM). The TM incorporates a number of design improvements over the MSS. It offers three visible spectrum bands, three in the near and middle infrared and one thermal infrared band. The improved spectral resolution on the TM system is complemented by an enhanced spatial resolution of just 30 metres.

The success of Landsat prompted other national space agencies to attempt to emulate the US achievement. In 1978 the French government announced the *Système Pour l'Observation de la Terre* (SPOT) programme; the first SPOT satellite was launched in 1986. Like Landsat, SPOT has a circular, near polar, sun-synchronous orbit. With coverage of almost the whole globe every twenty-six days, SPOT uses two high resolution visible (HRV) imaging systems, which offer two operation modes. They provide for either panchromatic (i.e. black and white) imagery with 10 metre spatial resolution or colour infrared multispectral imagery with 20 metre resolution. Both modes have already demonstrated their advantages over the Landsat sensors for many detailed applications.

There have been a number of other unmanned satellite programmes which have carried sensors of interest to geographers. For a few weeks in mid-1978, the short-lived SEASAT-1 gave us some insight into the potential of synthetic aperture radar (SAR) systems for spaceborne radar remote sensing. It gave weight to similar programmes being formulated for spaceborne radar in Europe, Canada and Japan, all of which are scheduling launches in the early 1990s.

Shortly after SEASAT's failure in space, NIMBUS-7 was launched. This satellite carried the Coastal Zone Colour Scanner (CZCS) which, with six bands operating in the visible, near infrared and thermal regions, was designed specifically to measure the colour and temperature of oceanic coastal zones. SEASAT and the CZCS together demonstrated the importance of satellite remote sensing in exploring the unmapped oceans of the world.

It was imagery from the hand-held cameras used by the Apollo astronauts which sparked the development of many of the large programmes described above. These did not totally eclipse interest in imagery acquired by crews on earth orbital satellites. In 1973, SKYLAB was launched. This was the first US space workshop, and its astronauts acquired over 30,000 images of the earth, which demonstrated the complementary nature of photography and digital imaging from space. The Space Shuttle has continued this tradition of space photography. In addition, the Shuttle has carried an imaging radar system on a number of missions.

Modern remote sensing is the lineal descendant of air photo reconnaissance, and, for all the current interest in imagery and data acquired from space, there has been no loss of interest in the value of photographs and other data derived from aircraft. The standard black and white or colour vertical aerial photograph still remains the most commonly used remote sensing product for many routine geographical applications, and in the United Kingdom, as in other developed nations, there are an increasing number of private sector agencies which recognize the commercial opportunities in maintaining up to date high quality coverage of large parts of the country at a variety of scales and formats. For many parts of Britain, more specialist imagery is also available, including false colour infrared photography, thermal linescans and airborne radar.

Applications

Remote sensing has not obviated the need for fieldwork in geography. Indeed, the continuing need to calibrate new data sources and to 'ground truth' imagery means that remote sensing has reinforced the need for field investigations in geography. But as a precursor to detailed ground surveys, and as an adjunct to other data collection techniques, remote sensing has an obvious contribution in geographical enquiry. Not least, it should not be forgotten that for large tracts of the planet's surface an up-to-date Landsat TM or SPOT image may be easily the most detailed map available. In many of the world's deserts, for example, published conventional maps cannot begin to compare, in terms of detail or timeliness, with recent satellite data products.

But remote sensing does not just compete with and augment existing geographical data collection systems. In many instances, it has facilitated the collection of wholly new categories of information. There is simply no other way of collecting broad-scale synoptic data about macroscale changes on and around the surface of our plant (plate III.5.2). Thus it was NIMBUS satellite data which first alerted scientists to the changing patterns of atmospheric ozone in high latitudes. Elsewhere, remote sensing has afforded insights through its imaginative exploitation of differing areas of the electromagnetic spetrum. Thus the Space Shuttle radar system in November 1981 first drew the scientific community's attention to ancient drainage systems and associated abandoned pastoral sites in the now hyperarid lands of the eastern Sahara. And much of what we now know about the morphology, distribution and aeolian dynamics of both present and former desert sand seas is derived from a combination of high-altitude reconnaissance aircraft imagery, Landsat and SPOT data, and other remote sensing sources.

Table III.5.2 summarizes a range of successful geographical applications of remote sensing. This wealth of applications has been made possible by several factors which commend the use of remote sensing.

1 *Homogeneous and comprehensive data coverage* at a variety of scales, allowing international comparative studies. It is also now becoming possible, for the first time, to develop longitudinal data sets for monitoring long-term changes in tropical forest cover. Additional benefits are associated with the spatially continuous nature of much remotely sensed data.
2 *Digital formats* allow for the near instantaneous worldwide trans-

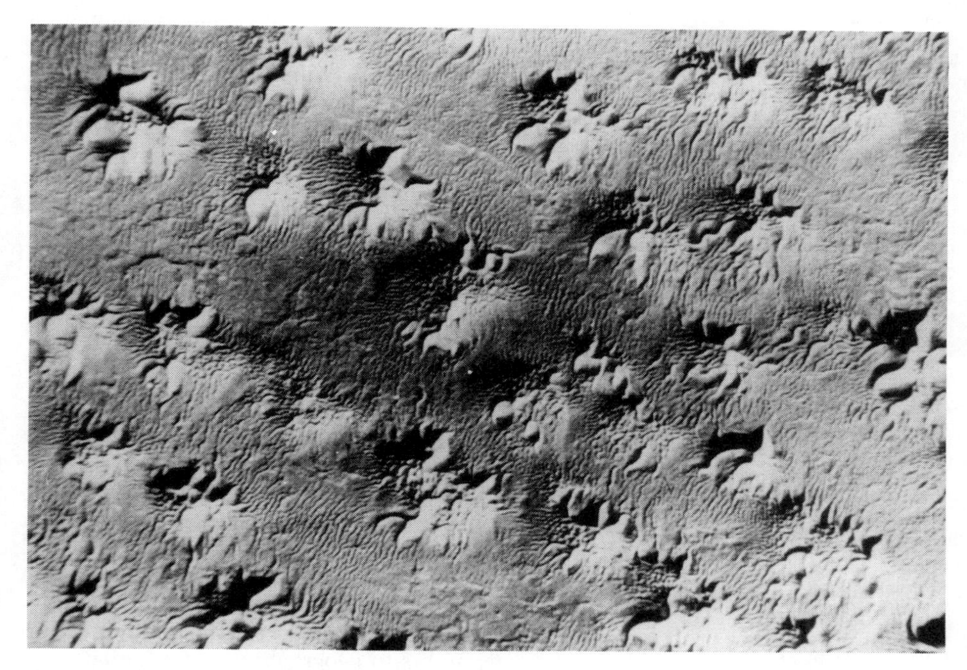

Plate III.5.2 A satellite image of star-shaped, or pyramidal, dunes from the Sahara. Remote sensing permits an overview of inaccessible areas.

mission of data, its processing by computer and its immediate reformatting in a variety of usable products.

3 *Frequency* of data collection is a major advantage. Although the present earth orbital repeat cycles of satellites such as Landsat (sixteen days) or SPOT (twenty-six days) may be too long to permit effective monitoring of ephemeral events such as short volcanic eruptions or desert dust storms, other satellites such as the NOAA meteorological satellites or METEOSAT offer much better temporal resolutions, albeit at the cost of worse spatial resolution.

4 *Cost* is a major factor in all environmental monitoring programes. Although there have been moves towards more commercial pricing of space remote sensing products, and prices rose steeply during the 1980s, satellite imagery frequently still offers a low cost way of accessing up to date spatial data.

These manifest advantages are tempered by a number of disadvantages and problems inherent in geographical applications of remote sensing. In a discipline that places considerable emphasis on studying the interactions between people and landscape, remote sensing will often suffer from the fact that it focuses on the morphological, the mappable and the concrete on the surface of the earth. It cannot easily be used to highlight the economic, social or political processes that underlie so many questions of resource management.

Other problems derive from the variable nature of the interactions between electromagnetic radiation, the atmosphere and the earth's surface. The atmosphere is rarely an ideal propagation medium for radiation, and signals can

Table III.5.2 Some successful geographical applications of remote sensing

	Status (1991)	
	Demonstrated experimentally	Routinely operational
Atmospheric composition monitoring	Yes	Yes
Weather forecasting	Yes	Yes
Hurricane tracking	Yes	Yes
Sea state forecasting	Yes	No
Sea and iceberg monitoring	Yes	Yes
Fisheries (shoal tracking)	Yes	No
Oil spill monitoring	Yes	Yes
Flood extent monitoring	Yes	Yes
Flood damage assessment	Yes	Yes
Drought monitoring	Yes	Yes
Vegetation mapping	Yes	Yes
Vegetation stress monitoring	Yes	Yes
Forest and range management	Yes	Yes
Crop yield studies	Yes	No
Wildlife studies	Yes	[a]
Soil mapping	Yes	[a]
Geological mapping	Yes	Yes
Geomorphological mapping	Yes	Yes
Coastal processes monitoring	Yes	Yes
Natural hazard assessment	Yes	Yes
Environment impact assessment	Yes	Yes
Mineral resource assessment	Yes	[a]
Urban growth monitoring	Yes	Yes
Population censuses	Yes	No
Archaeological heritage mapping	Yes	[a]

[a]Although remote sensing may now be routinely used for this purpose, operational or other constraints may severely limit the utility of the technique under certain circumstances.

be attenuated or even lost completely through adverse atmospheric conditions. Often haze and cloud preclude high quality sensing in the visible and near infrared regions of the spectrum. At longer wavelengths, we still do not know enough about the nature of interactions at the ground to interpret unequivocally the signals received by sensors; thus radar imagery remains very difficult to interpret.

A third and final potential problem is closely associated with one of remote sensing's principal merits. The digital nature of the data can also create its own difficulties. Image processing is a skilled and specialist enterprise, challenging geographers to acquire a range of new skills for which they may not necessarily be well equipped. Despite the increasing availability of appropriate facilities, a certain amount of initial training is essential to derive the best from satellite data.

Remote sensing thus offers a new and exciting technique to geographers. It has helped renew the discipline's concern for macroscale, continent-wide and global problems. Its full potential can only really be exploited when used in conjunction with other techniques described elsewhere in this volume. Thus the rich arrays of spatial data afforded by remote sensing provide major inputs to geographical information systems (see chapter III.7) and the display of maps derived from remotely sensed data benefits considerably from developments in computer-assisted cartography (see chapter III.6).

Getting Started

The term remote sensing is less than twenty years old. But in that brief period, remote sensing has emerged as a new subdiscipline in its own right, sustaining its own professional bodies, periodicals and a flourishing range of newsletters and magazines. For geographers, *Geographical Abstracts: Physical Geography* offers an overview, six times a year, of newly published scientific literature in the field. There are six major remote sensing journals; the following, in particular, are available in many university, polytechnic, and college libraries: *International Journal of Remote Sensing, Photogrammetric Engineering and Remote Sensing, Remote Sensing of*

Environment, and the *IEEE Transactions on Geoscience and Remote Sensing*.

There is now a bewildering array of undergraduate textbooks on remote sensing. Those by Barrett and Curtis (1982), Curran (1985), Harris (1987) and Lillesand and Kiefer (1987) are particularly recommended. In addition, the two-volume *Manual of Remote Sensing* (Colwall et al., 1983) provides an encyclopaedic overview on the theory, instrumentation and techniques of image acquisition and of the interpretation and applications of remote sensing.

In the United Kingdom, many universities now house collections of both digital satellite data and derived photographic products. The National Remote Sensing Centre (NRSC) provides a useful first point of contact for those seeking to acquire satellite imagery. The Centre, which publishes a regular newsletter, can be contacted at NRSC, Space Department, Royal Aircraft Establishment, Farnborough, Hampshire GU14 6TD.

Data from US satellites can be ordered through NRSC or by contacting NOAA directly at National Earth Satellite Service, EROS Data Center, Sioux Falls, South Dakota 57198, USA. The principal distribution centre for SPOT imagery is SPOT Image, 18 Avenue Edouard-Belin, F31055, Toulouse Cedex, France.

In the UK most district, county and regional councils continue to maintain comprehensive libraries of air photographs of their areas. The Ordnance Survey's Air Photo Cover Group markets photographs for the whole UK. It can be contacted at Ordnance Survey, Romsey Road, Maybush, Southampton, Hampshire SO9 4DH. Among the growing number of commercial companies who retail photographs are J. A. F. Photographic (Mitcham), Clyde Surveys (Maidenhead), Meridian Air Maps (Lancing) and BKS Surveys (Coleraine). Finally, many universities maintain collections of regional or even national importance; the Cambridge collection in particular is outstanding. It can be contacted through the Committee for Aerial Photography, University of Cambridge, Mond Building, Free School Lane, Cambridge CB2 3RF.

References

Barrett, E. C. and Curtis, L. F. 1982: *Introduction to Environmental Remote Sensing*, 2nd edn. London: Chapman & Hall.

Colwall, R. N., Simonett, D. S. and Estes, J. E. (eds) 1983: *Manual of Remote Sensing*, 2nd edn. Falls Church, VA: American Society of Photogrammetry.

Curran, P. J. 1985: *Principles of Remote Sensing*. Harlow: Longman.

Estes, J. E., Jensen, J. R. and Simonett, D. S. 1980: Impacts of remote sensing on US geography. *Remote Sensing of Environment*, 10, 43–80.

Harris, R. 1987: *Satellite Remote Sensing: an introduction*. London: Routledge & Kegan Paul.

Lillesand, T. M. and Kiefer, R. W. 1987: *Remote Sensing and Image Interpretation*, 2nd edn. New York: Wiley.

III.6
Cartography
Roger W. Anson

What is Cartography?

A well-known and oft-quoted maxim suggests that 'a picture is worth a thousand words', but if one accepts this statement it should also be acknowledged that, to an informed user, an effectively compiled and presented map has an even greater communicative value! The *Concise Oxford Dictionary* explains the term 'map' simply as a 'representation (usually on a plane surface) of the earth's surface, showing physical and political features etc., or of the heavens', but T. A. Magerison provides a rather more erudite explanation. He states that:

> *A map is the simplest, most elegant and informative way of presenting data which vary across a surface. It is a two dimensional model which the human mind recognises and comprehends with pictorial clarity, but yet provides quantitative as well as qualitative information.*
>
> *(Margerison, 1976, p. 3)*

The most recently promulgated definition, formulated in 1989 by a Working Group of the International Cartographic Association (ICA) and very much a subject of debate, suggests that a map is

> *a holistic representation and intellectual abstraction of geographical reality, intended to be communicated for a purpose or purposes, transforming relevant geographical data into an end product which is visual, digital or tactile.*
>
> *(Taylor, 1989)*

Although it cannot claim to be the oldest of the world's professions, map making was being actively practised as early as 3800 BC. However, probably the earliest formally recorded use of maps to communicate spatial relationships and illustrate geographical features occurs in the works of Claudius Ptolemy, a brilliant Greek astronomer and mathematician, who lived and worked in Alexandria during the second century AD (plate III.6.1). One of his monumental volumes, entitled the *Geographia*, contains an account of the principles of map making, a description of map projection methods, instruction on how to draw a world map and subsequently divide it to permit the larger scale representation of smaller regions, and a list of some 8000 place names together with their geographical co-ordinates. Although existing copies of the work are now accepted as being incomplete, the original version is also thought to have contained a series of maps which were still acceptable, in a redrawn form, as authoritative sources of geographical information some 1200 years later.

Not until the time of the Renaissance voyages of discovery was Ptolemy's geographical scholarship seriously challenged, but by 1620 thousands of miles of previously uncharted coastlines had been surveyed and recorded by European shipmasters, and the spatial arrangement of the continents (with the exception of Australasia and Antarctica) had been effectively fixed and mapped.

The second half of the eighteenth century witnessed the introduction, in Europe, of formally organized and

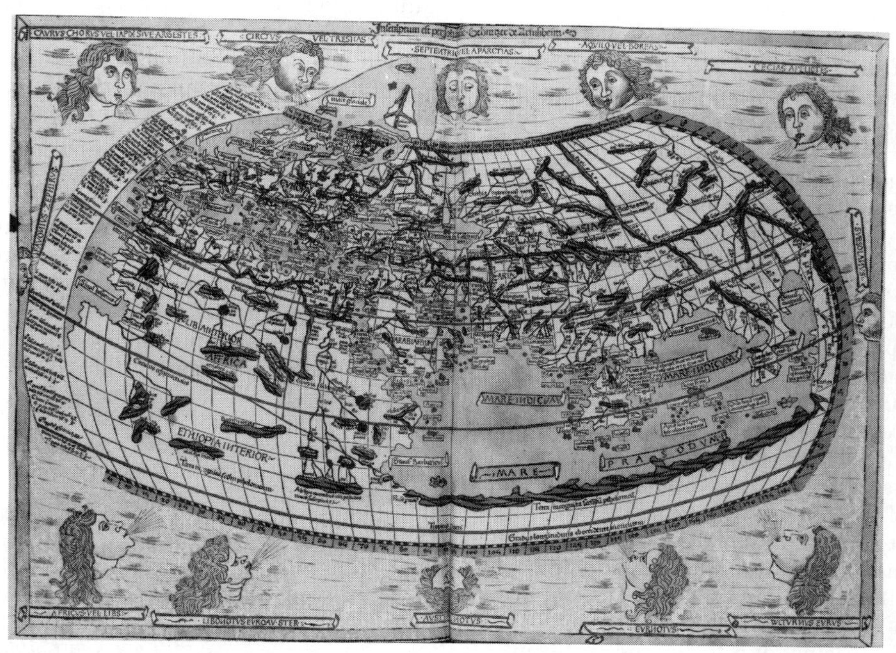

Plate III.6.1 Ptolemy's map of the World. Auct. P.1.5. Folio 27.

governmentally funded 'national surveys' (the Ordnance Survey of Great Britain celebrates its 200th anniversary in 1991), and the subsequent printing and publication of multiple copies of mapping series produced at consistent scales. These documents, particularly those at large scales (1:2500, 1:1250 and 1:500), exhibited a greater degree of positional accuracy and significantly more detail than was formerly possible and, as a result of parallel developments in production technology and with the contemporaneous onset of the Industrial Revolution, became indispensable aids to the work of geographers, planners, engineers and administrators. At this time, however, maps reached only a comparatively small number of literate educated people – today they are an essential part of everyday life and it is indeed difficult not to come into contact with them during a normal social routine.

A quick perusal of a current popular newspaper results in the observing, if not the reading and understanding, of small monochrome maps serving to locate and explain significant contemporary events. Similarly, the relevant or expected state of the nation's weather is regularly demonstrated to us by the use of meteorological charts appearing on television or in the press, and the so-called 'Green Revolution' and ever-increasing environmental concern have resulted in the recent publication of a spate of map-based illustrations.

In spite of a long and rich history, which is well documented in a number of excellent publications by experts (Thrower, 1972; Crone, 1978; Wallis and Robinson, 1987; and others), use of the term 'cartography' did not become current until 1839. It now seems that a major challenge confronts both professional map makers and those responsible for the education and training of students involved in an attempted understanding and depiction of the geographical environment in that they have, necessarily, to redefine the scope of a discipline in flux!

Little more than a quarter of a century ago cartography, although beginning to be appreciated as an academic discipline in its own right, was generally considered to be a pursuit rooted in convention with the majority of its practitioners performing operations and applying skills essentially similar to those used by their counterparts a generation earlier. Today it is in a stage of transition from being an analogue to a digitally based profession and little of its increasingly complex field, or those of associated disciplines which contribute to effective professional map making, has remained unaffected by almost constant change. This rapid rate of development shows no signs of decreasing, and techniques or even points of view which were taken for granted in the 1960s are being, or have been, superseded by alternative methods and ideas.

As a result, the generally accepted explanation of cartography, as proposed by the ICA and published in the *Multilingual Dictionary of Technical Terms in Cartography* (Meynen, 1973), is being constantly queried in terms of its relevance to the future of the subject. It states that it is

> *the art, science and technology of making maps, together with their study as scientific documents and works of art. In this context maps may be regarded as including all types of maps, plans, charts and globes representing the earth or any heavenly body at any scale.*
>
> *(Meynen, 1973)*

This elucidation of the *raison d'être* of cartography, published less than twenty years ago, indicates the breadth of the subject as it was then understood, but it is now considered inappropriate as a reflection of present-day activities and future intentions. It totally neglects the existence of electronic-based information technology and computer-generated products such as digital databases or ephemeral maps displayed on screen; the day to day use of maps and the importance of user feedback; the communications potential of maps; and their increasing distribution to ever more knowledgeable and geographically aware consumers! A more representative, but less easily understandable, 'definition' was proposed by Guptill and Starr in 1984, when they suggested that cartography is

> *an information transfer process that is centred about a spatial data base which can be considered, in itself, a multi-faceted model of geographical reality. Such a spatial data base then serves as the central core of an entire sequence of cartographic processes, receiving data inputs and dispersing various types of information products.*
>
> *(Guptill and Starr, 1984)*

Neither of these contributions is considered to be internationally acceptable and in consequence an ICA Working Party was established with, amongst other terms of reference, the intention of formulating an acceptable definition of the subject. The presently tabled proposal states that cartography is

> *the organisation and communication of geographically related information in either graphic or digital form. It can include all stages from data acquisition to presentation and use.*
>
> *(Taylor, 1989)*

Doubtless this statement will require further refining before achieving general acceptability within the cartographic community, but from it, and the explanations cited earlier, the reader must be aware that the subject is in the midst of a technological revolution which is affecting not only methods of map preparation but also the types of products generated.

The Use and Abuse of Maps

Cartography and maps have long occupied a place in the geographer's affections, but the last quarter of a century has witnessed an evident cooling of this relationship. Significant changes in the scope and approach to geography, most especially resulting from the so-called 'quantitative revolution', have tended to reduce the importance of standard topographic mapping as a tool of use to the geographer. Simultaneously developments within cartography, in particular the initiation and development of research relating to the previously neglected communication paradigm, have resulted in the emergence of what was earlier thought of as a practical technical subject as an academic discipline in its own right. It could be argued that previously map making was dependent on an association with geography in order to give it scientific respectability, but this is no longer the case.

However, in spite of these seemingly divisive trends and an apparent bifurcation into different areas of interest and development, it is still generally accepted that the map retains a central function within the geographical milieu and that an awareness of the principles and practices of cartography remains a key factor in a geographer's education. The majority of geography or environmental studies schools or departments within universities or institutes of higher education, throughout the world, have courses relating to their students' informed application of cartographic techniques. These may be of different durations and degrees of complexity, but most last for either one term or a semester. They tend to be included as a part of first year studies in the United Kingdom, and in the second year in North America, and are either compulsory courses or strongly recommended for students majoring in geography. The reason for their continued inclusion in college programmes is to provide a basic knowledge of both the spatial and informational aspects of maps, and also to encourage the student to practise the skills appropriate to their successful construction. By so doing it is considered that students should be able to function more efficiently and effectively as geographers (Greenhood, 1964).

As is suggested in *Perception and Maps: human factors in map design and interpretation* (Board and Taylor, 1977), it might realistically be assumed that, based on a study of the relatively small amount of published research on the design and ease of interpretation of detail from maps (i.e. their ability to communicate), these subjects are sufficiently understood as not to require investigation. After all, geographers – of all types, specializations and levels of learning – have employed maps in many roles for almost as long as there have been geographers. Their very familiarity with the map, as a medium for the communication of concepts and spatial relationships, has inevitably led to the assumption that the process whereby an efficient portrayal is created has been fully understood. However, the falsity of this belief is demonstrated by the very uneven record of monochrome cartographic illustrations associated with geographical texts. In the worst cases many maps are either illegible or only partly legible, probably because the original idea for the illustration is incapable of execution within the normal constraints imposed by format, scale, projection, available production facilities etc. It seems that, to some extent, many geographers only pay lip service to the map as a communications tool and do not really believe in them any more than they do in tables of statistics – unless they work to their advantage! Another probability is that geographers have been conditioned by almost two centuries of reliance on national mapping agencies that provide them with topographic maps that do not entirely suit their particular purposes and, as a consequence, are prepared to accept second best in personally generated mapping over which, theoretically, they have more control. It is also possible that not all geographers are born map readers or 'graphicate' – a term coined by Balchin and explained as the 'communication of relationships that cannot be successfully communicated by words or mathematical notation alone, but require maps, pictures, graphs, diagrams, etc.' (Balchin, 1972).

Today the availability of a range of well written, illustrated and instructive textbooks prepared by authors experienced in both geography and cartography does much to help improve the situation alluded to above. Special mention should be made of the individual efforts made by Dickinson (1973), Greenhood (1964), Keates (1989) and Monmonier (1985); the collective writings of Robinson, Sale, Morrison and Muehrcke (1984); plus the contributions made by the ICA in publishing multi-author works such as *Basic Cartography for Students and Technicians* (volumes 1 (Ormeling, 1984) and 2 (Anson, 1988)) and the *Compendium of Cartographic Techniques* (Curran, 1988). All these sources provide information, in English, of immediate use to the student intending to communicate geographical knowledge in a graphic form by using either manual or computer-assisted methods.

Map Preparation

In the ideal situation geography students should be able to demonstrate an awareness of, and familiarity with, cartographic principles and practices. These abilities should relate not only to traditional applications using manual methods of representation (Kanazawa, 1984), but also to the potential afforded by new and developing technology which is relevant to the graphical depiction of spatial data (Monmonier, 1985; Kadmon, 1988).

The consequence of efforts using either of the approaches should be the generation of maps and diagrams from previously compiled data. The data may be the result of personal researches and the collection of information in the form of existing map material; the recording of observations made in the field; detailed interpretation from aerial photography or remotely sensed imagery using analytical photogrammatic techniques; or statistical listings which have been either manually produced or computer generated. Potentially useful data must be interrogated with respect to their relevance, age, accuracy and reliability of origin, and may be either discarded or maintained depending on the subject and intended purpose of the proposed map (Robinson et al., 1984; Rouleau, 1984; Spiess, 1988).

The subsequent compilation process has as its objectives the assembling and location (in terms of proper planimetric position, projection system and employed scale) of the diversity of geographical data which it is intended to incorporate within a prescribed format (Spiess, 1988; Keates, 1989). Normally geographers are involved in the generation of 'thematic' or 'special purpose' as opposed to 'general' topographic maps (Robinson et al., 1984; Lehmann and Ogrissek, 1988). They are particularly concerned with the display of spatial variations of a single entity, or the relationship between phenomena, against a background or base consisting of the graticule, coastline, hydrology and relevant administrative subdivisions of the area being considered. These details may have been plotted with respect to the particular projection employed, abstracted from existing map material, or recalled from a computer-held database or geographical information system (see chapter III.7). Having generated a background, symbolized thematic data can be located against this to form a predominant layer of information which should appear to stand above the basic geography of the area.

An essential prerequisite to the preparation of a 'compilation' (a hand-drawn version of the intended end-product) is to work from larger to smaller scales (Spiess, 1988; Keates, 1989). The rationale for this is that all but the largest scale maps show data that have been generalized to suit, or be accurate for, a specific scale of presentation. Consequently, if compilation takes place from a small to a large scale, inappropriate over-generalization may become built in. Methods used to position base details, in the manual case, may be either photographic, mechanical or purely by eye, and will result in the generation of a work-sheet which is normally prepared on a translucent material such as tracing paper or polyester plastic (Kanazawa, 1984; Keates, 1989). *Basic Cartography for Students and Technicians*, volumes 1 (Ormeling, 1984) and 2 (Anson, 1988), includes numerous explanatory graphics relating to both good and bad compilation practices, and the other texts cited are also well illustrated with instructional drawings.

Thematic data displayed against the geographical base are, of necessity, the result of generalization which is practised by the compiler (Robinson et al., 1984; Balodis, 1988). Four main elements contribute to this process. First, it is essential to undertake the 'simplification' of facts by identifying the most important features and eliminating unwanted detail and, second, 'classification' should be employed in order to scale and group data. The resultant material must then be the subject of graphic coding with respect to its essential characteristics, comparative significance and relative position during a process referred to as 'symbolization' and involving the use of point, line and area devices to represent either qualitative or quantitative information. Finally, graphic 'induction' must be undertaken by the application of the logical process of inference. These elements are subject to a number of specific controls such as the 'objective' or purpose of the map, the intended 'scale' of the presentation, the 'graphic limits' of the system of communication employed (e.g. is detail to appear in monochrome or is colour available?) and the 'quality of available data'. Consideration of both the elements and the controls of generalization is relevant whether the map is to be produced by either manual or computer-assisted means, and can be related to either a tangible output on paper or a screen-displayed version of the geographical information. Assuming the former, decisions have to be made with respect to the specification of included information prior to the fair drawing of the detail (Spiess, 1988).

The names, relative importance, character and extent of geographical features can be illustrated by the informed selection and use of lettering styles and sizes. Care must always be taken to maintain legibility and to position type in such a way that it can be easily related to the feature to which it refers. All line work (e.g. that depicting the graticule, coastline, rivers, administrative boundaries, roads or railways) must be carefully and consistently drawn, and its width and nature suggesting the importance of the feature detailed. Similarly the design of both point and area symbols must be carefully specified to ensure a clear visual distinction between the nominal, ordinal and interval classes of data to be represented on the map (Dickinson, 1973; Spiess, 1988).

If colour is available it enables the introduction of additional visual variables, and can contribute significantly to the clarity, legibility, contrast structure and communications effectiveness of the eventual map. This is further assisted by permitting the inclusion of more detail, the colour coding of related data and the enhancement of visual interest (Robinson et al., 1984; Spiess, 1988; Keates, 1989). However, if the intended end-product is to be printed, the use of colour adds significantly to production costs, and its application must be carefully considered as certain conventions have developed which should be taken into account during the overall design process (Curran, 1988). The employment of colour generates aesthetic reactions and has particular human connotations which are psychologically or physiologically based. Thus red is typically thought of as representing warmth or danger, whilst blue suggests cold or wetness, and these factors can be important in maintaining effective communications and user understanding of a map. Unfortunately it is not, as yet, possible to programme aesthetic considerations when producing maps either on a screen or as computer print-outs.

Fuller descriptions of the compilation, generalization, specification and map construction processes mentioned above are included in textbooks written by the authors mentioned earlier, and at least some of these should be found in most libraries.

It is hoped that, from the statements made here, the reader has gained some appreciation of the value of cartographic awareness to the geographer. The ever-increasing availability of quantitative details relating to a host of topics has provided a wealth of new and up-to-date data, but the significance of much of this would be better understood if the inherent spatial relationships were depicted in the form of a map. Although the application of geographical information systems based on the computer's capability for the storage of vast quantities of alphanumeric information in an easily retrievable form is becoming increasingly important, much of the graphic output is decidedly second rate compared with that generated by a competent and knowledgeable cartographer. However, it must also be admitted that manual production methods are very much slower! The construction of an authoritative well-designed map which communicates included facts effectively to its user is a very satisfying experience, and it is to be hoped that geographers will realize the value and increase their compilation of such specialist graphics in the future.

References

Anson, R. W. (ed.) 1988: *Basic Cartography for Students and Technicians*, vol. 2. Barking: International Cartographic Association and Elsevier Applied Science.

Balchin, W. G. V. 1972: 'Graphicacy'. *Geography*, 57, 185–95.

Balodis, M. J. 1988: Generalisation. In R. W. Anson (ed.), *Basic Cartography for Students and Technicians*, vol. 2, Barking: International Cartographic Association and Elsevier Applied Science, 71–84.

Board, C. and Taylor, R. M. 1977: Perception and maps: human factors in map design and interpretation. *Transactions Institute of British Geographers, New Series*, 2, 19–36.

Crone, G. R. 1978: *Maps and their Makers*. Folkestone: Dawson.

Curran, P. J. (ed.) 1988: *Compendium of Cartographic Techniques*. Barking: International Cartographic Association and Elsevier Applied Science.

Dickinson, G. C. 1973: *Statistical Mapping and the Presentation of Statistics*, 2nd edn. London: Edward Arnold.

Greenhood, D. 1964: *Mapping*, 3rd edn. Chicago, IL: University of Chicago Press.

Guptill, S. C. and Starr, L. E. 1984: The future of cartography in the information age. In L. E. Starr (ed.), *Commission C Computer Assisted Cartography Research and Development Report*, Toronto: International Cartographic Association, 1–15.

Kadmon, N. 1988: Computer-assisted cartography. In R. W. Anson (ed.), *Basic Cartography for Students and Technicians*, vol. 2, Barking: International Cartographic Association and Elsevier Applied Science, 105–38.

Kanazawa, K. 1984: Techniques of map drawing and lettering: In F. J. Ormeling (ed.), *Basic Cartography for Students and Technicians*, vol. 1, Toronto: International Cartographic Association, 112–80.

Keates, J. S. 1989: *Cartographic Design and Production*, 2nd edn. Harlow: Longman.

Lehmann, E. and Ogrissek, R. 1988: Thematic cartography. In R. W. Anson (ed.), *Basic Cartography for Students and Technicians*, vol. 2, Barking: International Cartographic Association and Elsevier Applied Science, 85–103.

Margerison, T. A. 1976: *Computers and the Renaissance of Cartography*. London: NERC Experimental Cartography Unit.

Meynen, E. 1973: *Multilingual Dictionary of Technical Terms in Cartography*. Stuttgart: International Cartographic Association and Franz Steiner Verlag.

Monmonier, M. S. 1985: *Technological Transition in Cartography*. Madison, WI: University of Wisconsin Press.

Ormeling, F. J. (ed.) 1984: *Basic Cartography for Students and Technicians*, vol. 1. Toronto: International Cartographic Association.

Robinson, A. H., Sale, R. D., Morrison, J. L. and Muehrcke, P. L. 1984: *Elements of Cartography*, 5th edn. Chichester: Wiley.

Rouleau, B. 1984: Theory of cartographic expression and design. In F. J. Ormeling (ed.), *Basic Cartography for Students and Technicians*, vol. 1, Toronto: International Cartographic Association, 81–111.

Spiess, E. 1988: Map compilation. In R. W. Anson (ed.), *Basic Cartography for Students and Technicians*, vol. 2, Barking: International Cartographic Association and Elsevier Applied Science, 23–69.

Taylor, D. R. F. 1989: Opening Address by ICA President. In R. W. Anson and B. V. Gutsell (eds), *ICA Newsletter 14*, Toronto: International Cartographic Association, 5–6.

Thrower, N. J. W. 1972: *Maps and Man*. Englewood Cliffs, NJ: Prentice Hall.

Wallis, H. M. and Robinson, A. H. 1987: *Cartographic Innovations – An International Handbook of Mapping Terms to 1990*. Tring: Map Collector Publications (1982) in association with the International Cartographic Association.

III.7
Geographical Information Systems
Jonathan Raper

Introduction

Within contemporary geography one of the major forces for change in the way geography is studied is the geographical information system (GIS). GISs are computer systems for handling all types of spatial information and have been described as the 'biggest step forward since the introduction of the map'. This article will look at the basis for this claim, and will profile the rapid technological changes which have led to the development of GISs.

Any examination of how GISs are used and what they can contribute to today's geography must start with an analysis of the traditional concept of a map. Although it is customary for geographers to communicate their observations and analyses of spatial phenomena through maps, the paper map is, by definition, a static record of a particular geography at a given time and place. This limitation is reflected in the way in which maps are normally collected or made available, i.e. as thematic sets in atlases, in definitive map series such as 'topography' and as illustrations for scientific works. This process of map making and publishing is governed by cartographers, who have applied standards to map content and presentation.

However, as geographers have come to make maps by computer a shift in the perception of the map has become apparent. Maps when stored in the computer have taken on a dynamic form: their content, symbolization and the area covered can now be changed rapidly and easily. It can be argued therefore that maps are better thought of as databases, or collections of information linked by their spatial relationships. In this conception a *map* can be seen simply as one unique expression of a spatial database showing, for example, the area by area values of a spatially varying parameter where the categories chosen reflect a geographical model of some kind. Maps showing different data sets for the same area can also be overlaid on top of each other to generate many new integrated maps. The graph in figure III.7.1 shows how the number of such integrated maps grows rapidly as the number of initial maps is increased: thus from twenty original maps, 190 unique pairs of maps can be formed! Maps can also be made dynamic when spatial phenomena are mapped or projected through time – the resultant set of maps can be thought of as a model, or a spatial

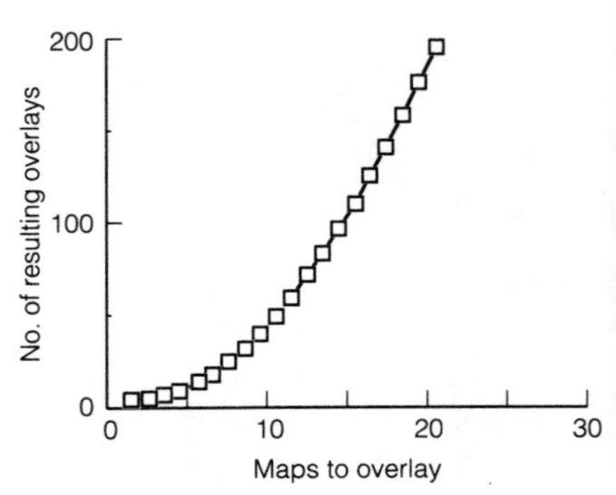

Figure III.7.1 Graph showing how the number of integrated maps increases rapidly as the number of initial maps increases.

and temporal database. Such a geographical database containing multiple spatially referenced data sets and allied with a set of spatial tools is the basis of a GIS.

Although the realization of GISs powerful enough to carry out spatial analysis has been restricted to the 1980s, the developments which have led to modern GISs go back two decades. Experiments in the early 1970s at the Experimental Cartography Unit of the Royal College of Art and the Ordnance Survey in the UK and the Canada GIS in North America were originally aimed at the automation of the mapping process, primarily to cut the costs of map production and to speed up the updating of topographic maps and plans. These early digital maps were made by converting conventional maps to sets of co-ordinates describing points and lines on the paper map. However, the scope of these systems was restricted to the scaling, reprojection and plotting out of the maps.

The realization of the GIS in the 1980s, however, has meant a fundamental change in the techiques used. This has been achieved by the use of systems which logically connect points, lines and areas to represent map features and which are tightly linked to a geographical database storing all the associated information about the features. These techniques have made it possible to query geographical databases rapidly, and to integrate fully different maps incorporating spatial and non-spatial data for the same area. With the parallel development of computer hardware making available ever faster computers at lower real prices, processing the (typically) large size of most spatial data sets is now within reach of all geographers, not just an elite of specialists with powerful computers.

Since these new techniques and systems have wrought a profound change in the ability of the geographer to handle spatial and spatially related data, many new horizons have opened for research using GISs. The emergence of a dynamic new technology has reinvigorated geographical analysis after a decade of methodological consolidation. The challenge of GISs is to build spatial databases for the key geographical questions, and to probe spatial relationships using these newly available tools for powerful spatial analysis.

The Development of Geographical Information Systems

GISs are integrated spatial data handling systems which have been developed for a range of purposes. A number of different definitions of GISs have been published, each reflecting a different perspective on their application. Three recent definitions illustrate this point, defining a GIS as

- 'a set of tools for collecting, storing, retrieving at will, transforming and displaying spatial data from the real world for a particular set of circumstances' (Burrough, 1986);
- 'an information technology which stores, analyses, and displays both spatial and non-spatial data' (Parker, 1988);
- 'a decision support system involving the integration of spatially referenced data in a problem solving environment' (Cowan, 1988).

The last definition by Cowan is the most applied, and best describes how GISs are being used in many fields of geography.

The evolution of GISs also reflects a range of purposes and applications. The functions of a comprehensive GIS can be traced to a range of early spatial data handling systems which evolved in several related fields such as computer-aided design (CAD), database management systems (DBMS), automated cartography (CARTO) and image processing (IM PROC). A comprehensive GIS today can be defined as the union of the functionality of each of these systems to form a true multi-function, multi-purpose spatial data handling system. Figure III.7.2 can be used to visualize this development, and to chart the origin and strengths of particular software packages sold as GISs. Depending on the functionality offered, most packages can be plotted on this diagram: many will plot in the overlap of CAD and CARTO, but only a few can actually be placed in the GIS category when defined in this way.

At the same time as the analytical functions of GIS software have been developing, new models have been defined to describe fundamental geographical features. Green and Rhind (1986) defined a geographical 'data model' as:

an abstraction of the real world . . . which incorporates only those properties thought to be relevant to the task(s) in hand.

The data models used by GIS are usually 'phenomenon based' and use the features in, or derived from, the geography at hand to specify the model used to represent them. Thus, point, line and

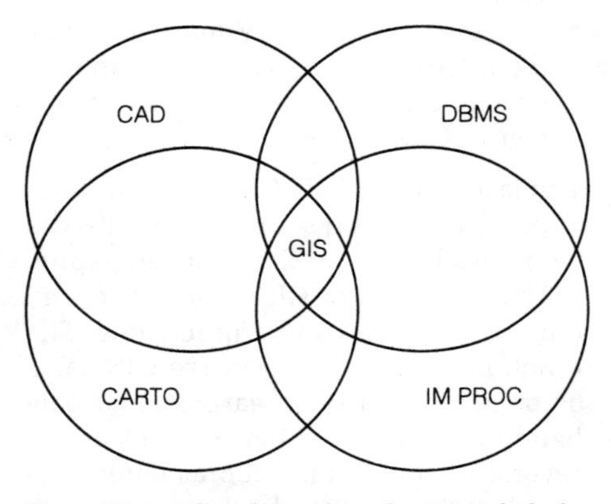

Figure III.7.2 The relationship of a GIS to linked data handling fields.

area units are usually used as the basic spatial building blocks for, respectively, specific locations, boundaries or routes, and zones or distributions.

However, to render this model of specific geographical features, a translation of these items into computer form must be made: this is known as a 'data structure'. A basic limitation for a data structure is that computer systems can only store and handle characters and numbers, not spatial objects such as lines. Hence spatial data structures can only be created from spatial referencing systems based on position finding or the use of co-ordinates (figure III.7.3). Two distinct approaches are used to represent points, lines and areas by spatial reference systems, which are known as 'vector' and 'raster' data structures.

Vector data structures use the idea of co-ordinate fixing as in navigation. A point can be described by its distance along each of two axes of measurement, e.g. north and east, while a line can be described by the shortest straight line distance between two such points – a 'vector' (see figure III.7.3). Areas can be represented by any number of lines joined together in a ring and arranged to reflect the specific shape of the area. In all cases the only information needed is the co-ordinates in numbers, which can easily be stored in the computer and recalled in the right order when needed.

Raster data structures use a different approach altogether, using tiny building blocks to build up the shape of a feature on a grid. Using such a grid of very small squares (a 'raster'), a point, line or area can be approximated by sets of squares or 'pixels', the resolution of the representation being directly controlled by the size of the grid squares. In a raster data structure the computer keeps a record of which pixels are switched 'on' and therefore form part of one of the geographical features.

Using one or other of these data structures allows the effective storage of

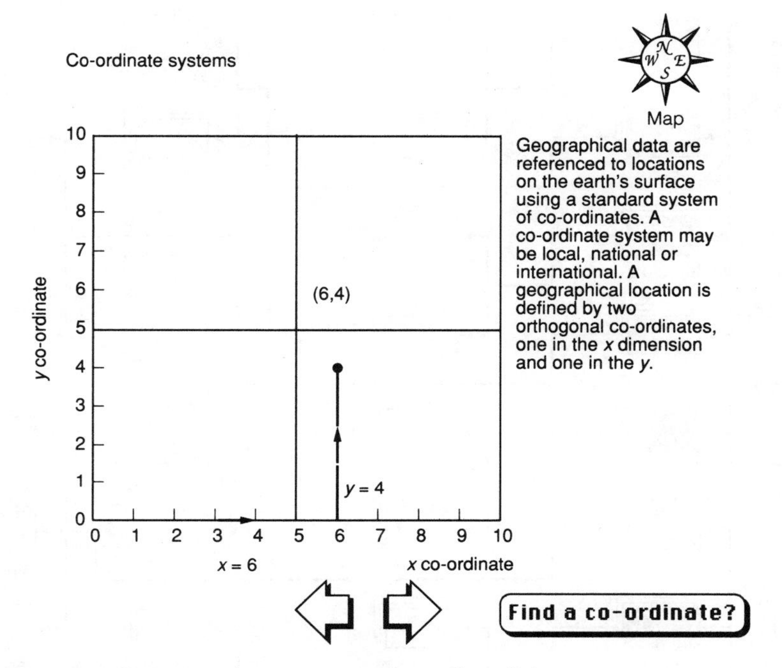

Figure III.7.3 A guide to vector data structures and co-ordinate fixing.

spatial objects in the computer. To complete the geographical database requires the addition of a unique identifier to all the points, lines and areas represented so that they all can be linked to a name or other non-spatial data stored alongside the spatial data. The completed database is an integrated spatial and non-spatial information system: however, its potential can only be unlocked by applying spatial analysis tools.

The Key Activities in Geographical Information Systems

The functions available in a typical GIS are based around the general principles of spatial data handling, and are easily framed by the sequence of operations carried out in the creation and analysis of a geographical database. These operations can be summarized using the scheme described in Raper and Green (1989) for the Geographical Information System Tutor (GIST), as shown in figure III.7.4. The functions are described in the following sections.

Data capture is divided into sections dealing with vector and raster methods. Vector data capture is most commonly achieved by the manual tracing or 'digitizing' of the map features with an electronic recording device called a cursor. To 'capture' the data the operator moves the cursor over a map sheet and traces all point, line and area features. These details are then stored as co-ordinates which make up the basic building blocks of the geographical database. Scanning is used to capture raster data: here a light sensor is moved over the map in a grid pattern, recording all the map features inked onto the map as pixels switched 'on'. The map is then

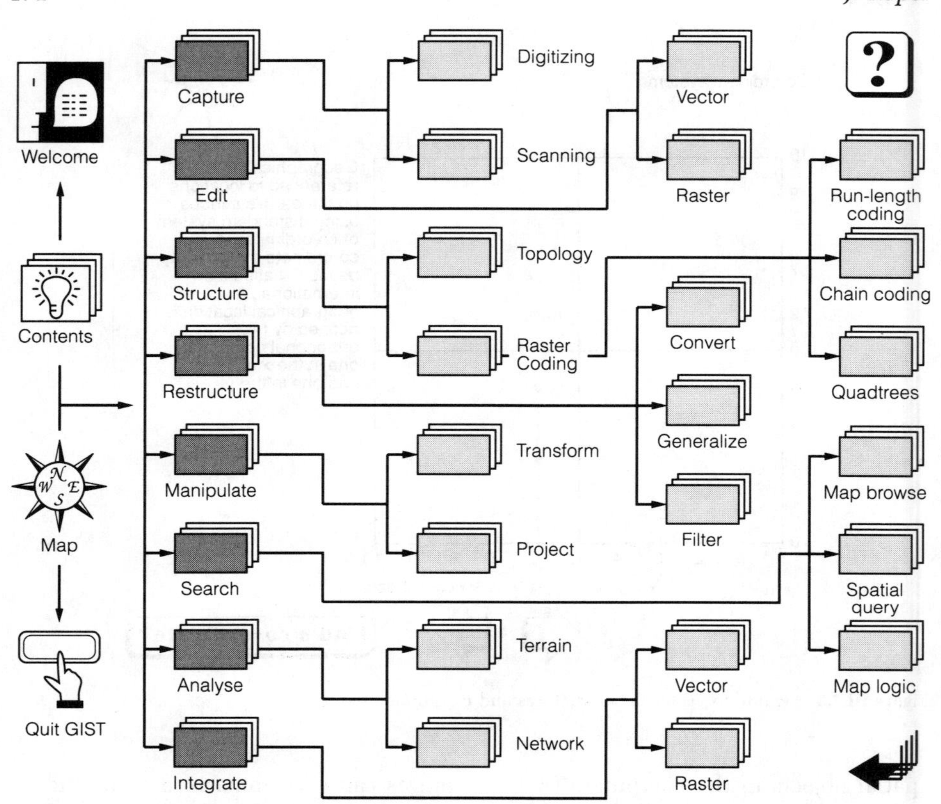

Figure III.7.4 The sequence of operations carried out in the creation and analysis of a geographical database.

represented by a matrix of pixels whose locations are stored as row and column numbers. In the course of both forms of data capture, associated text data can also be entered alongside the graphic data in a database, and are usually linked to the spatial data by the sharing of a unique identifying code.

Editing is also carried out on raster and vector data by distinct methods. Vector data editing seeks to correct the errors associated with digitizing, which include misplaced points and line overshoots or undershoots. Raster editing, however, aims to correct the image produced by scanning the graphic features on a map, for example, whose inks may show contrast variability and where ink smudges on the paper may be recorded accidentally. Thus, typically, raster editing involves gap removal, line thinning and stray pixel removal operations.

The process of *structuring* digital map data (storing the data in a form suitable for rapid spatial retrieval) is also distinct in the case of vector and raster data. Structuring vector data requires the establishment and storage of relations between the points, lines and areas. This step is extremely important as it allows the map database to be queried efficiently for spatial interrelationships. One of the common forms of structuring is based on the concept of 'topology', which defines the location of geographical phenomena relative to each other but is independent of distance or direction.

Typically, the connectivity and adjacency relations expressed in the map are stored in tables showing the lines which connect and the areas which are next to each other. Raster structuring is based on coding techniques to reduce the number of row and column addresses stored for the pixels, for example by coding the length of full rows and columns or using chains wrapped around features. Another distinct raster structure commonly used is called the quadtree: this is produced by repeatedly subdividing the grid into four parts, continuing the subdivision of the new quadrants only if they are *not* completely full or completely empty. Most raster structuring is aimed at the efficient storage of areas – boundaries between areas are inherent in the discontinuity between them.

The techniques of *restructuring* are used to change the level of map detail or convert between vector and raster data types. To reduce the number of points needed to represent a line in vector data a co-ordinate thinning algorithm such as the Douglas–Peucker technique can be used whilst maintaining basic line shape. In raster data filters can be applied to 'generalize' the shapes of areas or highlight edges. Conversion of spatial data structure from raster to vector and vector to raster can also be achieved using, respectively, raster generation from points and lines, or line or area tracing to create vector data.

Map *manipulation* describes the transformations which can be carried out on the spatial data, such as rotation, translation, scale change and warping. These functions are often used in the process of revising or adding to a map. A special type of transformation from a sphere to a flat plane is used to show real-world features with their geometrically correct interrelationships and in their correct location: this is called a map 'projection'. Several fundamental techniques of map projection are used, based on cylinders, cones and circular planes as a distinct transfer mechanism between a sphere and the planar grid of a map.

Specifying the *search* or retrieval of information from a GIS is one of the most important functions of a GIS. Retrieval of data can be done by querying either the graphical map data or the database of associated text information. Since the graphic and text data are linked in a GIS, a search of one can yield new selections in the other. Graphic searches use oblong windows, circles or buffers to search the graphic data and select the identifier for each map feature included; text searches use specified criteria to search the categories of text data stored so as to select the same identifiers stored within the geographical database. Each search can then be used to select items by identifier in the other type of data by using the selected identifiers as indexes. These searches can be made more sophisticated by using Boolean logical modifiers to combine search criteria such as AND, OR and NOT.

Spatial *analysis* can also be performed by a GIS, although usually the graphic data will need to be topologically structured. An example of a spatial analysis would be finding the shortest path through a network such as a road system. The building of terrain models relies on spatial analysis to assign estimates to the surface to be modelled: once this is done isolines (contours) can be drawn and isometric models can be generated to create a true three-dimensional perspective on the scene.

Finally, map *integration* is the combination of two (or more) similar scale maps of the same geographic area to create a new map. Vector-based integration involves the overlay of two sets of vectors, the location of their true intersections by geometric operations, and the creation of *new* lines and areas. Even with modern GISs this process is slow and difficult, whereas the same process carried out for raster data is relatively quick. Raster integration involves the overlay of two maps (therefore grids) with the same resolution to achieve a map union. However, neither map is retained as the pixel values for the new raster must be the *logical* union of the original maps. Note also that the text data associated with the spatial data for each map must also be integrated: all the new map features must have related text data from *each* original map.

The Applications of Geographical Information System

The functions associated with a GIS which are described above are best illustrated by some examples of the application of GISs. Typically, no GIS project will need to utilize all these functions: they should be considered as a flexible tool-box. Hence to carry out a project requires a choice of GIS which has the appropriate tools, access to the graphic and text data and a suitable computer system.

One typical application of GIS use concerns the identification of the population density around rail stations in order to plan train services. The graphic data in figure III.7.5 consist of a point at the centre of all the population census enumeration districts (EDs) for the county of Kent to the southeast of London, along with the railway lines and stations along them. The text data or attributes of the census EDs stored is the total population for that area, and for the railway stations is the name of the place served. In this project carried out for British Rail by the South East Regional Research Laboratory, the population within 1, 2, 5 and 10 km of the railway stations was required, in order to estimate the number of potential commuters likely to use each station.

To carry out this study the data were captured by digitizing appropriate maps, editing and structuring the data, and by storing the appropriate population and station name data. To establish the numbers living within the specified distances, circular search 'buffers' were placed around the stations. Starting with the smallest distance, the population of all the geographical centres of the EDs ('centroids') within the search circle were progressively added together to reach a total for each search radius. The respective population totals were then entered as new text data for the rail station in the attribute database. Finally, the attribute database was searched for all stations with catchment populations greater than planning targets for each level of train service, and the stations were graphically selected on the map to illustrate their spatial distribution across Kent.

In another example the incidence of soil erosion on the Greek island of Crete was to be investigated. This project was undertaken as part of the European Community Coordinated Information on the Environment project (CORINE) which is bringing together maps and text data to make better information available in the management of the environment. In this project maps of a wide variety of environmental factors were to be brought together to aid in the estimation of soil erosion potential (figure III.7.6.). As in the previous example the data were captured by digitizing appropriate maps, editing and structuring the data, and by storing the values for the different areas for which climate, soil erosivity and slope were stored. In this project the data were stored in vector form, and maps of the same scale were overlaid to produce a new map of estimated soil erosion. In this case the areas in all the maps were given values to reflect the parameter being mapped, rather than a name label. Thus, the overlay of the maps was by multiplication of the values from the areas on each map to produce a new map with factor scores for all the

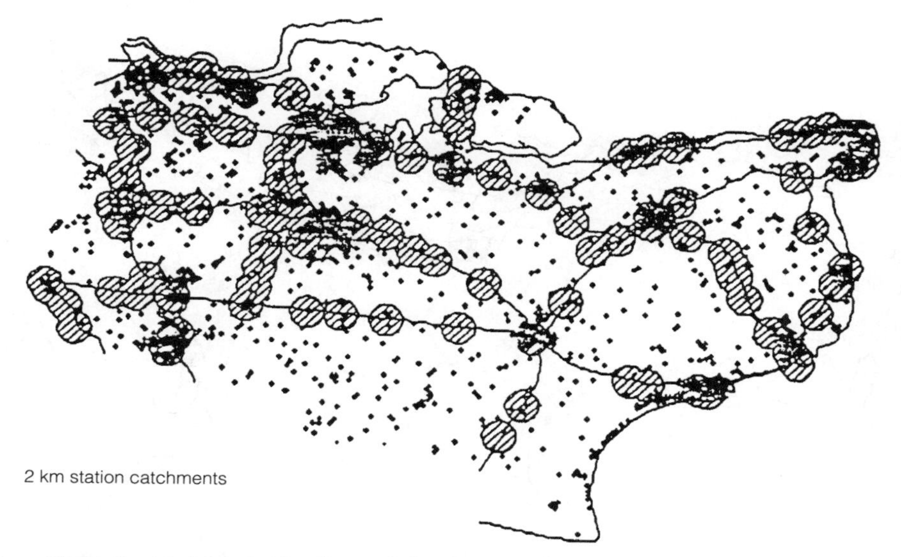

Figure III.7.5 Graphical data to identify population density within 2 km of railway stations in a portion of southeast England.

new areas. Finally all the new areas produced by overlaying and intersecting the three original maps were 'dissolved' together when adjacent areas had the same value. The speed with which this operation can be carried out enables much better use to be made of the existing information, which may not be useful on its own.

These two examples of a GIS in use indicate typical applications of GISs to geographical problems which would have been difficult and time consuming to complete manually. This improvement in the productivity of mapping and the handling of spatial data, however, has not gone unnoticed by the large institutional users of geographical information. In response to the growing interest in such techniques the government of the UK set up an enquiry in 1985 into the 'Handling of Geographical Information' which was chaired by Lord Chorley. The report of this enquiry was published in 1987 after a wide consultation exercise and made the following recommendations which are a model for the national implementation of GISs.

1 Costs – these systems are still expensive, but computer hardware costs are falling fast, and so investment returns can still be made.
2 Data availability – the UK central mapping agency (Ordnance Survey) should speed up data capture, and government departments should make data more readily available.
3 Linking data – the government should decide on standard spatial units (e.g. post codes) or locational reference systems, and develop transfer formats for data exchange between systems.
4 Awareness – improve awareness and training by further investment.
5 Research and development – streamline the approach to research funding, and allocate more funds.
6 Coordination – set up a centre for geographic information and research.

The greatest take-up of GISs amongst those organizations handling geographical data on a wide scale has been in the utilities – the providers of water, electricity, gas and telephone services to industry and domestic users.

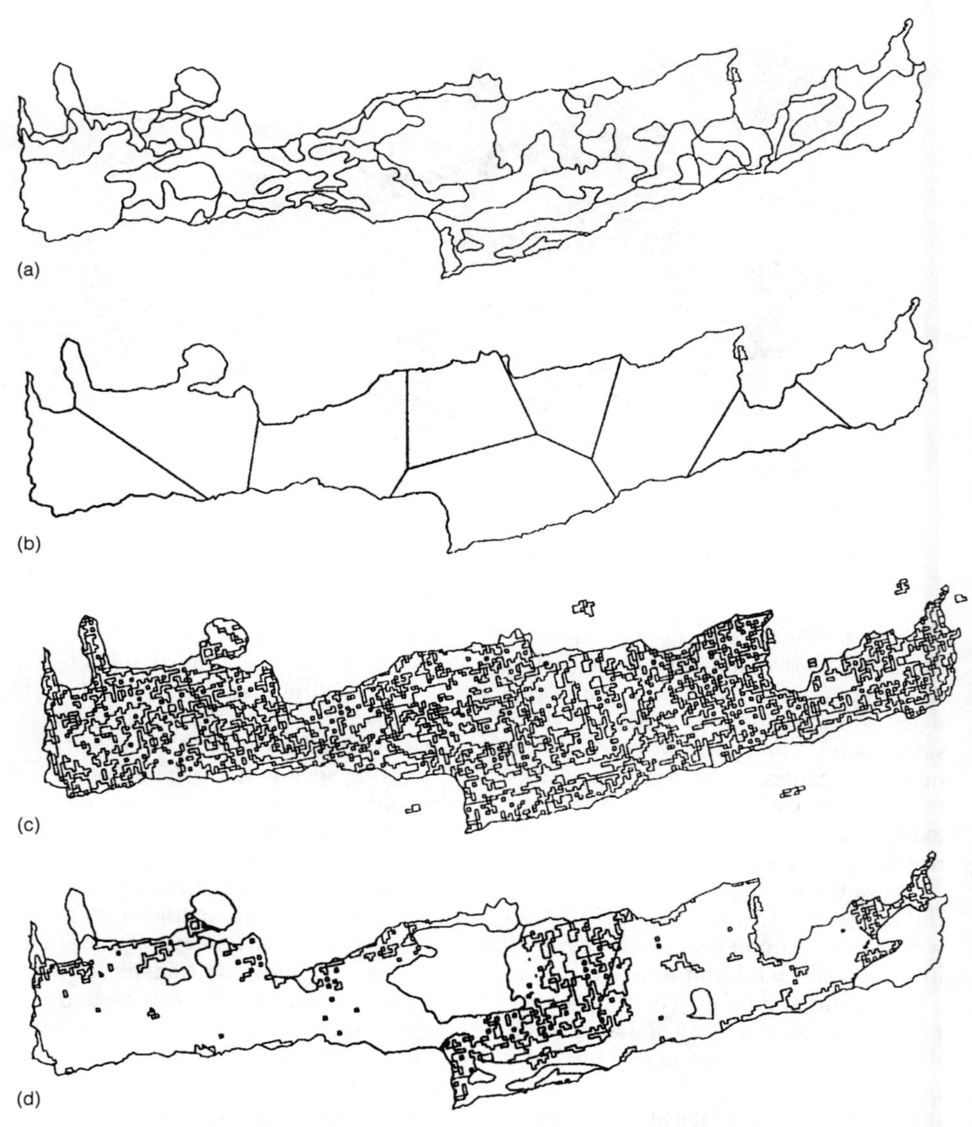

Figure III.7.6 Maps of a wide variety of environmental factors to aid in the estimation of soil erosion potential on Crete: (a) soil erodibility index/soil quality index; (b) climatic erodibility index/climatic quality index; (c) slope; (d) potential soil erosion risk.

The utilities require very large scale maps to identify the location of their equipment above or below the ground, and typically use maps at a scale of 1:1250 or 1:2500. These maps show individual buildings and are used to plan roadworks or prevent accidents to other services. Local authorities also use GISs to plan street cleaning programmes and to map the location of elderly people needing special health care and school catchment areas. Other users such as large superstores planning new store locations use smaller scale maps, perhaps 1:10,000 or 1:25,000, and may reference all their information to post codes rather than real-world co-ordinates.

The rapid increase in the capabilities of GISs and their increasing use in the activities described above have meant that GIS is both stimulating to study and potentially offers a good job opportunity for those with appropriate skills. At present such skills are in short supply worldwide: thus GISs offer an unrivalled opportunity for geographers to pursue a vocation in technical development or to enter business or administration where the computer handling of geographical information has rapidly become a key skill.

Research Frontiers

However, as in any rapidly growing area of technology and knowledge many questions remain unanswered, and some key aspects of GISs remain undeveloped or underdeveloped. Research programmes in GISs are now receiving priority funding in many countries, in particular in the following areas:

- new algorithms or programs for vectorization, overlay, generalization and spatial analysis;
- better interfaces (or spatial languages) to GISs to improve the accessibility of GISs as use expands;
- the design and implementation of spatial database structures which are efficient and integrate data sets;
- the development of models for the networking of spatial and non-spatial data in multiple locations;
- improved techniques for three-dimensional GISs and the modelling of surfaces and solids.

Much of this research is of direct interest to commercial vendors of GISs, and therefore much of the work is being funded collaboratively – an added benefit for those who wish to further their studies through research, but with a potential entry into business at a later date.

A New Geography?

It can now be argued that the spatial theory and techniques generated by GISs have created a new and vigorous subdiscipline within modern geography. However, GISs will also undoubtedly change geography as a whole: the analysis and integration of spatial data which can now be carried out has changed the terms upon which much geographical research can be done. The current period is one of profound change and is characterized by the *installation* and *configuration* of systems by all those handling geographical data: the future will see the *application* and *use* of the systems on a much wider scale. If the challenge of GIS development is to be met then geographers must get involved in the design and operation of many of those systems.

Acknowledgements

The author would like to thank his co-author of GISTutor for permission to use examples from the program in this paper. Applications examples appear by permission of the South East Regional Research Laboratory, funded by the UK Economic and Social Research Council, and the Directorate General XI of the European Commission. The author would also like to thank his colleagues David Rhind and Helen Mounsey and other speakers in the Birkbeck College Short Course Programme in GIS for stimulating discussions on the definition and application of GISs. Finally, thanks also to Ann Wilkes for preparing the maps of Crete.

References

Burrough, P. A. 1986: *Principles of Geographic Information Systems for Land Resources Evaluation*. Oxford: Clarendon Press.

Cowan, D. J. 1988: GIS versus CAD versus DBMS, what are the differences? *Photogrammetric Engineering and Remote Sensing*, 54, 1551–5.

Green, N. P. A. and Rhind, D. W. 1986: Spatial data structures for geographic information systems. Conceptual design of a geographic information system for the Natural Environment Research Council, Report 2.

Parker, H. D. 1988: The unique qualities of a geographic information system: a commentary. *Photogrammetric Engineering and Remote Sensing*, 54, 1547–9.

Raper, J. F. and Green, N. P. A. 1989: The development of a tutor for geographic information systems. *British Journal of Educational Technology*, 20, 164–72.

III.8
Laboratory Work
Heather A. Viles

'Geographers in the Laboratory?! Surely Not!'

Many non-geographers (and quite a few geographers as well!) find it difficult to imagine what physical geographers get up to in the laboratory. 'Surely physical geography is a field science?', they say, or, 'Of course, geographers aren't proper scientists, so they can't possibly do any serious laboratory work'. However surprising it may seem, geographers can and do carry out laboratory work and it forms an important part of most physical geography projects. Geography undergraduates will normally be expected to participate in some organized laboratory classes, but for many students this is their first and last experience of laboratory work. There are many opportunities, however, for using laboratory techniques as part of a dissertation.

Two main types of laboratory work are carried out by physical geographers, i.e. *analysis* of samples collected in the field (where one is asking such questions as what is it made of? and how old is it?) and *experimentation* (where one is finding out how samples behave under certain conditions). For both these types of laboratory work there are two main approaches, i.e. *low-tech* and *high-tech* methods. *Low-tech* methods use simple, cheap apparatus but may take a long time, whereas *high-tech* methods use expensive equipment but are usually quick and accurate. Most geography laboratories have a wide range of *low-tech* equipment, and some are also well equipped with *high-tech* instruments.

In this chapter I review some of the major laboratory techniques used by physical geographers for the analysis and experimental study of sediments, rocks, solutions and organisms. In reality, most projects will involve several techniques on different types of samples and the final section of this chapter presents some examples of laboratory-based projects suitable for undergraduates.

First a few words of caution are necessary. *Safety* in the laboratory is of paramount importance, involving taking care of yourselves and other people, equipment and samples. All laboratories have safety regulations which should be read before any laboratory work is started. Some general rules are as follows: always obey what the laboratory technicians say; if in doubt about anything, ask; follow all instructions carefully; own up immediately if anything goes wrong or if you break any equipment. Finally, it is most important to look after your samples properly. After all the hard work involved with collecting samples in the field it is a tragedy if they get lost, damaged or otherwise mixed up. You cannot do science without samples; and you cannot do good science without good samples.

Before starting on any of the laboratory techniques suggested below it is important to consider what you are going to do with the results. Why are you using the techniques? What information are you going to get from these techniques? How accurate do you need your results to be? What statistical analyses are you going to carry out on the data collected?

How are you going to display the results? If you have a clear idea of what you are trying to do and why, you are more likely to get useful results with minimum fuss.

There are several useful general introductions to laboratory work including Goudie (1990), and two excellent chapters on laboratory techniques and microscopy in Haynes (1982).

Methods of Sediment Analysis and Experimentation

Sediments studied by physical geographers include river and lake sediments; beach material; loess; dune sands; moraines; and marsh muds. Analysis of sediments involves investigating their physical, chemical and biological make-up. So, for example, the variation in these parameters down a core obtained from lake sediments might be studied, or their variation across a number of samples from dune crest and interdune areas. A suggested scheme for the analysis of sediment samples is shown in figure III.8.1.

Physical characteristics of interest include the size and shape of particles and the porosity of sediment. Initial observations may also be made of colour (using standard colour charts, e.g. Munsell Soil Colour Charts) and any stratification. Low-tech grain size analyses involve sieving and settling tube methods. Wet sieving is used to separate sand from mud; in dry sieving, particle size distributions within the sand fraction are analysed, and hydrometer or pipette settling methods are used to produce size distributions for silt- and clay-sized sediment. There are standard methods for graphing the results, and estimates of distribution statistics can also be produced, e.g. mean, standard deviation, skewness and kurtosis. Sample preparation before use of any of these techniques is normally necessary to remove organic material and any other binding material. Folk (1974) provides an excellent handbook for such procedures.

Various high-tech methods are available for rapid and accurate determination of particle size within the silt–mud size range, including the Coulter Counter (described in Goudie, 1990) and the CILAS granulometer.

Grain shape can be assessed by comparing grains with standard charts of roundness and sphericity. More complex equipment is available in some laboratories to investigate grain 'rollability' (a multivariate measure of grain shape). For grains less than 1 mm in diameter, microscopes provide a useful tool for studying shape and composition of the grains and their surface features. In microscope work samples need mounting on microscope slides and then point-counting techniques are usually used to gain a representative sample (200–500 grains per slide is usual).

High-tech methods for studying sediment shape and composition involve scanning electron microscopy (SEM) and energy-dispersive analysis of X-rays (EDAX) (for samples of prepared grains), X-ray fluorescence and X-ray diffraction (often powdered samples) and atomic absorption spectrophotometry (AAS) (for samples of dissolved grains). SEM methods also provide useful information on grain surface textures as described in Krinsley et al. (1973). More information on these techniques, which are used on a wide range of geographical samples, is shown in table III.8.1.

Laboratory experiments on sediments include studies of grain movements in flumes or wind tunnels; experimental weathering of quartz grains in an environmental cabinet; and studies of water and solute movements through a soil column (see table III.8.2). Such studies normally require some high-tech equipment and all need careful research design. Two invaluable sources of refer-

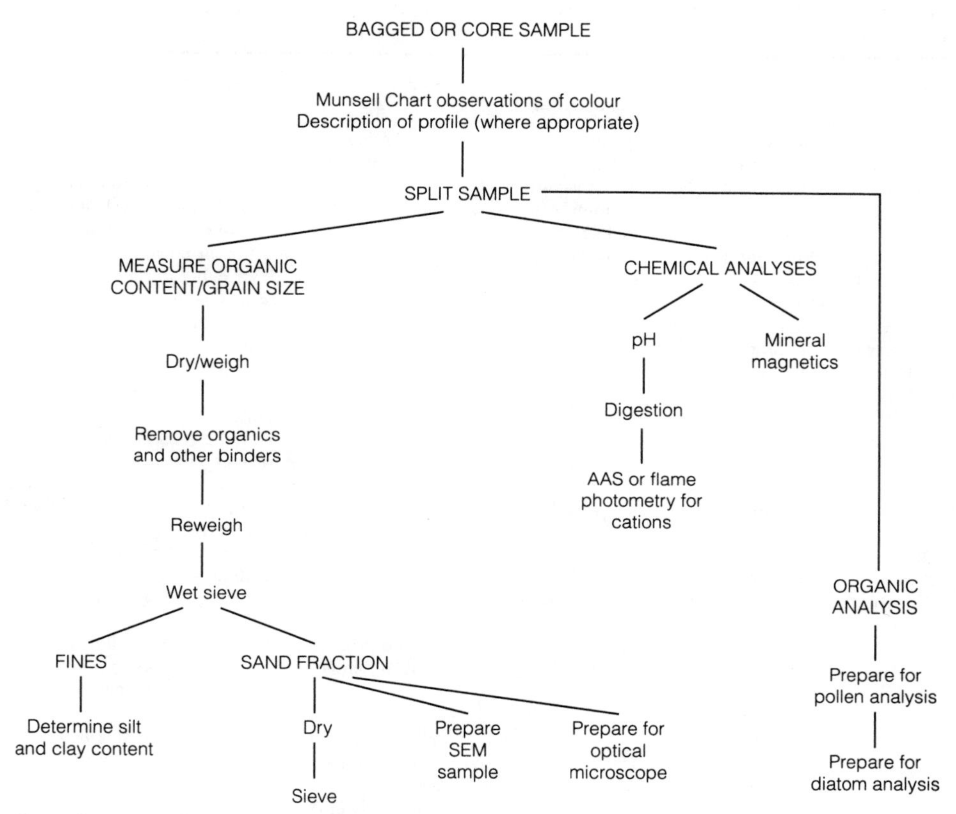

Figure III.8.1 A sediment analysis scheme.

ence for such studies are Yoxall (1983) and Allen (1985).

Analysis and Experimentation on Rock Samples

Many of the methods used to investigate rock samples are similar to those used in sediment studies, although the consolidated nature of rocks permits a whole additional suite of techniques. Johnson and Maxwell (1989) discuss preparation and analysis for investigation of various rock parameters. Hutchinson (1974) is a useful general guide to available techniques. A suggested scheme for the analysis of rock samples is presented in figure III.8.2.

As with sediments there are several simple preliminary descriptions which can usefully be made of rock samples, including photographs of hand specimens (remember to include a scale), observations of colour and surface microtopography. Physical characteristics of interest include porosity, insoluble residue amount and composition, and various parameters of rock strength, all of which are described in Goudie (1990).

Standard petrographical techniques are used to provide more detailed descriptions of rock characteristics such as fabric and mineralogy. All of these

Table III.8.1 A summary of high-tech laboratory equipment used by physical geographers

Name	Function	Uses
SEM (scanning electron microscope)	Microscope using electron beam instead of light. Provides high magnifications and depth of focus	Any small robust samples capable of withstanding vacuum, e.g. rock fragments, sand grains, diatoms
EDAX (energy-dispersive analysis of X-rays	Used in association with SEM to provide qualitative and quantitative analyses of elements present, at specific points on sample and overall	As above
XRD (X-ray diffraction)	Provides data on minerals present in suitable powdered samples as the degree and intensity of diffraction is related to chemical composition and crystal structure	Powdered samples usually used obtained from rocks, soils etc. Clay minerals identified by XRD
XRF (X-ray fluorescence)	Provides data on elements present when samples bombarded with X-rays	Many possible sample forms can be used and very low concentrations detected. Suitable for soils, rocks and water samples
AAS (atomic absorption spectrophotometry)	Provides data on concentration of metal elements in a solution, e.g. Ca, Mg, Fe, Na, K	Reliable and accurate method suitable for many samples once made into solution
AA (auto-analyser)	Provides data on concentration of anions in a solution, e.g. nitrate, chloride, sulphate	Reliable and accurate method for use on any dissolved samples

Table III.8.2 Suggestions for laboratory experimental studies

Topic	Approach
1 *Weathering*	
Salt weathering	Rock cubes placed in salt solutions in environmental cabinet/laboratory; before and after observations of rocks and debris produced
Freeze–thaw weathering	Rock cubes in water subjected to freeze–thaw cycles in environmental cabinet; before and after observations of rocks and debris
Biological weathering	*Either* growth of fungal cultures on rock cubes (before and after observations, SEM) *or* rock cubes placed in various organic acids (before and after observations, SEM)
2 *Flow/sediment movements*	
Dune formation	Using sand in wind tunnel, with different wind patterns/velocities used to produce barchans etc.
Bedforms in alluvial channels	Using different sediment sizes and water flows in a flume to produce various bedforms in a flume
Waves and sediment movement on beaches	Using sand in wave tank to produce beach profiles under varied wave conditions
3 *Soil/water movements*	
Infiltration	Use rainfall simulator/sprinkler to rain on soil block; monitor infiltration capacity under different conditions
Splash erosion	Use rainfall simulator/sprinkler to rain on soil under different rain intensities; monitor changing runoff and erosion
Rill development	Use sediment in a tilting flume to investigate the production of rills under different conditions

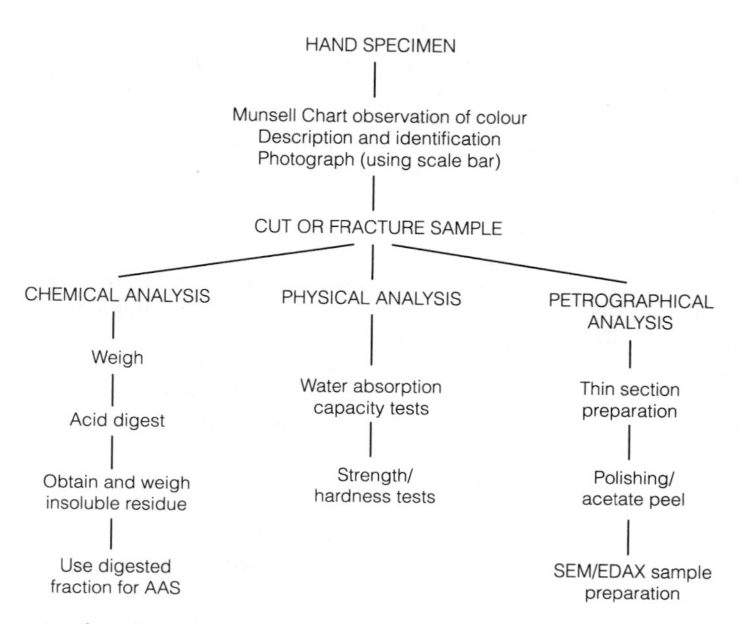

Figure III.8.2 A rock analysis scheme.

involve microscopes. The simplest method is the acetate peel which uses an impression from a cut, acid-etched face (sometimes stained to identify certain minerals). This peel is then observed using a petrographical microscope. More complicated to produce are thin sections, which are polished slivers of rock about 30 μm thick, which are also observed under the microscope and can also be stained. SEM/EDAX techniques provide detailed information on rock structure and composition. Rock samples for SEM are easy to prepare (use hammer and cold chisel to obtain suitable fragments and then mount on SEM stubs) but require skill to analyse. Various books are available to help identify rock samples microscopically, e.g. Adams et al. (1984) and Welton (1984).

Various low- and high-tech methods are available to identify the elemental and mineralogical composition of rock samples (often in dissolved or digested form). Samples need careful pretreatment for such studies. Titrations are commonly used to investigate calcium and magnesium contents, and are more fully described below.

Experimental weathering studies are frequently carried out on rock (and sediment) samples with 'before' and 'after' comparative observations of selected parameters such as weight, surface texture etc. Such studies are often reported in the literature and many are suitable for undergraduate project work (see table III.8.2).

Water Samples: Analysis and Experimentation

Samples of polluted river water, rainfall and glacial meltwater, among many others, are frequently analysed by geographers for their physical, chemical and biological characteristics. Water samples must be carefully treated so as to minimize disturbance and contamination, and analyses should be carried out as quickly as possible after sample collection. In many cases, field methods for chemical analysis of water are available, e.g. for pH, conductivity, calcium and total hardness. These methods are quick and easy to carry out, but accuracy is often low. For many analyses water samples require careful pretreatment to remove particulate matter etc. A suggested scheme for water sample analysis is shown in figure III.8.3. Mackereth et al. (1978) give a useful, simple introduction to water analysis.

Low-tech methods of water chemistry analysis primarily involve titration for determination of the concentrations of specific elements, once pH, Eh (a measure of the oxidation–reduction potential), conductivity, suspended solid load and total dissolved solids have been measured. Bassett et al. (1978) provide a useful review of techniques. Titrations involve the addition, drop by drop, of a standard solution to the sample under investigation; the solution reacts with the element to be determined quantitatively, producing an obvious end-point (usually a colour change). Titrations can be time consuming and require care and patience, but are commonly used for calcium, magnesium and chlorine determinations. Selective ion electrodes are available for the determination of common cations and are also useful for anions such as nitrate and chloride. High-tech methods include flame photometry, AAS and AutoAnalyser equipment (see table III.8.1).

Laboratory experimentation using water often involves a flume, but rainfall simulators may also be used. Studies frequently involve water and sediment movements and/or interactions. Some examples are given in table III.8.2.

Plant/Organic Samples: Analysis and Experimentation

Laboratory studies on organisms often involve identification of the species present, enumeration of individuals and analysis of their chemistry (see figure III.8.4). Common organisms and organic remains studied by physical geographers include diatoms and other algae, lichens,

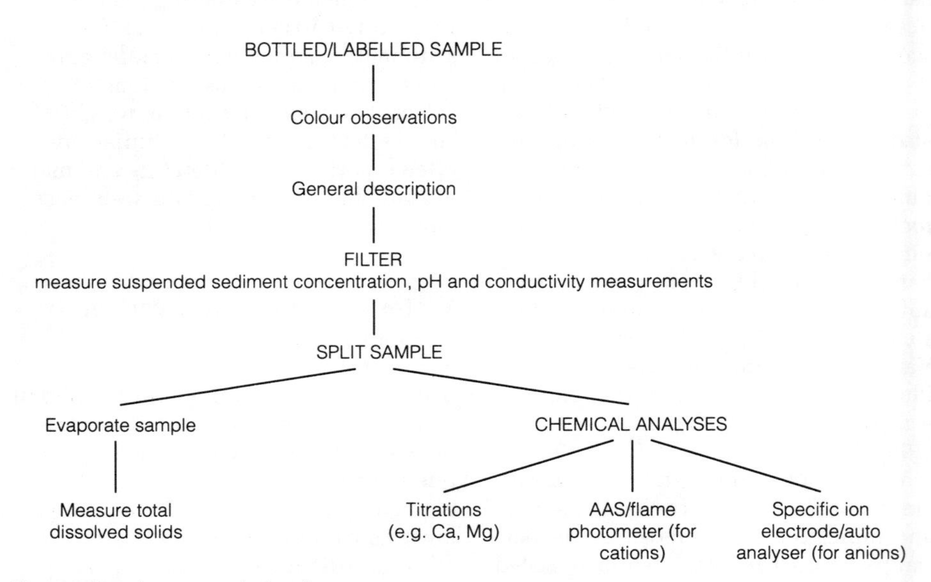

Figure III.8.3 A water analysis scheme.

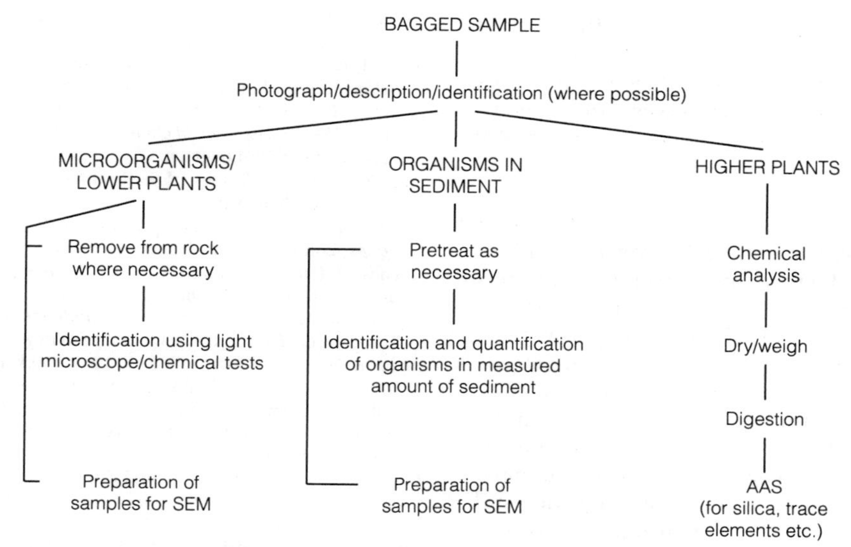

Figure III.8.4 A scheme for plant/organics analysis.

grasses and crops, ostracods, molluscs, foraminifera and pollen grains. These may be obtained from soils, lake cores, water or rock samples.

In studying small organisms and organic remains microscopic techniques are invaluable. Sample preparation is always important and methods vary depending on the organisms involved. Good guides are available for the identification of most organisms and organic material, e.g. Moore and Webb (1978). Counting techniques can be used to obtain an estimate of the number of individuals of different species present on properly prepared microscope slides. SEM may aid identification of some species, e.g. diatoms, and is also useful in biological weathering studies to provide a detailed view of organism–rock relationships.

Higher plants and lichens can be subjected to chemical analyses in the laboratory. Samples need careful pretreatment before standard chemical analysis techniques can be carried out. Common chemical analyses of plant material include investigation of the silica contents of grasses, trace element concentrations in crops, and the contents of lead and other pollutants in lichens.

Experimental laboratory studies involving organic material are rarely carried out in geography laboratories, but cultures of some microorganisms can be obtained and used in biological weathering experiments, following standard biological techniques. Plants can similarly be used in many geomorphological and hydrological laboratory experiments, e.g. splash erosion simulations around crops.

Laboratory-based Projects/Dissertations

As the preceding sections have illustrated, a huge range of laboratory techniques is available to physical geographers. Laboratory analyses of material collected in the field can provide a valuable adjunct to field-based studies. Laboratory work can also form the basis of a project where experiments are used to simulate naturally occurring processes or events. Table III.8.2 gives some examples of such studies.

The obvious advantage of laboratory studies is that processes can be monitored under controlled conditions. With careful planning and experimental design it is possible to avoid many of the problems that may plague field-based projects (e.g. no rain over the study period when you are trying to investigate storm hydrology). The other major advantage of laboratory experiments is that they can 'speed up' natural processes – so you can investigate phenomena which otherwise would take several years to study in the field. Of course, different geography departments have very different laboratory facilities and expertise, and so you must investigate the possibilities open to you in your department before getting too excited over an idea for a project.

References

Adams, A. E., Mackenzie, W. S. and Guilford, C. 1984: *Atlas of Sedimentary Rocks under the Microscope*. London: Longman.

Allen, J. R. L. 1985: *Experiments in Physical Sedimentology*. London: Allen & Unwin.

Bassett, J., Dennell, R. C., Jeffery, E. H. and Mendham, J. 1978: *Vogel's Textbook of Quantitative Inorganic Analysis*. London: Longman.

Folk, R. L. 1974: *Petrology of Sedimentary Rocks*. Austin, TX: Hemphill.

Goudie, A. S. (ed.) 1990: *Geomorphological Techniques*. London: Unwin Hyman.

Haynes, R. (ed.) 1982: *Environmental Science Methods*. London: Chapman and Hall.

Hutchinson, C. S. 1974: *Laboratory Handbook of Petrographic Techniques*. Chichester: Wiley.

Johnson, W. M. and Maxwell, J. A. 1989: *Rock and Mineral Analysis*, 2nd edn. Chichester: Wiley.

Krinsley, D. H. and Doornkamp, J. C. 1973: *Atlas of Quartz Sand Surface Textures*. Oxford: Oxford University Press.

Mackereth, F. J. H., Heron, J. and Talling, J. F. 1978: *Water Analysis*, Freshwater Biological Association Publication 36.

Moore, P. D. and Webb, J. A. 1978: *An Illustrated Guide to Pollen Analysis*. London: Hodder & Stoughton.

Welton, J. E. 1984: *SEM Petrology Atlas*. Tulsa, OK: American Association of Petroleum Geologists.

Yoxall, W. H. 1983: *Dynamic Models in Earth-Science Instruction*. Cambridge: Cambridge University Press.

III.9
Physical Geography Fieldwork
Heather A. Viles

How Not To Do It

My first experience of geography fieldwork was on the standard school field excursion to the Lake District of northern England. I seem to remember spending several hours in a stream with leaky waders, feeling cold, wet and grumpy. Exactly what we were attempting to do and why escapes me now. I probably didn't even have much idea at the time.

My second experience of active fieldwork was rather more memorable, although it still involved getting wet and cold. Having decided to do my undergraduate dissertation on the Essex salt marshes in southeast England, I agreed on a plan of campaign with my tutor and set off carrying one expensive piece of surveying equipment, an unwieldy tripod and a staff. My first mistake was to be too timid: I was too scared to show my ignorance and ask how the said piece of equipment worked, and too shy to ask the laboratory technician for other useful bits of apparatus (e.g. filter papers – I ended up having to buy a home brewing kit instead). Luckily, I found a good book on surveying, but I was never totally confident.

My next mistake was to ask my hugely attractive next door neighbour from home to be my assistant. Having said yes initially he then proceeded to let me down, probably realizing the limited attractions of spending three weeks on some muddy marshes with me. Luckily, I could do much of the work on my own. Unluckily, I couldn't drive at the time and so had to spend an hour each way every day on the bus and then walk a long way, with heavy equipment, to my field sites. The surveying work simply couldn't be done on my own, so finally I managed to persuade my Mum into being my staff holder (and provider of sandwiches and lifts there and back). We even spent her birthday on the marshes, including a celebration lunch picnic on a Martello tower.

Although willing, my Mum had her limitations as a field assistant. She spent much of the time gazing out across the estuary watching the Thames barges sailing up and down. The longer she gazed, the more the surveying staff she was holding slipped away from the vertical, and the more difficult it became to attract her attention. Not surprisingly, she wasn't very keen on total immersion in mud, so we had to negotiate a complicated course around the marsh creeks, which made surveying in a straight line almost impossible.

My Mum finally put her foot down when we came to a particularly muddy and smelly marsh area, and my Dad was then pressed into service. He proved more athletic, but rather less reliable. At one key moment he dropped the surveying staff into a deep creek, and we wasted a lot of time and energy rescuing and cleaning it.

Much of my time alone on the marshes was spent placing filter papers on the marsh surface and removing them after a suitable time interval. The papers were designed to collect sediment and provide an estimate of the rate of accretion. This proved to be a tricky operation, during which I became thoroughly accreted in mud. What I also hadn't bargained for was the fact that the papers tended to get eaten or otherwise removed. More disastrously, I soon discovered that the marshes in this area are dominated by erosion rather than accretion. Thus much of my fieldwork was rather wasted and in the middle of the field period I had to change the thrust of my project.

So where did I go wrong? Obviously I needed much more careful planning of experimental design (see chapter III.2). I hadn't seriously considered the logistical problems I would face, nor even acquainted myself properly with the equipment I was to use. All I learnt was how not to do it. Let us now look at how to tackle field research properly.

How To Do It

Once you have chosen a particular topic worth investigating (not an easy task in itself) there are some important preliminary decisions to take. First, you need to consider the research design of the overall project carefully (see chapter III.2) and assess the scale and length of field study that is desirable. Second, you must choose your field site and ensure that you have any necessary permission for access. Third, you must determine whether you need any assistants for carrying out any of the fieldwork, and finally, you must consider safety. Fieldwork safety, especially if you are working in an isolated area on your own, is of paramount importance and must be taken seriously. Table III.9.1 sets out some general safety considerations.

Careful preparation before fieldwork ensures that you collect useful data to make into a good project. It is useful to try out all your equipment and techniques in a pilot study before you start with the main project to iron out any problems before you get out in the field

Table III.9.1 Fieldwork safety

1 Before setting out

Discuss likely safety problems or risks and check equipment with your supervisor or tutor before starting work

Plan work carefully, taking into consideration experience, terrain and weather. Do not overestimate what you can do

Never go out into the field without leaving word, and preferably a map showing expected location and time of return

2 Clothing

Wear sensible clothing for the conditions, e.g. warm clothes and brightly coloured waterproof jacket for average UK weather; long-sleeved shirts, wide-brimmed has and loose cool trousers for the tropics

Wear solid, non-slip soled shoes or boots, e.g. training shoes, walking boots

Wear a safety helmet (all geography departments have them) when working on quarries, cliffs, scree slopes

3 In the field

Do not work alone without adequate experience for the conditions

Know what to do in an emergency (e.g. accident, illness, bad weather, darkness)

Carry at all times a small first-aid kit, some emergency food (e.g. glucose tablets), a survival bag, whistle, map, compass, watch, torch, water

Know the international distress signal, i.e.
(a) six whistle blasts, torch flashes or waves of a light-coloured cloth;
(b) one-minute pause;
(c) another three blasts (flashes, waves) at twenty-second intervals

4 *In certain areas*

On the coast: get local information on tides and currents; avoid getting trapped by the tide on sand banks or at the base of steep cliffs

In the desert: carry water at all times; learn desert survival techniques

On mountains/in caves: learn the relevant safety codes, and in particular the causes and effects of exposure

Overseas: obtain all necessary vaccinations, take malaria pills and an AIDS kit (containing disposable syringes etc.) where advised (especially if you are travelling to some isolated place with few facilities).

What do physical geographers actually do in the field? In the following sections I give some examples of the different techniques commonly used. It is obviously impossible (and would be extremely tedious) to give an exhaustive list, and so I have been rather selective. I also include some examples of fieldwork-based dissertation topics, showing how a combination of field techniques may be used to answer specific questions. Useful books providing information on fieldwork techniques include Gardiner and Dackombe (1983), Goudie (1990), Berglund (1986) and Moore and Chapman (1986). A very simple but useful introduction to physical geography fieldwork is provided by Hanwell and Newson (1973). For all fieldwork you need a good field notebook and pencil or pen (although portable computers are becoming increasingly used, and a cassette recorder may also be useful), and you should consider keeping copies of all field notes in case of any mix up.

Surveying, site description and sampling

One of the major tools of the geographer is the map, or perhaps nowadays the geographical information system (see chapter III.7), and a sound description of the field study site is a useful prelude to any project. Such description may involve mapping, photography, field sketches and written observations (e.g. soil profile descriptions).

Surveying techniques (plate III.9.1) necessary for producing maps and topographic profiles are commonly taught in geography courses, but there are several useful texts available, e.g. Pugh (1975) and Ritchie et al. (1988). Surveying equipment ranges from the simple and cheap (Abney levels, tapes, poles and clinometers) to the expensive but accurate, e.g. automatic levels and electronic distance measurers (EDMs). Before beginning surveying you must ask yourself the following basic questions. Why are you producing the survey? How accurate must it be? What area should be covered? How detailed does it need to be? For an accurate survey you should start and finish at properly surveyed benchmarks. A proper base map is also useful. For most dissertations a series of surveyed profiles will suffice; it is unlikely you will have the time or resources to produce a full topographical map! Parameters usually measured in topographic surveys are altitude, slope angle and horizontal distance between two points.

Many measurements and analyses can be produced easily in the field without the need to collect samples for subsequent laboratory analysis. Often, however, it is useful to take both field and laboratory measurements of the same parameter to check for accuracy and consistency. You should devise a workable sampling strategy before starting work and, where possible, produce forms for logging data to make things easier. Laboratory analyses can provide detailed information for field projects, but require proper sampling, transport and storage of the samples to be analysed. Some laboratory techniques require very precise types and sizes of samples which much be confirmed before field sampling. There are some general considerations applicable to all samples. First, all samples must be labelled clearly, accurately and permanently. Each sample must also have adequate documentation, e.g. time and date of collection, temperature when collected. Duplicate or back-up samples are also a good idea, and may be vital for some studies.

Historical data collection

For many physical geography projects a historical dimension provides useful

Plate III.9.1 Fieldwork is a fundamental part of geography and most courses involve a practical fieldwork element in which skills like surveying can be developed.

comparative information. For example, short-term studies of pollution within a drainage basin may be compared with longer-term records of pollution and changes in land-use practice in the area. Similarly, studies of short-term coastal dune changes may be set in context with long-term records culled from maps and air photographs. The book by Hooke and Kain (1982) is a must for any student contemplating such interesting investigations.

Fundamentally, historical records come in five main categories: pictorial representations (maps, photos, paintings); written records (which may be direct or indirect, where indirect records indicate variation in some linked parameter); oral records (e.g. recollections of local people); statistical sources (e.g. meteorological and hydrological records); and proxy records in the landscape (e.g. moraines which tell some of the history of deglaciation). This final category of record can be investigated using simple geomorphological and ecological techniques, but the other records are found elsewhere (e.g. public record offices; archives) and require different treatment.

Most historical records have gaps and many are difficult to interpret and may at times provide a confusing record of the subject under study, but they can provide the background to, or basis of, some fascinating investigations, as shown by the examples given in Hooke and Kain (1982).

Geomorphological and pedological measurements

Geomorphological mapping gives detail on slopes and landforms. The standard symbols used in geomorphological maps are shown in Gardiner and Dackombe (1983, ch. 2) and range from simple descriptors of slope profiles to symbols for individual landforms such as pingos, stone stripes, clints and grykes.

Descriptions may also be required of more limited areas, e.g. sections or soil profiles, and here also standard techniques and symbols are available. An accurate base diagram is again necessary. Sedimentary sequence symbols are given in Gardiner and Dackombe (1983, pp. 34–8), as are sample soil descriptions (pp. 94–105).

Many geomorphological measurements can be made easily in the field. Geomorphometry is a general term covering many landform measurements (see Goudie, 1990, ch. 2). Geomorphometry has been used on stream networks, closed depressions in karst landscapes, and sand dunes, amongst many examples. Subsequent processing of such geomorphometric data can produce useful insights into the nature of the landscape. Commonly measured geomorphometrical parameters of specific landforms include width, depth, height, length and shape. Geomorphometrical measurements may also be produced of the landscape viewed as a continuous surface and in this case the common variables are altitude, slope (gradient and aspect) and convexity.

On exposed rock surfaces several techniques can be used to collect information on hardness and topography. Rock mass strength is assessed in the field using the Schmidt hammer (a rebound hammer device whose use is reviewed by Day and Goudie, 1977). Precautions and procedures for using the Schmidt hammer are given in Gardiner and Dackombe (1983, pp. 91–3). Measurements of weathering rind thickness (using ruler or calipers) are easily obtained and can be used to provide information on long-term weathering rates. Rock surface microtopography is measured using a number of simple devices. Rillenkarren (solutional flutes on limestone), for example, have been measured using a Carpenter's profile gauge, which is cheap and simple to use. Similar profiling techniques can be used to measure glacial striations, cave scallops and a range of solutional micro-forms.

Some particle size, shape and distribution measurements are best carried out in the field, at least where large particles are involved. In till fabric analysis, for example, data on the shape, size and orientation of particles within glacial till deposits are collected. Clasts on alluvial fans may be measured for size and shape using a ruler or calipers, as may beach pebbles or armoured mud balls.

Sample-taking of rocks and sediments for subsequent laboratory analysis is straightforward, but care must be taken to ensure that representative samples of adequate size are collected, clearly labelled, carefully transported and stored properly.

Rock samples may be taken with a geological hammer which must be used with caution. Safety glasses should always be worn. A cold chisel is useful for taking small, accurately placed pieces. Samples must be large enough to permit a range of analyses, but obviously small enough to be carried easily! Rock samples, especially of crumbly material, must be carefully packed (e.g. in newspaper inside a plastic bag) to prevent damage in transit. Rocks may be labelled themselves and also on the sample bags, and notes should be taken of the exact position of the sample *in situ*.

Unstratified bulk sediment samples (e.g. dune sands) may be sampled in carefully numbered plastic bags, ensuring that enough sediment is available for all analyses required. Stratified samples, or samples where position in the profile is important, need more careful collection. Core techniques are often used, where the sample is collected in a pipe, or tube, from which the core is then extruded and cut up in the laboratory.

Weathering, erosion, transport and

deposition in some areas occur fast enough to monitor directly. Mass movements, for example, can be monitored using arrays of pegs. Ice heave (a diurnal process) can easily be monitored on periglacial soils using photography, pegs, paint marking and other simple techniques. Bank erosion along alluvial channels or salt marsh creeks is similarly easy to measure using such techniques.

Sedimentation rates on salt marsh surfaces, dunes, beaches etc. can also be monitored using filter papers, marked layers, pegs and other simple methods. Beach sediment movements are often recorded using tagged or dyed grains.

Hydrological measurements

Water in channel flow, surface flow, ponds, pools, lakes and seas can be analysed easily in the field. For channelled flow, velocity, discharge and stage can all be measured or estimated using simple equipment (see Gardiner and Dackombe 1983, ch. 7). Current meters or floats may be used to measure velocity. For both methods, averaging of results is necessary. Discharge may be estimated from velocity and area calculations, where cross-sectional area is measured with chains and poles.

Ice flow requires rather different measurement techniques. Often, the position of marker stakes in the ice is monitored over a period of time to detect the slow ice movements. Coastal waters provide scope for measuring wave height, periodicity and angle as well as velocity.

Water chemistry analyses which should be carried out in the field include temperature, pH, dissolved oxygen content and specific conductance. Simple field kits are also available in most geography departments for measuring calcium and magnesium concentration and the total hardness of water (useful for many karst water studies).

There is a very useful section in Goudie (1990) on water sampling for solute studies which is equally relevant for all water chemistry studies. Sampling frequency, spatial distribution and any special preparation necessary for bottled samples are all covered in that chapter and will not be repeated here. A summary of basic water sampling considerations is given in table III.9.2. Unique problems arise when sampling ice and snow, as discussed in Goudie (1990).

Most hydrological process monitoring involves repeated measurements of the variables listed above (a) over an area at one time, or (b) at a point over time, or (c) over an area over time. So, for example, sampling of surface runoff, subsurface flow and channel flow at various points within a drainage basin might be undertaken to investigate changes in water quality; or water quality variations might be monitored at a point in a stream over a number of storm events, or a combination of both methods might be used.

Hydrological monitoring in karst areas

Table III.9.2 Tips for water sampling

1 Is your sample location representative in space (laterally and vertically?)
2 Is your sample representative in time?
3 The larger the bottle the less chance of contamination there is from the bottle
4 Glass bottles break and contaminate samples with silica
5 Polythene bottles allow some gas exchange and solute absorption onto the bottle walls
6 Do not store for long
7 Samples susceptible to chemical precipitation, algal effects etc. need to be treated, e.g. 5 ml of chloroform per litre
8 Completely fill bottle to minimize gas exchange
9 Shake samples as little as possible
10 Filter as quickly as possible to remove sediment
11 Does your bottle have a stopper?
12 For complete analysis you may need one litre of sample
13 Before filling your bottle where possible slosh water of the same type around it to 'clear' it out

can also take the form of water tracing (often using dyes) to investigate patterns of subsurface water flow in unexplored karst drainage systems.

Meteorological and climatological measurements

The weather can be monitored using simple cheap equipment. Rainfall, temperature (ground and air), relative humidity and wind (speed and direction) are commonly recorded using standard equipment available in all geography departments. More specialized equipment is available to monitor atmospheric composition, e.g. smoke, sulphur dioxide and nitrous oxides found in polluted air. It is vital to select a suitable site for monitoring in order to obtain representative data for the area under study. For example, if you are studying air flow over a dune field, you should monitor wind speed and direction at a number of dune crest and interdune sites within the dune field. You must also make decisions as to when in the day, and how frequently, to take meteorological measurements.

Micrometeorological measurements over a period of weeks can make good dissertations (e.g. forest climate studies, weather patterns and sediment movement on coastal dunes), and pollution monitoring over such a period is also feasible (e.g. short-term variations in air pollution related to weather conditions). For all such studies monitoring simply involves repeated measurements of the parameters listed above. Careful consideration must be given to sampling frequency.

Ecological and biogeographical measurements

Studies of vegetation, and to a lesser extent animals, make good geography dissertations. The first observations necessary for most such projects are to identify the species present. There are simple floras and handbooks available for most types of vegetation to help with such identification. Simple description of vegetation communities can use characteristics of physiognomy (involving height, colour, luxuriance, leaf size and shape) or floristics (involving species composition, abundance of each species and species richness). Moore and Chapman (1983, ch. 9) provide a good introduction to the different methods available.

Several sampling techniques are routinely used in vegetation surveys, with quadrats often providing the basic sampling unit. It is important to consider whether quadrats are appropriate for any particular investigation and if so the size, shape and number required. Quadrats may be sampled in a random, stratified or regular manner or across particular transects. Greig-Smith (1983) provides a good introduction to sampling strategies. Data collected in vegetation studies can be analysed in several ways to elucidate associations between species, and how patterns of species composition vary with environmental conditions. Classification and mapping of different habitats is often the final goal of such studies.

Finally, lichenometry merits a mention, as this technique has important uses in many geomorphological studies. Lichens are slow-growing primitive plants, made up of an algal and a fungal component, which can be identified using visual, microscopic and chemical characteristics. Lichens, especially crustose species growing on rock surfaces, can be measured and a growth curve can be obtained which permits the calculation of the age of the lichen (and, by inference, the age of the surface on which it is growing). Various introductions to this useful technique have been written (e.g. Innes, 1985) and it has been used for dating moraines, river channel changes and archaeological material (e.g.

stone circles). Suitable lichens can be measured easily using simple techniques (rulers or calipers), but there has been some debate about how best to measure non-circular lichens and about which lichens to sample on any surface.

Vegetation samples for laboratory analyses must be packed carefully to ensure that they do not decay in transit. For example, you may wish to measure the concentrations of various elements in plants and associated soils. The exact precautions depend on the analyses to be carried out. In some cases, air drying and storage in paper is adequate. Samples should be analysed as quickly as possible after collection. Avoid using sealed plastic bags for transporting plant material because of respiration effects.

Some ecological processes occur fast enough for short-term monitoring. Colonization of substrates by coastal algae or marine invertebrates often proceeds very quickly and can provide useful information for biological weathering studies. Nutrient cycling can also be monitored, although such studies often require specialized techniques. Some animal ecology processes are suitable for short-term study, e.g. bird migrations, predation, termite movements and their impact on sediment movement.

Ecological field experiments are also possible involving, for example, defoliation of test patches followed by observations of recolonization, or exclusion of grazing animals from test vegetation patches.

Fieldwork-based Projects in Physical Geography

There are many excellent worthwhile topics which can be successfully carried out by undergraduates using some of the techniques set out in the preceding sections. Table III.9.3 gives a selection of recent, prize-winning physical geography dissertations from Oxford University which included at least some fieldwork component. This selection is obviously biased, and to some extent reflects the interests of the lecturers in the Oxford department, but it nevertheless gives an indication of the range of feasible topics.

Table III.9.3 A selection of fieldwork-based dissertations in physical geography

1 Some aspects of coastal landslips and cliff falls at Portland
2 Mass movement phenomena of the Trotternish Peninsula, Isle of Skye
3 The morphology and drainage patterns of the salt marsh at Brean, near Weston super Mare
4 A study of some of the factors influencing the distribution of vegetation on industrial waste tips in the Wigan area of Lancashire
5 The chemical pollution of the river Cam to the southwest of Cambridge: a study of the spatial and temporal variations in water quality
6 Alluvial fans in the upper Simeto Valley, Sicily: geomorphology and history
7 Landsliding and other large-scale mass movements on the escarpment of the Cotswold Hills
8 The glacial deposits of Wensleydale, Yorkshire
9 A study of nitrate concentrations in Slapton Ley, South Devon
10 Lichenometric dating of moraines around Solheimajokull, South Iceland
11 Dirt ogives on the Mer de Glace, Chamonix, France
12 Salt weathering on alluvial fans in Northern Chile: two case studies in Antofagasta province
13 A study of channel and sediment dynamics as indicators of urbanization in East Cheshire

If you plan ahead, attempt something that is possible and worth doing, and use the right equipment, then fieldwork should be rewarding and provide useful data upon which to base a dissertation. So, don't do what I did! THINK AHEAD!

References

Berglund, B. E. (ed.) 1986: *Handbook of Holocene Palaeoecology and Palaeohydrology*. Chichester: Wiley.

Day, M. J. and Goudie, A. S. 1977: Field assessment of rock hardness using the Schmidt Test hammer. *British Geomorphological Research Group Technical Bulletin*, 18, 19–29.

Gardiner, V. and Dackombe, R. V. 1983: *Geomorphological Field Manual.* London: Allen & Unwin.
Goudie, A. S. (ed.) 1990: *Geomorphological Techniques,* 2nd edn. London: Unwin Hyman.
Greig-Smith, P. 1983: *Quantitative Plant Ecology,* 3rd edn. Oxford: Blackwell Scientific.
Hanwell, J. D. and Newson, M. D. 1973: *Techniques in Physical Geography.* London: Macmillan.
Hooke, J. M. and Kain, R. J. P. 1982: *Historical Change in the Physical Environment.* London: Butterworth Scientific.
Innes, J. L. 1985: Lichenometry. *Progress in Physical Geography,* 9, 187–254.
Moore, P. D. and Chapman, S. B. (eds) 1986: *Methods in Plant Ecology,* 2nd edn. Oxford: Blackwell Scientific.
Pugh, J. C. 1975: *Surveying for Field Scientists.* London: Methuen.
Ritchie, W., Wood, M., Wright, R. and Tait, D. 1988: *Surveying and Mapping for Field Scientists.* Harlow: Longman.

III.10
Questionnaire Surveys
Gary Bridge

'I Know! I'll Do a Questionnaire!'

It's close to midnight and you're in a sweat. Tomorrow you have to hand in the proposal for your undergraduate dissertation. You know that you want to do something on the increasing socio-economic status of many inner London neighbourhoods (a process known as gentrification) but you're not sure how to do it. You must prove to the examiner that you have done some actual field-work. Then it comes to you in a flash of inspiration (or perhaps because you can't be bothered thinking about it any more) – 'Of course! I'll do a questionnaire!'. You collapse into bed for a contented night's sleep.

So far you have made two crucial errors. First of all you have only defined an area of research and not a specific research problem. What is it that you want to know about gentrification? What is your research question? For example you might ask why gentrification occurs where it does, or who is doing the gentrifying and why, or what are the feelings of working-class residents about the social change taking place in a neighbourhood. Each of these questions would require a different research method and different sources of information.

The second crucial error, and this often applies even where the research problem is well defined, is the assumption that a questionnaire is the best method available. That is because it has come to be associated with so-called 'hard' social science involving large surveys and statistical analysis. All too often the questionnaire is seen as the cure-all for the problem of doing field-work. However, a questionnaire is only effective when it is the *most appropriate method* of providing the information needed to address a *well-defined research problem*. Let us say, for example, that you are interested in finding out whether gentrification is a back-to-the-city movement of the suburbanized middle class or a within-city movement of middle-class residents who are choosing not tc suburbanize. You will have to know where gentrifiers have come from (i.e. their previous addresses). A questionnaire survey of residents in the Docklands in London might seem like the obvious way of getting the information. However, other sources of information may be available. For example, you may be allowed access to local estate agents' records which will list purchasers' previous addresses. This would be a simpler and less costly way of getting the required information.

Even when it has been established that a questionnaire survey *is* the most appropriate method of gaining the information to address the specific research question, success is not guaranteed. The success or failure of a questionnaire survey is determined by three things: (a) the sampling theory, i.e. are you asking the right people; (b) questionnaire design (wording of the questions, layout of the questionnaire etc.) and (c) analysis and interpretation of the results.

'A survey is a method of collecting information directly from people about their feelings, motivations, plans, beliefs, and personal, educational, and financial background' (Fink and Kosecoff, 1985). Surveys can take the form of questionnaires or interviews. Questionnaires are distinguished from interviews by the fact that they are either self-administered (i.e. filled in by the respondents themselves, as in a postal questionnaire) or filled in by a researcher (in person or on the telephone) with no prompting or interaction with the respondent other than to ask the questions themselves. Interviews, in contrast, whether formal or informal, involve more dialogue between interviewer and respondent. Interviews tend to delve more deeply into people's attitudes, beliefs and feelings (see chapter III.11). They usually involve qualitative analysis of the information gained (often in the form of case studies) whereas the information gained from questionnaires is usually subjected to quantitative analysis involving statistics.

Sampling

A questionnaire survey starts with the definition of the population of interest and procedures for contacting a sample of that population and ends with the analysis of the data from the questionnaires.

When conducting a questionnaire survey it is seldom possible to question all the members of the population of interest. In the example above it would be too expensive to question all the residents of the Docklands. A sample of the population must be taken. To do this you must have a clear notion of the population of interest. Unless you know that, the sample is meaningless. Ask yourself 'Who should be asked and how do I contact them?' What is the sampling frame that is appropriate to the population you are interested in (e.g. electoral rolls, telephone directories, trade directories, tax registers)? Does the sampling frame fairly represent the study population? For example, in Britain the electoral register is the traditional way of sampling residents in a neighbourhood but it only records those residents who are over the age of 18 who have registered to vote. The most appropriate way of selecting households or individuals from your sampling frame is determined by sampling theory. The aim of sampling theory is to avoid bias and ensure that your sample is as representative of the total population of interest as possible. There are a number of sampling methods, and by consulting the textbooks you will be able to decide which is the most appropriate for your study. The size of the sample is also important. It is necessary to allow for non-responses, especially in postal questionnaires. The general rule on sample size is 'the bigger the better', and the upper limit is likely to be set by practicality, e.g. how much postage you can afford, how many streets you are willing to walk. For most statistical tests the minimum sample size is thirty.

All these questions can be resolved by applying common sense to your particular research question and by consulting the textbooks (especially Dixon and Leach, 1978, for an introductory guide; Fink and Kosecoff, 1985, for a straightforward account; Moser and Kalton, 1971).

Finally, if you are doing a residential questionnaire, which member of the household is to answer the questions? If you are doing such a questionnaire in person, unannounced, pay a visit to the local police station before you start, so that they know you are in the area, and always carry identification.

Analysis and Interpretation

You will have to follow the procedures laid out in the textbooks (especially Wrigley, 1985; Clark and Hosking, 1986) to help you process, analyse and interpret results of the survey. You might have a fantastic questionnaire, with a high response rate, and then ruin your study with poor analysis. Do not necessarily leap for the most sophisticated software. Statistics must be used critically. Within limits there is probably a statistical test that will do anything you can think of. The problem is knowing what it is you are looking for rather than knowing about the statistics. When you have got the problem straight then you can look up the appropriate statistics in a textbook.

Questionnaire Design: the Difficult Middle Bit

As I have argued, sampling and analysis are dependent on getting the research problem straight and then referring to the appropriate textbooks. The most difficult part of a questionnaire survey is stage (b), collecting the information, and so this stage will occupy the rest of the discussion.

The form of your questionnaire will differ according to whether it is postal or interview. Deciding between the two will probably be determined by the nature of your study and practical limitations. For example, if potential respondents are scattered all over the country, then a postal questionnaire is the only practical method. If you are in a position to choose between postal and interview questionnaires bear in mind the following pros and cons of each method.

Postal questionnaire: pros and cons

Pros

1 It cuts down on travelling time and legwork.
2 There is no interviewer bias.
3 It gives the respondent time to answer difficult questions.
4 It is good for personal and embarrassing questions.

Cons

1 The questions must be easily understandable and unambiguous – this requires a lot of work in the design of the questionnaire (see below).
2 The answers cannot be re-checked with the respondent.
3 There is no respondent spontaneity.
4 The respondents can see all the questions before answering, which gives them an insight into the line of your questioning and they may therefore tailor their responses to fit your reasoning, so biasing the responses.
5 Who is answering? Even if you are specific in your instructions about who should answer, you can never be sure they have been followed.
6 Supplementary observational data are not available. If you are actually there you can see what the respondents look like, how they respond to the questions and what their environment is like. This is useful contextual information for the survey.
7 There is a low response rate and waste of resources. Response rates for postal questionnaires tend to be low. A response rate of 30–40 per cent from a survey of residents in an ordinary neighbourhood is considered good. Rates may be higher if you are surveying a particular interest group (e.g. other geography students) or if there is something in it for the respondents (e.g. you are using the information to promote their grievances or offering gifts for responding). This last possibility is an unlikely one for a geography student.

To achieve the absolute minimum requirement for statistical analysis, a sample size of thirty, you would need to send out at least a hundred questionnaires, given average response rates. That means at least 200 postage stamps – two for each respondent, one for the outgoing questionnaire and another on the self-addressed envelope enclosed with the questionnaire so that the respondent can send the completed questionnaire back to you. You may also want to send out reminders after a couple of weeks or so. It is easy to see how the costs mount up. If you can hand deliver or collect all or some of the questionnaires this will help reduce costs but of course will be a drain on your time. It is important to give people sufficient time to fill in the questionnaire before sending reminders. You should allow them at least one weekend.

As well as the self-addressed envelope the postal questionnaire should be accompanied by a covering letter from your academic institution. This letter should explain who you are, the purpose of the survey, how they have been selected for the survey and the reason they are being approached. A guarantee of confidentiality is also essential. Don't be officious but equally don't be apolc-getic. An example of a covering letter is given in figure III.10.1. It is not a formula to be followed rigidly but the tone is important.

Good layout of the questionnaire and the ordering of questions is essential, especially for a postal questionnaire, and this will be discussed later.

Questionnaires administered by the researcher: pros and cons

Pros

1 There is a higher response rate than for postal questionnaires.
2 Fewer resources are needed, provided that travel is minimal.
3 You may get positive feedback on the design of the questionnaire, so that you can adapt the design as you go.
4 It provides the chance to clarify the questions.

Cons

1 It is time consuming.
2 It may be inconvenient for the respondent.
3 You may invade a person's privacy.
4 There is the problem of you, the interviewer: commercial polling organizations rarely employ students since they are the last people likely to get a sympathetic response. Don't let this last point put you off. Good interviewing depends on the personality of the interviewer as much as age or sex. It also depends on a well-constructed quesionnaire which, as already mentioned, is crucial for postal questionnaires too. So we now turn to the issue of questionnaire design.

Questionnaire Design: Committing Yourself to Paper

There are no hard-and-fast rules for questionnaire design. It will vary according to the nature of the topic and the people who are to be canvassed. Thus a well-crafted questionnaire is a product of a clearly defined research objective, a sensitivity to the potential respondents, trial and error (using a pilot survey or, if that is not possible, by passing the questionnaire around friends, family and colleagues to make sure that they understand the questions in the same way that you do) and, the most valuable commodity of all, *common sense.*

Although there are no golden rules for questionnaire design there are some handy hints based on the past experiences (and mistakes) of other researchers.

Asking the right questions

Questions of content are made much easier if you have thought about your objectives and about the final method of analysis. One of the biggest mistakes made in dissertations is the reliance upon meaningless questions. How do

UNIVERSITY OF OXFORD

SCHOOL OF GEOGRAPHY

School of Geography
Mansfield Road
Oxford OX1 3TB
England

Tel: (0865) 271919
Telex: 83147 VIA.OR.G
Fax: (0865) 270708
(attn: School of Geography)

Direct line:

30th March, 1990

Dear Resident,

I am a Geography student at Oxford University and am currently doing research on the social changes occurring in the Sands End area.

The views and experiences of local residents, such as yourself, are a crucial part of the research. Your address is one of a number that have been chosen on a chance basis. It would be of great help if any one member of your household, aged 18 or over, could spend a few moments filling in the brief questionnaire enclosed.

All the information you give will remain anonymous and confidential. It will be covered by the Data Protection Act.

A stamped/addressed envelope is provided for you to return the completed questionnaire by post.

I hope you can find the time to help me with my research.

Yours faithfully,

JANET address: GEOGMAIL @ UK.AC.OXFORD.VAX

Figure III.10.1 Specimen letter of introduction.

the terms in the question relate to the abstract categories of your analysis? Have you included all the necessary questions and are all the questions necessary?

Asking the right questions in the right way

There are four main considerations here: (a) the type of answer required, (b) the words themselves, (c) bias, and (d) ambiguity.

Type of answer required Is it fact, opinion or attitude? Different question formats will be appropriate for the different types of answer needed. Closed questions (like multiple choice questions in an exam) are often suitable for factual questions (see figure III.10.2). Closed questions have the advantages that they are quicker for the respondent to fill in, are more precise and are easier to analyse. They are also easier to code. Coding means giving a number to each of the possible responses so that the answers can be fed into the computer. Instructions on coding can be found in any survey textbook.

Attitudes can sometimes be recorded using a closed question format in the form of a rating scale. Rating scales come in various guises (e.g. nominal, ordinal, interval, graphic, comparative, additive) depending on the sophistication of the information required. The tenure question (Figure III.10.3) is an example of a nominal or categorical rating scale. Again this is textbook stuff. It is important to note that the form of the answers, whether categorical (yes/no), continuous (age) and scaled (using diagrams), has relevance for the type of statistics that can be used. An example of a rating scale to capture attitudinal information is given in figure III.10.3.

Sometimes questions have no obvious answers or you may want the respondents to answer in their own words. In this case open-ended questions must be used. These can always be coded afterwards. An example of an open-ended question is found in figure III.10.4.

The words themselves The second element of good questioning is using the right words. Keep them simple. Use everyday words that have immediate meaning to people and avoid jargon and specialized words (unless you are surveying a specialized group of people where those words have specific and acknowledged meanings). Even apparently simple terms like 'friend' and

4a Is your home ... ? (please tick)

☐ owner occupied

☐ privately rented (furnished)

☐ privately rented (unfurnished)

☐ rented from the council

☐ rented from a housing association

☐ other (please state) ....

Figure III.10.2 An example of a closed question.

15 Do you have a feeling of community, living here in Sands End? (please tick)

a strong feeling of community

some feeling of community

no feeling of community

don't know

Thank you very much for your help in this study.

****

Figure III.10.3 An example of a rating scale to capture attitudinal information.

'community' may mean quite different things to different people. In general, do not use a complicated word where a simple one will do; for example 'live' is better than 'reside'.

Bias Certain names, places or phrases are emotionally charged and they can unfairly influence questionnaire responses. For example a questionnaire of geography students might ask:

1 Would you attend a lecture given by Dr Spock?
2 Would you attend an 8 a.m. lecture given by Dr Spock?
3 Would you attend an 8 a.m. lecture given by Dr Spock, the expert on soil profiles?

Options 2 and 3 add more information, but they may also bias the answers.

Another source of bias is when you as researcher are unaware of your own position on a topic. You need to check for this by showing the questionnaire to your friends and family, and people you know less well, to get a range of reactions to the questions. A fairly blatant example of researcher bias might be the question, 'in what ways do you think yuppies have ruined the neighbourhood?'

Bias may also be introduced by asking questions that are too personal. Asking the respondents 'How much do you earn?' may bias answers upwards, or, at worst, put respondents off altogether. Often alternative formats can be used to cope with questions that are too personal. In this case listing a number of income brackets (£000–£10,000; £11,000–£20, 000) for the respondent to indicate which band he or she falls into would be a more sensitive way of asking the question.

Ambiguity Ambiguous questions are usually ones that contain more than one thought. For example, 'do you think that the local government should cut its education or sanitation programmes?' This question is vague as well as ambiguous. It contains two thoughts (cutting education and sanitation). It is also not clear whether it is asking

This last section asks about your opinions of Sands End as a place to live.

13 What are the advantages and disadvantages of Sands End as a place to live? Please make your answers as full as possible

advantages

....

....

....

....

disadvantages

....

....

....

....

TURN OVER

Figure III.10.4 An example of an open-ended question.

whether education in general or sanitation in particular should be cut. The easiest way to avoid ambiguous questions is to follow the rule of 'one thought per question'. More than one thought requires additional questions.

Asking the right questions, in the right way, and in the right order

Ordering of questions is important, especially for a postal questionnaire in which you need to grab potential respondents' attention and keep it without putting them off. Easy factual questions usually come first. This helps respondents relax into the questionnaire and, often, helps you establish who they are. More complicated material should come later. Sensitive questions should be put towards, but not at, the end. Questions should follow a logical order. Where you have to change a line of questioning or where you are asking background questions that respondents might feel are unrelated to the central topic, explain briefly why you are doing this. In general proceed from the most familiar to the least.

Questionnaire Layout: You've Got the Look

Good layout is essential for a postal questionnaire. It must be clear which questions are to be answered and how they are to be answered. Do not try to save paper. Again common sense is the best guide. Questionnaires that are cramped, with poorly defined sections, are likely to be binned. One such poorly laid out questionnaire (figure III.10.5) and one with a better layout (figure III.10.6) are given as examples. Questions are in bold type and the level of detail required in each answer is directly specified. Lead the respondents by the hand with linking instructions. Rarely will all questions apply to each respondent. Filters must be used. Filters are used in

Q21d How often do you see this person? (Please circle)

daily / weekly / monthly / yearly / rarely /

Q21e Where do you usually meet?

....

Q22a Think of your favourite evening's entertainment. Who (other than your wife/husband or partner) would you most like to be with you on such an evening?

....

Q22b Which of the following terms best describes this person? (please circle)

relative / co-worker / neighbour / friend /
acquaintance / member of same organization /

Q22c Where does this person live? (please circle)

Sands End / elsewhere in Fulham /
elsewhere in London / elsewhere in UK / abroad /

Q22d How often do you meet this person? (please circle)

daily / weekly / monthly / yearly / rarely /

Q22e Where do you usually meet this person?

....

Q23 What are the advantages and disadvantages of Sands End as a place to live? (please make your answers as full as possible)

advantages

....

....

....

Figure III.10.5 An example of a poorly laid-out questionnaire.

8e Where does s/he work?

work establishment | location (street/district/country)

....

---

9 Do you have any children?

yes ☐ -> continue Q9 no ☐ -> go to Q10

9b How old is/are your child/children?

....

10a Do any members of your family, friends or relatives live in Sands End, apart from those who live with you?

yes ☐ -> continue Q10 no ☐ -> go to Q11

10b For those people connected to you who live in Sands End but not in your home, please state the type of the relationship in each case (e.g. brother, cousin, friend).

....

....

---

11 Which of the following 2 statements below comes closest to your opinion? Please tick.

The most important job for the government is to make certain every person has a decent steady job and standard of living. ☐

The most important job for the government is to make certain that there are good opportunities for each person to get ahead on their own. ☐

TURN OVER

Figure III.10.6 An example of a well laid-out questionnaire.

figure III.10.6. Arrows are also useful in guiding respondents through filters. However, they can only operate successfully if the questions they are directing the respondent to are actually on the same page.

Underlining sections, or even single questions, gives a compartmentalized visual image that looks tidy. It also gives the respondents a greater sense of accomplishment as they complete each section. Don't forget to ask respondents to turn over the page. There's nothing more frustrating than receiving a carefully completed questionnaire with the back page blank because the respondent did not realize that he/she had not finished. Number each page for the same reason.

Length Up to a certain point the length of a postal questionnaire does not seem to be a serious deterrent. It is whether the questionnaire interests the respondent and looks good that counts. It should be as short as reasonably possible, all other things being equal. This requires precise questioning and relevance at all times (with the exception of dummy questions used to soften up respondents for difficult questions). It is possible to be too short: through peremptory questioning or trying to cram questions into too small a space. This will only irritate respondents. Precision and relevance are the watchwords.

Length may appear to be less of a problem for interview questionnaires but you will probably find that you will often be asked 'How long will this take?'. Don't understate the length. If you feel panicky and are afraid of losing a respondent say that you can end the questioning whenever he/she wishes. Respondents rarely cut you dead half way through unless they really have to go or unless, of course, you have offended them or bored them rigid.

And Finally...

Questionnaire surveys should be enjoyable. You are conducting original research and discovering more about people. You will probably be pleasantly surprised by how co-operative they are. And remember, the best questionnaire asks precise questions, of the right people!

References

Clark, W. A. V. and Hosking, P. 1986: *Statistical Methods for Geographers*. New York: Wiley.

Dixon, C. J. and Leach, B. 1978: *Questionnaires and Interviews in Geographic Research*, Concepts and Techniques in Modern Geography 18. Norwich: GeoAbstracts.

Fink, A. and Kosecoff, J. 1985: *How to Conduct Surveys: A Step by Step Guide*. London: Sage.

Moser, C. and Kalton, G. 1971: *Survey Methods in Social Investigation*. London: Heinemann.

Wrigley, N. 1985: *Categorical Data Analysis for Geographers and Environmental Scientists*. London: Longman.

III.11
The Art of Interviewing

Jacquelin Burgess

Street Corner Society, first published in 1943, is one of the classic urban ethnographies. In the Appendix, William Whyte discusses some of the problems he encountered in doing the field research, including difficulties in getting people to talk to him about sensitive issues. His key informant, Doc, gave him some timely advice.

> *Go easy on that 'who', 'what', 'why', 'when', 'where', stuff, Bill. You ask those questions, and people will clam up on you. If people accept you, you can just hang around, and you'll learn the answers in the long run without ever having to ask the questions.*
>
> *(Whyte, 1955, p. 303)*

Doc was right, of course, but the problem facing undergraduates who are required to carry out research for a dissertation or extended project is that of time. You simply do not have enough time to hang around in the field, working on being accepted by the group you want to study. So I write this chapter in the expectation that you have probably already left things just a little late – and we shall indeed consider some of the basic *why, who, what, how* and *then what* questions of interviewing.

Most human geography courses are still heavily biased towards the quantitative methods of social science. I want to persuade you that qualitative research is an equally useful and important way of doing geographical research and to dispel some of the myths which still exist among geography students that qualitative methods are 'too subjective', lacking in rigour, and something which can be done in the pub for the price of a couple of pints. Qualitative field research has a long and respectable tradition in sociology and anthropology. Within geography it is most closely associated with so-called 'humanistic' approaches and the 'new' cultural geography (see Cosgrove and Jackson, 1987; Eyles and Smith, 1988). Both are characterized by an emphasis on people as creative human beings who act in the world on the basis of their subjective understanding of the society and structures within which they live out their lives. If that sounds a bit of a mouthful, what it means in practice is a commitment to understanding peoples' experiences through listening to the ways in which they describe and account for aspects of their lives and activities. Not surprisingly, therefore, qualitative methods provide more complex interpretations of feelings and actions than do quantitative studies. The data are usually linguistic rather than statistical, contextual rather than cut out from everyday life; the researcher is engaged with the informants rather than separated from them as in a questionnaire survey.

What do you need to be a good interviewer? In the jargon of American sociology, 'successful interviewing is not unlike carrying on an unthreatening, self-controlled, supportive, polite and cordial interaction in everyday life' (Lofland, 1971, p. 90). In plain English, if you like talking to people – you don't go around shouting at them, putting them down, ridiculing them or not listening to what they say – you have the potential to become a good interviewer. The art of interviewing is to be able to conduct a conversation in such a way that the persons you are talking to are able freely to express their opinions and feelings while, at the same time, enabling you to meet your own research objectives. It is not uncommon to come away from a fascinating conversation with someone about life, the universe and everything, to use Douglas Adams' phrase, only to realize that you still don't know whether that individual really does know that the answer is 42! So the goal is to achieve an end result which satisfies both of you – informant and researcher. What follows is some practical advice on how to achieve that result.

Why Use Interviews as the Basis for Dissertation Research

In order to reach your decision, you must consider the following points:

- the nature of your research proposal;
- alternative methods of achieving your objectives;
- the constraints which will affect the ways in which you can achieve those objectives.

If your research proposal is concerned with aspects of human geography which require that you make interpretations of the feelings, values, motivations and constraints which contribute to our understanding of people's behaviour, then interviewing may be an appropriate research technique. You might, for example, want to understand how different groups within a locality which has lost its main source of paid work make adjustments to their changed circumstances. Or perhaps you are interested in the decision-making strategies of planners or recreation managers to encourage people to use country parks as a way of relieving pressure on fragile upland ecosystems. Maybe you want to explore the geographical experiences of groups such as women or children or ethnic minorities or handicapped people and discover the extent to which their needs and interests are currently being met in the local services provided. Whatever the topic, you need to consider whether interviews would provide the best way of achieving your objectives. Would a questionnaire survey be more appropriate? The answer will be no, if you want to concentrate on how individuals describe and account for their own experiences; if you want to understand the complexities of the problem rather than reducing it to a set of key explanatory variables; if your aim is to undertake a *case study* rather than a representative sample of a wider population.

Among the major constraints you will face will be those of *access* and *time*. Can you get hold of the people you want to talk to? It is often very difficult, for example, to gain access to members of élite groups, such as the managing directors of multinational companies or the very rich – one reason, perhaps, why most ethnographies have been undertaken with the poor or other marginalized groups in society. Time is also important. If you are interested in understanding why Alton Towers is such an attractive tourist destination, it is not really a good idea to carry out your field research in January. The best kind of qualitative research builds over time with the researcher moving backwards and forwards from the field to the interpretation of data and back into the field for more interviews, informed by the experiences of what has gone before. Bear in mind that the transcription and interpretation of interview data takes much longer than for statistical data; writing up is a more creative and interactive process than that associated with quantitative analysis.

What Kind of Interviews Will You Conduct

Having done the background reading and formulated your research objectives, you will now need to decide:

- who you will interview;
- how you will make contact;
- what kind of interview you will conduct.

The selection of interviewees will be determined theoretically by the nature of your project and practically by the relations you are able to establish in the field. Let us take one of the examples I have already mentioned. You want to study the impact of economic restructuring in the coal industry on a local community. On a theoretical level, you know that different groups within the community will be differentially affected by closure of the coalmine. So you will need tc interview members of those different groups: male and female workers who have been made redundant; those who have accepted transfers to pits in other places; retired miners who have lived in the community for many years; women and men in family groups who are affected by the closure; young people without prospects of future employment. So your aim will be to adopt a strategy of *theoretical sampling*, interviewing individuals from those different groups of interests. Your choice of informants is made on this basis rather than on random sampling from a whole population. The number of people you interview is less important than the quality of information you gain from your interviewees. As you proceed, you may well find that new ideas and issues are emerging from the field research and that the crux of the problem seems to be the way in which the decision to close the pit was communicated to local people. You will then need to interview some of the key decision-makers in British Coal and the local authority. Deciding when to stop becomes the key issue – and will be based both on practicalities of time and an intuitive judgement that you have heard the range of stories that people within the community have to tell about their experiences and explanations of what is happening to them.

Normally, you should first contact your interviewee by letter, by telephone or through a personal introduction to make an appointment. With locality research, the best idea is to make an initial contact with someone who might be able to provide you with other introductions – these people are often members of community-based organizations, for example. This kind of research technique is often described as *snowballing* – inevitably you find that you get passed on and the list of 'people you really should talk to' grows at quite an alarming rate. Say who you are and who recommended that you talk to the interviewee. Say what you are doing; try to make it interesting to the other person – why should someone want to give up their time to talk to a geography student – but don't make any rash promises about being able to change the world as a result of your dissertation findings. You will also need to think about problems of confidentiality – people will need to be reassured about what will happen to the information they might give you. It is also polite to write and thank the person after the interview has been completed.

Qualitative researchers normally make a distinction between *formal* and *informal* interviews (see Plummer, 1983, pp. 93–8; Burgess, 1984, pp. 101–22). Formal interviews most closely resemble questionnaire-based interviews in that the researcher has a clear agenda of issues that he or she wishes to cover. These will usually be written down beforehand, not as set questions, but as important topics to be discussed.

However, formal interviews differ from questionnaires in one very important respect. The sequence of topics covered in the interview is determined through the interaction of researcher and informant. The important thing is to go with the flow of the interview, being flexible in the order in which topics are discussed but making sure that, by the end of the meeting, you have met all your objectives. Formal interviews give you the security of knowing that you have covered the range of issues with all your informants. By contrast, informal interviews much more closely resemble ordinary conversations. The aim is to discover how individuals describe and make associations between different kinds of ideas and experiences. These interviews tend to be much longer and more tangential to the problem – but full of insights into the life and personality of the person you are talking to.

How to Conduct a Successful Interview

We need to think about two issues here:

- the interpersonal skills you need to conduct the interview;
- the recording of information.

Interviewing skills should be learned and practised before you go into the field. They are as much part of the geographical repertoire of research techniques as learning how to use a depth-integrating sampler. Taking interpersonal skills first – the fundamental goal is to create a rapport between yourself and the interviewee. If she or he likes you, finds talking to you a pleasant and interesting experience and trusts you, then the interview will go well. You can do several things to ease the transition from stranger to friendly acquaintance. Prepare yourself for the interview by being sensitive to the expectations of the person you will be meeting. It is not a very good idea to arrive for an interview with the chief planning officer of a local authority in designer-torn jeans and dirty trainers. Neither would it really be appropriate to interview adolescents in the neighbourhood gang in your most formal clothes. Communication covers much more than the language we use in talking to one another. Non-verbal communication through our body language is just as powerful. Think about your posture in the interview. Are you sitting hunched up with arms and legs crossed, clutching your notebook and pencil as if your life depended on them? Your nervousness will communicate itself to your interviewee who may begin to wonder what is wrong with him/her – or you. Be relaxed in your posture but continue to convey interest and attention. Make lots of eye contact with the other person – acknowledge that they exist – but also be careful that it does not get out of hand and you end up interrogating them with piercing stares or inviting them to bed! Practise a range of facial expressions in the mirror or on a friend and see how you express interest, pleasure, confusion and uncertainty. See how you take the initiative in asking questions or changing the topic of conversation.

In the interview itself, you need to think about two related issues: the ways in which you phrase your questions and the pace of the interview. It is possible to ask questions in different ways which will give rather different kinds of answers. *Closed questions* with the familiar 'what, where, when, how often, how much, who, why' require that the informant give you pieces of information and often leave all the initiative with you. *Open questions* such as 'tell me about...' and 'in what ways do you feel...' are invitations to encourage communication. A good interviewer will use both kinds of phrasings. Perhaps even more importantly, a good interviewer will learn to *listen* not only to what he or she is being told but also to how it is being said and what lies underneath the remarks. Listening closely enables you to handle the different kinds of silences which arise in conversations. A few years ago, I conducted interviews with economic development officers. Playing the tape-recordings afterwards, I was dismayed by the number of times I jumped in with the next question rather than giving the interviewee time to develop his or her point. The majority of inexperienced interviewers find silences difficult to deal with. It is a reflection of their own anxieties about the interview. Try to identify what kind of silence you are dealing with and then respond according. For example:

- Is it a thoughtful silence?, in which case make the appropriate 'mm's and 'aah's to ease the interview on.
- Is it a stuck silence? The informant is having difficulty with the point you have asked, perhaps, in which case rephrase, recapitulate what they have just said or clarify with an example.
- Is it an embarrassed silence? Maybe something has been said or asked which should not have been. If you made the error, say that you hadn't realized that it would cause difficulties, apologize and move on to a safer topic. In general, be sensitive to the feelings of your interviewee and tactfully change the subject if necessary.

The other major problem is working out an appropriate method of recording the answers to your questions. Clearly, if you are frantically scribbling down everything that is being said to you, you are not going to be able to develop a good rapport with your informant – who, in turn, is likely to become more self-conscious and anxious about what you might be writing. Tape-recorders would seem to be the obvious solution – but beware! Many people will simply refuse to be 'on the record' or they will severely censor what they say to you. Others will become acutely embarrassed. I tried to record an interview with a miner in County Durham who had a broad Geordie accent. The session was going very badly – despite the fact that we had previously struck up a very good relationship. I stopped the tape and asked if it was a problem for him. 'Yes, it bloody well is,' he said. 'Your students in London will listen to me and think I'm just an ignorant nobody because I don't talk like them.'

Best practice is to use a reporter's notebook and jot down key words or phrases during the course of the interview – but no more than that. Once the interview has finished, *as soon as possible*, find a quiet spot and write down everything you can remember about the course of the interview. Review in your mind precisely how it went, the sequence in which topics came up and what was said. You will find that you can remember very much more than you think you can. Then, once you get back to base, type up a full transcript of the interview, preferably on a personal computer. Do not forget to include date, time, name of person, address and telephone number. Put in other details which will help you later to remember the interview. How did it go? Did you have any problems? What was the person like? This procedure will take a considerable amount of time but you will gain from it when you come to interpret the data and write your final repcrt.

And Then What?

The analysis of qualitative data is not easy and it is the most difficult aspect of the research procedure to describe succinctly. Good discussions of different approaches can be found in Jones (1985), Burgess (1984, pp. 166–84, and Strauss (1987). With your interview transcript, you will be able to distinguish factual information from opinions and feelings;

when things happened, who was involved, why certain decisions were made and by whom. Further, you will need to identify the key concepts which appear to underpin the stories you have been told. Do this for each interview once you have completed it, create indices of these different kinds of information, and write yourself memos about the ideas which arise while you are carrying out the interpretation. Use your discoveries to inform subsequent interviews in the field. As you will realize by now, qualitative research is a much more interactive process between field and desk than questionnaire surveys, where you collect all the data before undertaking the analysis. When you finally decide to draw a halt to your field programme, you will then need to classify and sort all your research material to draw out the central interpretive themes. These will emerge from a creative synthesis of your background reading, your interviews and any other data that you might have collected. The end product, hopefully, will be an interesting, lively and insightful project which is grounded in the realities of everyday life – an example of genuinely humane geography.

References

Burgess, R. G. 1984: *In the Field: an introduction to field research*. London: Allen & Unwin.

Cosgrove, D. and Jackson, P. 1987: New directions in cultural geography. *Area*, 19, 95–101.

Eyles, J. and Smith, D. M. (eds) 1988: *Qualitative Methods in Human Geography*. Oxford: Polity Press.

Jones, S. 1985: Depth interviewing/the analysis of depth interviews. In R. Walker (ed.), *Appliedw7 Qualitative Research*, Aldershot: Gower, 45–70.

Lofland, J. 1971: *Analyzing Social Settings: a guide to qualitative observation and analysis*. Belmont, CA: Wadsworth.

Plummer, K. 1983: *Documents of Life: an introduction to the problems and literature of a humanistic method*. London: Allen & Unwin.

Strauss, A. L. 1987: *Qualitative Analysis for Social Scientists*. Cambridge: Cambridge University Press.

Whyte, W. F. 1955: *Street Corner Society: the social structure of an Italian slum*, 2nd edn. Chicago, IL: University of Chicago Press.

III.12

Landscape Studies in Practice

Paul Coones

Context

This chapter is designed to be read in the light of the conceptual discussion embodied in chapter II.7 entitled 'Landscape geography'. It offers a brief survey of the content, approaches and sources relating to landscape geography, in such a way as may provide a preface and accompaniment to the pursuit of practical work in the field. The ideas contained in the other chapter are of importance here, too, because the underlying theory or purpose of landscape study cannot be taken to be what is sometimes called 'a given'. Certain specialists in landscape, especially in the historical disciplines, have tended to make unstated assumptions about their objectives, the modes of explanation involved, and their selection and use of evidence. In other words, it is impossible to give advice on 'how to do it' if the nature of 'it' is merely implied and not made clear.

There may be said to be three stages in this process, or three levels in the mode of explanation. Although usage varies, the terms philosophy, methodology and technique are often applied to them. First, what questions are being addressed? What is the fundamental objective of landscape study? How does it relate to other branches of knowledge? What are its theoretical underpinnings? Second, which approaches are being adopted? What is meant by 'looking' and 'seeing'? What methods are being used to provide a framework? The working concepts or structures to be employed might include the region, the value system of 'literary landscape' studies, or the empirical reconstruction of a past landscape. Third, and most directly, what techniques and tools are of use in the actual business of gathering the relevant information and processing it?

In this short chapter, landscape studies in Britain are reviewed from the various points of view embraced by geographers, within the general context of the geographical ideas previously outlined.

Subject Matter

The rarity of truly natural landscapes is often cited in support of the argument that the physical environment is irrelevant in the modern 'man-made' landscape. By the same token, the physical landscape is frequently dismissed as 'scenery' and excluded from landscape studies proper. Yet the totality of the geographical environment remains; natural elements are not mere static background but are employed in the creation of cultural landscapes. Human history is set within a period of intense environmental change (Goudie, 1983); many 'anthropogenic' processes are in fact accelerated physical processes (soil erosion, for example) (Goudie, 1990), and it is sometimes hard to distinguish the natural from the artificial. (The Norfolk Broads were thought to be natural lakes until Lambert et al. (1960) recognized them as flooded medieval peat-diggings.) Even slight natural changes can crucially affect the human use of marginal areas (Parry, 1978); the techniques of both physical and human geography, and the evidence produced by the field and the archive, need to be employed in understanding landscapes in which the agency of man and nature require joint consideration, notably those associated with coasts, rivers and wetlands. Especially important to the interpretation of landscape changes (such as settlement desertion) is the reconstruction of past environments (Evans, 1975; Smith, 1979; Simmons and Tooley, 1981), together with an accompanying knowledge of physical events in terms of magnitude and frequency, process rates, lag times, trigger mechanisms and response patterns. In such contexts, the human–physical divide, which is generally accepted without question as a basic structure in geographical enquiry, is shown to be both unrealistic and unworkable.

A useful guide to methods, sources and techniques for investigating historical change in the physical environment is provided by Hooke and Kain (1982). The following texts provide good starting-points for the examination of the different aspects of the natural landscape in Britain: geology (Trueman, 1971; Whittow, 1977); geomorphology (Goudie and Gardner, 1985; Brunsden et al., 1988); soils (Curtis et al., 1976); climate (Manley, 1971); vegetation (Pennington, 1974; Rackham, 1986, 1990). A survey of environmental hazards is presented by Perry (1981).

Analysis of the cultural landscape is usually attempted through specialization by topic, period, region or approach, and few authors since Hoskins have ventured an overall conspectus (Coones and Patten, 1986); collections of essays are a more common format (Reed, 1984; Woodell, 1985). Even the general surveys tend to place the emphasis upon an essentially chronological treatment of standard cultural themes viewed from the perspectives of archaeology and history (Reed, 1990). Of the host of individual subjects, those which have attracted most interest and research are associated with the rural landscape (see Aston, 1985), especially historic agrarian landscapes (Adams, 1976): fields (Taylor, 1975), hedges (Pollard et al., 1974), woodlands (Rackham, 1986, 1990), rural settlements (Roberts, 1977, 1987a; Rowley, 1978; Muir, 1980; Taylor, 1983) and place-names (Gelling, 1984; Field, 1987). Also popular are historic towns (Aston and Bond, 1976), transport and communications (Taylor, 1979) and specific categories such as prehistoric monuments (Dyer, 1982), medieval castles (Austin, 1984), moated sites (Aberg, 1978; Wilson, 1985) and ruins generally (Bailey, 1984). A comprehensive range of field antiquities is described by Wood (1979).

Industrial landscapes (Trinder, 1982) have been neglected, the fashion being for the nostalgic and arguably illusory images of pre-modern bucolic bliss (see the masterly discussion of English landscape tastes by Lowenthal and Prince (1965)). In addition to the pronounced leaning towards villages and fields, there is plenty on the planned landscapes of landed wealth – the country houses and estates (Clemenson, 1982), the aristocracy and the landscape parks (Prince, 1958; Cantor and Hatherly, 1979), the gardens (Jacques, 1983) and the model villages (Darley, 1975). The rural bias is particularly evident in the tendency for the story of the landscape and 'the history of the countryside' to be regarded as synonymous (Rackham, 1986; see also Aston, 1985; Scholes, 1985).

The 'Shire Archaeology' series contains several excellent titles on landscape-related topics, and Hodder and Stoughton's county volumes on 'The making of the English/Welsh landscape' (Hoskins and Millward, 1970 onwards) form one of a number of regional series produced by individual publishers.

Approaches

It will be plain by now that landscape has been studied in many different ways, both as a means and as an end in itself. It is important to be clear about the aim of any study being undertaken. As hinted above, practitioners of particular disciplines sometimes take the objective for granted, and this is apparent in the selection of evidence, methods of description and explanation, and the general emphases which characterize the result. Enthusiasm for a new technique, and the adoption of that technique as a bad master rather than as a good servant, can colour and even distort a whole piece of research. Unacknowledged and implied modes of thought can similarly determine the outcome. A notable case of this is the morphogenetic approach mentioned on p. 73, in which artefacts, relics and lineaments (often treated essentially as objects *in* the landscape) are examined in order to construct a story of landscape change, or, in practice, to trace the evolution of individual forms over time. Archaeologists, for example, naturally concentrate upon the kinds of artefact evidence with which they are familiar, sometimes using the term 'landscape' merely to imply a spatial scale greater than that of the traditional single site. Landscape thereby becomes simply a collection of sites or, in effect, one large 'site' comprising a scatter of artefacts.

Geographers, and some historians, have employed reconstructions of former landscapes and the evidence of relics in the present landscape as sources for broader purposes of enquiry and interpretation. By this means the role of human agency in altering past environments may be more fully revealed, and the changing relations between societies and environments illuminated. In this respect landscape is conceived and used as a record, perhaps 'the richest historical record we possess' (Hoskins, 1955). To pose the deceptively simple question, 'Why is this landscape as it is?', raises a host of issues relating to the nature of description and explanation, subject matter and evidence, the juxtaposition of chronology and chorology, the unique and the general, analysis and synthesis; ultimately, it may call upon the full breadth of skills encompassed by the geographer's trade.

In this connection, a further dimension is revealed. Increasingly apparent in recent years has been a group of related ideas which challenge this essentially phenomenal treatment of landscape as a means of realizing an empirically reconstructible past. Landscape may constitute a record, but it is a selective, distorted and potentially treacherous one, replete with enigmas and riddles. Moreover, landscapes are not made up of facts that 'speak for themselves'; they cannot be read directly like a book, not only because so many pages are missing, torn or stained, but also because they were simply not designed to be read in this way. The value systems both of peoples of the past and of modern observers interpose themselves and preclude any commonality of language (Lowenthal, 1975, 1985). A school of landscape study has grown up, connected with the 'new' cultural geography, which conceives of landscapes as repositories of human values – embodiments of the attitudes, ideals and beliefs of their creators, infused with cultural meaning, symbolism, imagery and ideology (Relph, 1981; Cosgrove, 1984; Penning-Rowsell and Lowenthal, 1986). Such work embraces non-material culture and has involved the interpretation of landscape as 'text' (Duncan and Duncan, 1988), exploring links with art, literature, aesthetics and the history of ideas, and with themes such as the emotional and experiential ties of people with specific places ('topophilia' and 'geopiety') (Tuan, 1974, 1977; Pocock, 1981). The interest is consequently as much in the urban and the ordinary (Meinig, 1979) as in the rural and the historical, 'culture' being the central issue (Cosgrove and Jackson, 1987).

Examination of these values and perceptions, and an attempt to assess the inspirations, motives and significance underlying the inherent subjectivity of such ways of 'seeing' landscapes, would all constitute worthwhile and legitimate subjects fcr individual projects. These could focus on particular landscapes, topics or people. Good examples to take as starting-points might include the following: for literature, Barrell (1972) on the poetry of John Clare, Barrell (1982) on Thomas Hardy's Wessex, and Hudson (1982) on Arnold Bennett; for architecture and its social meaning, Neale (1981) on Bath; for painting and the visual arts in general, Prince (1984) (and the references therein), Pugh (1990), Cosgrove (1979) on Ruskin, and Barrell's (1980) study of the rural poor in English painting. Some interesting techniques are discussed by Pocock (1982).

Sources

A grasp of this diversity of pcssible approaches and of their conceptual ramifications is an essential preliminary to an assessment of sources for landscape research. This is especially so with respect to the most important source of all – the landscape itself. The guides to the practical aspects of 'reading the landscape', either general in nature (Muir, 1981a, 1985) or organized from particular perspectives such as archaeology (Aston and Rowley, 1974; Taylor, 1974) and local history (Hoskins, 1982, 1984), can then be used as aids to fieldwork. It will be immediately apparent from a perusal of these books that people see different things in the landscape, depending on their interests, predilections, training and point of view. Indeed, the whole process of 'seeing', which lies at the heart of landscape study, has already been observed to be highly value laden. The geographer must perforce be aware of its multifaceted nature and be prepared to exploit the possibilities which it offers.

The traditional 'eye for country', perhaps the geographer's lost art, is a skill which can be developed with practice and experience. Acute observation, performed in association with the posing of a sequence of questions, constitutes a central element. The evidence provided by the landscape – the shapes, forms and patterns of its interrelated components – is assessed in turn in the light of information gleaned from libraries, record offices and other archives (Stephens, 1981; Riden, 1983, 1987; Richardson, 1986). In addition to documents of various kinds, the student needs to consult maps (Harley, 1972; Hindle, 1988; Smith, 1988) and aerial photographs (Wilson, 1982, 1987; Muir, 1983) which can reveal whole secret landscapes of 'ghost' forms not normally evident at ground level. A classic work of this kind is Frere and St Joseph's (1983) survey of Roman Britain from the air. The revelations continue (Whimster, 1989). Archaeologists, equipped with new techniques, methods of dating and broader conceptual frameworks are discovering and interpreting much buried evidence that is not manifested in the visible landscape. This is particularly the case with regard to earlier periods, for which documentary material is wellnigh non-existent and where morphologies, which in fact evolved subsequently, can seriously mislead (Roberts, 1987b). For more recent times, the rich topographical literature of Britain (Anderson, 1881) involves not only contemporary written descriptions but also drawings, prints and paintings (Barley, 1974; Russell, 1979). Current specialist journals such as *Landscape History*, *Landscape Research* and *The*

Local Historian contain articles on all kinds of new tools and strategies pertinent to landscape study, and comprise a useful fount of ideas, topics and themes for the individual to pursue.

Landscape research has been likened to detective work, in which a conundrum or puzzle is presented and clues of any kind and from whatever quarter are followed up in an attempt to unravel a sequence of events or account for a pattern (Muir, 1981b; Burton and May, 1986). The subject may be the reasons behind the desertion of a village; the outlasting of function by form in the suggestive disposition of a town's streets, centuries after the urban defences – in which lie the origins of the plan – were dismantled; a field system which embodies traces of its precursors right back to the prehistoric clearance from ancient woodland; a pool by a house, which represents the fishpond of the earlier medieval manor whose foundations are now hidden (but partly revealed in aerial photographs or other forms of remote sensing imagery); or it may be something so apparently unimportant as a sharp bend in a country lane, a routeway which turns out to be an enclosure road following an old track and respecting the boundaries of the medieval open field furlongs. This is the richness of the 'palimpsest' of the British landscape, in which the shapes and forms of successive layers are superimposed, modified and handed down, living on as remnants or preserved indirectly in other guises entirely (plate III.12.1).

Passing on to the realms of symbolism and ideology, other examples may be cited. What does a 'planned' landscape of the eighteenth century – a park and its associated model village, say – suggest about contemporary attitudes towards

Plate III.12.1 In a long-settled area like the British Isles, the landscape shows the imprint of many past human activities. In this picture Hadrian's Wall snakes along a ridge in northern England and looks across to a landscape that, under natural conditions, would be forested.

nature, the sources of aesthetic inspiration drawn upon by the designer, or about land, property, social structure and class relations? How can a related study of relevant personal documents, popular literature, paintings, fashion and life-styles throw light upon what is seen? Often the various approaches will complement each other. To take another instance, drawn from the same period, the 'improvers' attempted to overcome adverse physical conditions prevailing in parts of England – through drainage, clearance or new agricultural practices – and mould human circumstances in order to create landscapes which realized a complex set of ideals. Nature and people alike were to be tamed and ordered, and certain economic, social and aesthetic aims achieved at one and the same time. The landscape records the forms, but also bears witness to the meaning.

The materials and ideas which bear upon such questions are multifarious. Landscape study is a multidisciplinary activity, and therein lies one of its principal challenges and sources of fulfilment. It is a pursuit for which geographers are admirably equipped, and to which they may make a significant contribution.

References

Aberg, F. A. (ed.) 1978: *Medieval Moated Sites*, Research Report 17. London: Council for British Archaeology.

Adams, I. H. 1976: *Agrarian Landscape Terms: a glossary for historical geography*, IBG Special Publication 9. London: Institute of British Geographers.

Anderson, J. P. 1881: *The Book of British Topography....* London: Satchell; republished, Wakefield: EP Publishing, 1976.

Aston, M. 1985: *Interpreting the Landscape: landscape archaeology in local studies*. London: Batsford.

——and Bond, J. 1976: *The Landscape of Towns*. London: Dent; reprinted, Gloucester: Alan Sutton, 1987.

——and Rowley, T. 1974: *Landscape Archaeology: an introduction to fieldwork techniques on post-Roman landscapes*. Newton Abbot: David & Charles.

Austin, D. 1984: The castle and the landscape. *Landscape History*, 6, 69–81.

Bailey, B. 1984: *The National Trust Book of Ruins*. London: Weidenfeld.

Barley, M. W. 1974: *A Guide to British Topographical Collections*. Council for British Archaeology.

Barrell, J. 1972: *The Idea of Landscape and the Sense of Place 1730–1840: an approach to the poetry of John Clare*. Cambridge: Cambridge University Press.

——1980: *The Dark Side of the Landscape: the rural poor in English painting 1730–1840*. Cambridge: Cambridge University Press.

——1982: Geographies of Hardy's Wessex. *Journal of Historical Geography*, 8, 347–61.

Brunsden, D., Gardner, R., Goudie, A. and Jones, D. 1988: *Landshapes*. Newton Abbot and London: David & Charles with Channel 4.

Burton, A. and May, J. 1986: *Landscape Detective*. London: Allen & Unwin.

Cantor, L. M. and Hatherly, J. 1979: The medieval parks of England. *Geography*, 64, 71–85.

Clemenson, H. A. 1982: *English Country Houses and Landed Estates*. London: Croom Helm.

Coones, P. and Patten, J. 1986: *The Penguin Guide to the Landscape of England and Wales*. Harmondsworth: Penguin.

Cosgrove, D. 1979: John Ruskin and the geographical imagination. *Geographical Review*, 69, 43–62.

——1984: *Social Formation and Symbolic Landscape*. London: Croom Helm.

——and Daniels, S. (eds) 1988: *The Iconography of Landscape: essays on the symbolic representation, design and use of past environments*. Cambridge: Cambridge University Press.

——and Jackson, P. 1987: New directions in cultural geography. *Area*, 19, 95–101.

Curtis, L. F., Courtney, F. M. and Trudgill, S. T. 1976: *Soils in the British Isles*. London: Longman.

Darley, G. 1975: *Villages of Vision*. London: Architectural Press.

Duncan, J. and Duncan, N. 1988: (Re)reading the landscape. *Environment and Planning D: Society and Space*, 6, 117–26.

Dyer, J. 1982: *The Penguin Guide to Prehistoric England and Wales*. Harmondsworth: Penguin.

Evans, J. G. 1975: *The Environment of Early Man in the British Isles*. London: Elek.

Field, J. 1987: What to read on place-names in Britain. *The Local Historian*, 17, 396–404.

Frere, S. S. and St Joseph, J. K. S. 1983: *Roman Britain from the Air*. Cambridge: Cambridge University Press.

Gelling, M. 1984: *Place-names in the Landscape*. London: Dent.

Goudie, A. S. 1983: *Environmental Change*, 2nd edn. Oxford: Clarendon Press.

——1990: *The Human Impact on the Natural Environment*, 3rd edn. Oxford: Basil Blackwell.
——and Gardner, R. 1985: *Discovering Landscape in England and Wales*. London: Allen & Unwin.
Harley, J. B. 1972: *Maps for the Local Historian: a guide to the British sources*. London: Standing Conference for Local History and National Council of Social Service.
Hindle, B. P. 1988: *Maps for Local History*. London: Batsford.
Hooke, J. M. and Kain, R. J. P. 1982: *Historical Change in the Physical Environment: a guide to sources and techniques*. London: Butterworth.
Hoskins, W. G. 1955: *The Making of the English Landscape*. London: Hodder & Stoughton.
——1982: *Fieldwork in Local History*, 2nd edn. London: Faber & Faber.
——1984: *Local History in England*, 3rd edn. London: Longman.
——and Millward, R. (eds) 1970–: *The Making of the English/Welsh Landscape* (county volumes). London: Hodder & Stoughton.
Hudson, B. J. 1982: The geographical imagination of Arnold Bennett. *Transactions of the Institute of British Geographers*, New Series, 7, 365–79.
Jacques, D. 1983: *Georgian Gardens: the reign of nature*. London: Batsford.
Lambert, J. M., Jennings, J. N., Smith, C. T., Green, C. and Hutchinson, J. N. 1960: *The Making of the Broads: a reconsideration of their origin in the light of new evidence*, RGS Research Series 3. London: Royal Geographical Society.
Lowenthal, D. 1975: Past time, present place: landscape and memory. *Geographical Review*, 65, 1–36.
——1985: *The Past is a Foreign Country*. Cambridge: Cambridge University Press.
——and Prince, 1965: English landscape tastes. *Geographical Review*, 55, 186–222.
Manley, G. 1971: *Climate and the British Scene*, 5th impression. London: Collins.
Meinig, D. W. (ed.) 1979: *The Interpretation of Ordinary Landscapes: geographical essays*. New York and Oxford: Oxford University Press.
Muir, R. 1980: *The English Village*. London: Thames & Hudson.
——1981a: *Shell Guide to Reading the Landscape*. London: Michael Joseph.
——1981b: *Riddles in the British Landscape*. London: Thames & Hudson.
——1983: *History from the Air*. London: Michael Joseph.
——1985: *Shell Guide to Reading the Celtic Landscapes*. London: Michael Joseph.
Neale, R. S. 1981: *Bath 1650–1850, a Social History; or, A Valley of Pleasure yet a Sink of Iniquity*. London: Routledge & Kegan Paul.
Parry, M. L. 1978: *Climatic Change, Agriculture and Settlement*. Folkestone: Dawson.
Penning-Rowsell, E. C. and Lowenthal, D. (eds) 1986: *Landscape Meanings and Values*. London: Allen & Unwin.
Pennington, W. 1974: *The History of British Vegetation*, 2nd edn. London: English Universities Press.
Perry, A. H. 1981: *Environmental Hazards in the British Isles*. London: Allen & Unwin.
Pocock, D. C. D. (ed.) 1981: *Humanistic Geography and Literature: essays on the experience of place*. London: Croom Helm.
——1982: Valued landscape in memory: the view from Prebends' Bridge. *Transactions of the Institute of British Geographers*, New Series, 7, 354–64.
Pollard, E., Hooper, M. D. and Moore, N. W. 1974: *Hedges*. London: Collins.
Prince, H. C. 1958: Parkland in the English landscape. *Amateur Historian*, 3, 332–49.
——1984: Landscape through painting. *Geography*, 69, 3–18.
Pugh, S. (ed.) 1990: *Reading Landscape: Country–City–Capital*. Manchester: Manchester University Press.
Rackham, O. 1986: *The History of the Countryside*. London: Dent.
——1990: *Trees and Woodland in the British Landscape*, revised edn. London: Dent.
Reed, M. (ed.) 1984: *Discovering Past Landscapes*. London: Croom Helm.
——1990: *The Landscape of Britain from the Beginnings to 1914*. London: Routledge.
Relph, E. 1981: *Rational Landscapes and Humanistic Geography*. London: Croom Helm.
Richardson, J. 1986: *The Local Historian's Encyclopedia*, 2nd edn. New Barnet: Historical Publications.
Riden, P. 1983: *Local History: a Handbook for Beginners*. London: Batsford.
——1987: *Record Sources for Local History*. London: Batsford.
Roberts, B. K. 1977: *Rural Settlement in Britain*. Folkestone: Dawson.
——1987a: *The Making of the English Village: a study in historical geography*. Harlow: Longman.
——1987b: Landscape archaeology. In J. M. Wagstaff (ed.), *Landscape and Culture: geographical and archaeological perspectives*, Oxford: Basil Blackwell, 77–95.
Rowley, T. 1978: *Villages in the Landscape*. London: Dent; reprinted, Gloucester: Alan Sutton, 1987.
Russell, R. 1979: *Guide to British Topographical Prints*. Newton Abbot: David & Charles.
Scholes, R. 1985: *Understanding the Countryside*. Ashbourne: Moorland Publishing.
Shire Archaeology Series. Princes Risborough: Shire Publications.
Simmons, I. G. and Tooley, M. J. (eds) 1981: *The Environment in British Prehistory*. London: Duckworth.
Smith, D. 1988: *Maps and Plans for the Local Historian and Collector*. London: Batsford.

Smith, R. 1979: Environmental issues in landscape studies. *Landscape History*, 1, 16–28.

Stephens, W. B. 1981: *Sources for English Local History*, 2nd edn. Cambridge: Cambridge University Press.

Taylor, C. 1974: *Fieldwork in Medieval Archaeology*. London: Batsford.

——1975: *Fields in the English Landscape*. London: Dent; reprinted, Gloucester: Alan Sutton, 1987.

——1979: *Roads and Tracks of Britain*. London: Dent.

——1983: *Village and Farmstead: a history of rural settlement in England*. London: George Philip.

Trinder, B. 1982: *The Making of the Industrial Landscape*. London: Dent; reprinted, Gloucester: Alan Sutton, 1987.

Trueman, A. E. (revised J. B. Whittow and J. R. Hardy) 1971: *Geology and Scenery in England and Wales*. Harmondsworth: Penguin.

Tuan, Y. -F. 1974: *Topophilia: a study of environmental perception, attitudes and values*. Englewood Cliffs, NJ, and London: Prentice Hall.

——1977: *Space and Place: the perspective of experience*. London: Edward Arnold.

Whimster, R. 1989: *The Emerging Past: air photography and the buried landscape*. London: Royal Commission on Historical Monuments (England).

Whittow, J. B. 1977: *Geology and Scenery in Scotland*. Harmondsworth: Penguin.

Wilson, D. 1985: *Moated sites*. Princes Risborough: Shire Publications.

Wilson, D. R. 1982: *Air Photo Interpretation for Archaeologists*. London: Batsford.

——1987: Reading the palimpsest: landscape studies and air-photography. *Landscape History*, 9, 5–25.

Wood, E. S. 1979: *Collins Field Guide to Archaeology in Britain*, 5th edn. London: Collins.

Woodell, S. R. J. (ed.) 1985: *The English Landscape: past, present, and future*, Wolfson College Lectures 1983. Oxford: Oxford University Press.

III.13
Archives in Geographical Research
Michael Williams

The task of conveying some idea of the scope and range of material in archives in three continents in the short space available in this chapter is daunting. Archival collections are vast, sometimes very difficult to access, and growing in size and complexity all the time. All one can do is to draw on one's own experience and relate some impressions, hints and guidance based on thirty years' work in archives in Britain, Australia and the United States, primarily on matters of land settlement, landscape change and resource evaluation and conversion, from the medieval period to the present (Williams 1970, 1974, 1989). Hopefully these comments will serve as an introduction to the nature, variety and use of archives, and the three works cited above offer examples of their use in historical geographical research.

The Image and Utility of Archives

Archives are a place in which public records and historic documents are kept and preserved, and, by transference, the name also applies to the records or documents themselves. Unfortunately, the very name often conjures up the image of dustiness, fustiness and the dead past. To some, archives seem to have little relevance to the here-and-now of geographical research. At worst, they seem useless and antiquarian.

But if one can get over those prejudged impressions archives open up exciting possibilities to would-be researchers in geography. They are a rich depository of material that can add depth and context to 'present day' studies (at what interval or date, one might ask, does the present become the past?) and they provide fascinating glimpses of geographies long gone but which leave legacies in the landscape and society, and which are still intrinsically interesting and intellectually stimulating and demanding in themselves. In addition, because archives are the depository of public records, they often contain material from the recent past, as well as material that is old. Hopefully, as soon as official and private documents become non-working documents they are passed on to archives for storage and preservation. Thus, archives are a part of a dynamic and constantly updated continuity between a living present and a living past.

The Purpose of Archival Research

Some basic questions need to be asked at the outset of any foray into archive collections. What is the purpose of the research, what does one want to achieve and what does one want to look at?

One could, conceivably, have any one of the following three aims, or even all three intertwined:

1 to investigate all that is known about a particular portion of the world's surface through time, as in either the re-creation of a historical geography or the evolution of a cultural landscape of a region from the beginning of settlement;

2 to investigate the antecedents of a particular present-day geographical factor or feature that one needs to know more about, for example the work of Phillips (1989) on the practice of underdraining farmland in nineteenth-century England based on private estate records;
3 to investigate a particular source that would have a rich geographical yield. In the UK the work of H. C. Darby on the Domesday record of 1066 is the classic example of this approach (Darby, 1977); another example is the work of Kain and Prince (1987) on the Tithe surveys of the mid-nineteenth century.

In the first aim the search, to a certain extent, is random and has the element of serendipity. One combs indexes (if there are any), one dips into samples of record groups, and sometimes material is found that is important and relevant. But for the enormous amount of documentary dross that is sifted out of the intellectual pan there is often little gold at the end of the day, week or month. In the second the search is more focused on a particular location or topic or time (or all three), but there is still a high element of luck involved. In the third the search is totally committed to a particular set of records and auxiliary documents that lend meaning to them. Thus, such research is 'source led' and may seem, paradoxically enough, ahistorical in its lack of context by focusing on the present of some past time. In reality most research in archives combines all three approaches, in different proportions as the need arises.

It goes without saying that the ease of archival research is in some inverse relationship to the knowledge of the documents to be used. Aim 3 is much easier than aim 1. But in all cases it pays off handsomely to read all one can about a topic before venturing into the archives, so that the enormous investment of time, money and energy is not wasted, and one's confidence is sustained that results will appear at the end. The situation is analogous to fieldwork in physical geography. The geomorphologist or biogeographer does not go immediately to, say, the edge of a terminal moraine in the Canadian Rockies or Greenland, with all the expense and effort involved, without first having read all that has been written about previous fieldwork in the area and generally about morainic formation or lichen growth. Similarly the would-be human/historical geographer does not arrive in an archive with all the expense and effort involved without having first written to the archive for advice, perused the various guides and finding aids suggested below and attempted to read everything about the topic or region to be studied. And when you get to the archive enlist the help of the trained archivist – that is what they are there for – and your task will be lightened.

The Nature of Archives

Unlike printed books of which there are many copies, documents are unique; there is usually only one of them. That is why they are deposited in archives and guarded zealously. This has a number of implications. First, the intending researcher has to get permission, usually backed up by a recommendation from some responsible person, that he/she is a serious and fit person to handle such valuable material. Second, because documents are unique they are in one place only. You cannot borrow them but have to go physically to see and study them. This can be costly.

Working in archives can be time consuming and laborious. Because documents are old, fragile and sensitive to light and heat they cannot be photocopied. At best they can be photographed.

Therefore there is an immense labour of copying that tires one's wrist and ruins one's handwriting. Many archives insist that the copying is done in pencil in order to avoid blots of ink or damage to documents by the pressure from ballpoint pens. For those that can afford it the ideal is the laptop computer, totally portable and silent. Increasingly archives are getting their most used documents onto microfiche or microfilm. This saves handling them and also allows for cheaper copying, often by instant photocopying of individual frames from the film or fiche, by self-operated coin-fed machines.

Occasionally, one finds maps or charts that give one exciting glimpses of past situations, and they seem like pieces of gold in the dross. If not available the maps can sometimes be reconstructed from data with a locational component. However, the catch here is to find out whether the boundaries to which the data refer are the same as those of today, and unless one can be sure of finding the contemporary boundaries the data may be of little value.

Most countries rationalize the great volume of archival material in specialist depositories, usually divided as to whether they are national/federal or county/state. Occasionally specialist thematic collections have been formed that transcend this division, but they are rare, and the national/local division provides a ready way of categorizing archives in the three continents. What follows is a preliminary guide of what is available and how to find it.

United Kingdom

Two useful general publications are Adkins (1988) which is a guide to all the libraries and their addresses in the United Kingdom (but not county record offices) and Downs (1981) which is a bibliographic guide to guides, such as library catalogues, which can include archives.

National depositories

The records of central government departments (e.g. Foreign Office, Home Office, Department of Trade, Ministry of Agriculture, Fisheries and Food etc.) for England and Wales from the eleventh to twentieth centuries are kept in the new Public Record Office (PRO), Ruskin Avenue, Kew, Richmond, Surrey TW9 4DU. Medieval records (custumals, rent rolls, taxation returns), the records of the State Paper Office from the early sixteenth to late eighteenth centuries, records of the law courts and legal departments (Exchequer and Chancery), together with records of the Privy Council Office are held in the old PRO in Chancery Lane, London.

There are 84 miles of shelves of documents in the PRO and the range of material is immense. Consequently it is always advisable to write beforehand stating what you might want, having first looked up the *Guide to the Contents of the Public Record Office* (three volumes). It is as well to remember that a thirty-year confidentiality rule applies; therefore at the moment of writing no documents before August 1960 can be made available for public scrutiny.

The PRO also holds the census enumerators' books (CEBs), the books into which the census enumerators transferred the information they had collected on forms issued to all households. But in the interests of conservation only microfilm copies can be studied at Portugal Street, London WC2. Fortunately copies are widely available throughout the country and the best guide to their availability is Gibson (1988). The CEBs cannot be looked at before a hundred years has elapsed; hence only those for 1841–81 are available. Mills and Pearce (1989) is an excellent guide to their use.

In all cases the records that survive will only be as good as the original records made and will reflect the laxity or thoroughness of the culling of records before they were sent to the PRO.

Another national record depository is the British Library (Great Russell Street, London WC1B 3DG) which, in addition to its more than 9 million books, has a Manuscript Room with an extensive collection of charters, rolls, maps and plans and topographic drawings, as well as 87,000 volumes of private papers of individuals of literary, political, artistic or land-holding affiliations. In addition there is a Map Room, with over 1.6 million items (see chapter V.1). At Collingdale Avenue, London NW9 5HE, there is the British Library's collection of over half a million newspapers, for London from 1801 onwards, and English provincial, Scottish, Irish, Commonwealth and foreign newspapers from 1700 onwards. Currently 3000 titles are taken and kept. Often regarded with suspicion on the day they are printed, newspapers in time become legitimate historical documents reflecting the views and opinions of the day in a condensed and, of course, biased form.

In addition to the actual documentary material, there is the printed official material that is often housed in archives with manuscripts. This consists of printed versions of documents, e.g. publications of the Historical Manuscripts Commission, the Calendars of State and Domestic papers, the journals and debates of both Houses of Parliament (Hansard), local and private Acts. Above all there are the British Parliamentary Papers (BPP) which consist of annual reports of government departments and instrumentalities, tax returns, commissions of enquiry, and reports on everything from slavery to sewage schemes, from railway construction to agricultural conditions, and from urban health and factory production to education, to name but a few. They are an indispensable accompaniment to archival work and may almost be treated as if they are archives. A useful guide to the nineteenth-century material in these papers is Ford and Ford (1953) and the various updates.

Local depositories

Nearly every county also has its record office that concentrates on local material. For example, there are parochial records, including parish registers (Finlay, 198x); borough records (rates, housing and health reports); parliamentary enclosure awards and correspondence; estate records; records of local firms, circuit and county courts; local photographs, newspapers and pamphlets; and often very helpful research notes made by previous archivists and research workers on particular localities and topics. Many counties have their own individual guides, e.g. Avon (Moore, 1979), and for an excellent guide and introduction to all local history sources see Richardson's *The Local Historian's Encyclopedia* (1986).

Intermediate between these national and local records are other repositories of a more regional significance, for example the National Library of Scotland, George IV Bridge, Edinburgh EH1 1EW, and the Welsh National Library, Aberystwyth, Dyfed SY23 3BU, with extensive collections of manuscripts, maps, plans and drawings. Of national and regional importance are the Bodleian Library, Oxford, and the Cambridge University Library with medieval and literary records, and the John Rylands Library, Manchester, with transport and economic history records.

United States of America

Because of the size and regional variety of the continent the seemingly simple uniform system of archival holdings in the United Kingdom is replaced by thousands of depositories. According to the American Library Directory (1988–9) there are 34, 379 libraries in the United States and Canada – 30,717 in the United States, 113 in territories administered by it and 3649 in Canada, and many of these have archival collections. The libraries (plate III.13.1) are situated mainly in major cities (e.g. New York Public Library), universities (e.g. the Bancroft Library of Berkeley Campus of the University of California) or state capitals (e.g. the State Historical Society Library, Madison, Wisconsin).

Nevertheless the same twofold division of national and local depositories helps in sorting out what is available. Good guides for an initial foray are Hamer (1961) and *The National Union Catalog of Manuscript Collections* (1962 and updates).

National depositories

Records of Federal government departments and agencies, e.g. Agriculture, Census Bureau, Forest Service, Indian Affairs Bureau, Interior, Land Management Bureau, Soil Conservation Service, as well as all Congressional records are kept in the National Archives and Records Administration Building, 8th and Pennsylvania Avenue NW, Washington, DC 20540. Each agency's records are given a distinct Record Group Number, those above being RG 16, 29, 95, 75, 48, 49, and 116 respectively. A useful entrée into this vast collection of over 1 million cubic feet is National Archives, *Guide to the Records of the National Archives* (1984), and this publication and other finding aids and individual Record Group catalogues can

Plate III.13.1 The Library of Congress in Washington, DC, USA.

be purchased from The Publications Sales Branch, National Archives, Washington, DC 20408.

The Library of Congress at 10 1st Street SE, Washington, DC 20540, besides being the largest library in the world, like the British Library also has vast collections of private papers (some 50 million items). These are kept in the Manuscript Division in the James Madison Building, a purpose-built and superbly equipped manuscript reading centre. To take just one example, the Library has the papers and diaries of Gifford Pinchot, chief forester and leading conservationist during the first forty years of the century. Again, perusal of finding aids is essential, and useful in this respect are the Library's own *Manuscripts on Microfilm* (1975) and its older *Handbook of Manuscripts in the Library of Congress* (1918).

While the James Madison Building also has a large map collection in the basement, most of the older cartographic and air photographic material that emanates from government records and activities is kept in the Cartographic and Architectural Branch of the National Archives, 841 South Pickett Street, Alexandria, Virginia, on the southern outskirts of Washington, DC.

Among other national collections of particular significance to geographers is the land survey material at the Bureau of Land Management (BLM), Eastern States Office (there are local offices for other regional groupings of states) at 350 South Pickett Street, Alexandria, Virginia. The BLM office brings together a nearly complete collection of public land survey plats on microfilm, and sometimes the originals, from the various regional offices.

As in other countries the manuscript data of the federal census of the United States is of particular interest to geographers. The originals are kept in the National Archives (Record Group 29), but copies of returns as they relate to particular states are often kept in the respective State Historical Libraries. Another source is the Mormon Genealogical Centre, Salt Lake City, Utah, where all available census data for the United States (and incidentally for all other countries of the world for which birth data are available) have been microfilmed and stored. Copies of microfilms can be purchased on request.

Federal government publications, which include proceedings of both Houses of Congress, major investigations and hearings, the annual reports of departments, and special returns or reports requested by individual members and Congressional Committees of either House since the end of the eighteenth century, are now appearing at the rate of over 200 shelf-metres per annum. This vast store of printed material is often housed in archives and offers an indispensable background and framework for any work on official documentary material. They are analogous to the BPP. A *Monthly Catalog of United States Publications* lists the publications of the Government Printing Office (GPO) during the month, and consolidated catalogues appear in Lester, Favel and Lester (1980), Buchanan and Kanel (1975) and *United States Government Publications, Monthly Catalog, Decennial Cumulative Personal Author Index.* In Britain, the best collection of US official publications is in the library of the London School of Economics.

Local depositories

Just as each county has record offices in the United Kingdom, so does each state in the United States. It is difficult to generalize but official and unofficial (private) archives are often housed in the headquarters of the State Historical Society, usually located in the state capital, although sometimes state material may be located in a university library, such as the Bancroft Library of the Berkeley Campus of the University of California, which has one of the best collections on California. A good start in unravelling the riches of the local scene are the general guides noted above but also lists of accessions noted in the monthly or quarterly magazines and journals of the respective State Historical Societies, most of which should be found in any major university library.

Australia

National depositories

National records barely existed before 1901 because Australia as a country did not exist before that date. What records there were were stored in huts and basements of the first buildings of the raw 'bush capital' of Canberra from the time of the meeting of the first Parliament in 1927 until the opening of the Commonwealth National Library in 1968. Now Commonwealth archives and many private papers are accessible in the new and magnificent building of the National Library, Canberra, ACT 2600. Prior to that, a national collection of sorts was (and still is) in existence in the Mitchell Library, the manuscript section of the Sydney Public Library, in the state of New South Wales, which by dint of being the first and most important of the states settled has perhaps the greatest collection of manuscripts relating to the early settlement and exploration of the entire continent (State Library of New South Wales, 1967–8).

Useful guides are National Library, *Guide to Collections of Manuscripts relating to Australia*, which has been published in instalments since 1966, the seventeenth appearing in 1990. Series A–C (instalments 1–12) were published in hard-copy form, but the later instalments were only available on microfiche. The guide covers collections for some forty-nine repositories in Australia, but is incomplete. In addition there is a general finding aid in National Library, *Principal Collections in the National Library of Australia* (1978). Many collections are also described, in a discursive way, in Burmester, *National Library of Australia; guide to the collections* (1974–82).

Other useful finding aids to printed material are National Library, *Union List of Newspapers in Australian Libraries* (1959, 1960 with updates), and National Library, *Australian National Bibliography* (monthly and annually), which lists all publications relating to Australia and accessions to the Library.

Local depositories

Because of the importance of the individual states throughout the nineteenth century and the centralization of settlement in the capital cities, large local collections exist. Each state has its archive collection which is usually a branch of the State Public Library, the special case of the Mitchell Library in New South Wales already having been noted. The State Library of New South Wales published a good two-volume catalogue in 1967–8 entitled *Catalogue of Manuscripts of Australasia and the Pacific in the Mitchell Library, Sydney*, and each State Library has similar though not such thorough finding aids.

In what must have been the most bureaucratic of nineteenth-century societies, records seem to have been kept about every decision made and every activity undergone. Depending on the state concerned, much of this is printed in the State Parliamentary Papers (those of South Australia, Victoria and New South Wales perhaps being the fullest). Initially state government departments were cavalier about the records they preserved or destroyed until the state archives established a regular practice of vetting the process. Many departments, in particular lands and surveys departments, argue that records are working documents, and indeed they were until digital mapping and computer storage allowed them to relinquish the hand-written ledgers and dockets.

The result is that, despite some important gaps, state archives have an abundance of manuscript, cartographic and local topographical and biographical material on every conceivable aspect of land settlement and the creation of new geographies that is relatively easy to access and work on. Particularly interesting are the records of the state lands departments that are revealing about the settlement process and environmental evaluation, e.g. agricultural and pastoral lease documents, correspondence between settlers and government inspectors and surveyors, enquiries into new lines of roads and railways, the construction of ports and water supply systems. State archives will also hold state census material.

No overall guide has been published to all state records but, as mentioned before, each State Public Library in the state capital issues finding aids and other information, and occasionally guides have been published, as in South Australia (Crowley, 1966). Addresses can be found in the respective state *Year Book* or the *Commonwealth Year Book* (both annual publications).

Conclusion

For those who have the patience to master the intricacies of manuscript collections there are rich rewards. In practical terms, the answer to so many geographical questions lies in the antecedents of actions and landscapes. In addition, the possibility exists of building up authoritative databases of changing phenomena. Those who have the myopic view that only the present matters make a fundamental practical and philosophical error. The past can be made to speak clearly and authoritatively to us, and archive material is one of the ways in which that can be achieved.

References

Examples of Geographical Research Using Archives

Darby, H. C. (1977): *Domesday England.* Cambridge: Cambridge University Press.

Kain, R. J. P. and Prince, H. C. (1987): *The Tithe Surveys of England and Wales.* Cambridge: Cambridge University Press.

Phillips, A. D. M. (1989): *The Underdraining of Farmland in England during the Nineteenth Century.* Cambridge: Cambridge University Press.

Williams, M. (1970): *The Draining of the Somerset Levels.* Cambridge: Cambridge University Press.

——(1974): *The Making of the South Australian Landscape.* London: Academic Press.

——(1989): *Americans and their Forests: A historical geography.* New York: Cambridge University Press.

Finding Aids, United Kingdom

Adkins, R. T. (ed.) (1988): *Guide to Government Departments and Other Libraries.* London: British Library Science Reference and Information Services (published every 2 years).

Downs, R. B. (1981): *British and Irish Library Sources: a bibliographic guide.* London: Mansell.

Finlay, R. (1981): *Parish Registers: An introduction*, Historical Geography Research Series, 7. Available from Dr. C. Withers, Department of Geography, The Cheltenham and Gloucester College of Higher Education, Cheltenham, Glos. GL50 2RH.

Ford, P. and Ford, G. (1953): *Select List of British Parliamentary Papers, 1833–1899.* Oxford: Basil Blackwell.

Gibson, J. S. W. (1988): *Census Returns on Microfilm: a directory of holdings in Great Britain, Channel Islands, Isle of Man.* Publication of Federation of Family History Societies, available from compiler at Harts Cottage, Church Hanborough, Oxford OX7 2AB.

Mills, D. and Pearce, C. (1989): *People and Places in the Victorian Census: a review and bibliography of publications based substantially on the manuscript Census Enumerators' Books, 1841–1911*, Historical Geography Research Series, 23. Available from Dr C. Withers, Department of Geography, The Cheltenham and Gloucester College of Higher Education, Cheltenham, Glos. GL50 2RH.

Moore, J. S. (1979): *Avon Local History Handbook.* Chichester: Phillimore.

Public Record Office (1963): *Guide to the Contents of the Public Record Office*, 3 vols. London: HMSO.

Richardson, J. (1986): *The Local Historian's Encyclopedia.* New Barnet, Herts: Historical Publications.

Finding Aids, United States

American Library Directory (1988–9). New York and London: R. M. Bowker.

Buchanan, W. and Kanel, E. M. (comp.) (1975): *Cumulative Subject Index to the Monthly Catalog of the U.S. Government Publications, 1900–1971.*

Hamer, P. M. (1961): *A Guide to Archives and Manuscripts in the United States.* New Haven, CT: Yale University Press.

Lester, D. W., Favel, S. K., and Lester, L. E. (comp.) (1980): *Cumulative Title Index to United States Public Documents, 1788–1976.*

Library of Congress (1975): *Handbook of Manuscripts in the Library of Congress.* Washington, DC: GPO.

Library of Congress, Manuscript Division: *Manuscripts on Microfilm: a checklist of the holdings in the Manuscript Division*, compiled by Richard B. Bichel. Washington, DC: GPO.

National Archives (1984): *Guide to the Records of the National Archives.* Washington, DC: GPO.

The National Union Catalog of Manuscript Collections (1962 and updates), 12 vols. Washington, DC: Library of Congress.

United States Government Publications, Monthly Catalog, Decennial Cumulative Personal Author Index.

Finding Aids, Australia

Burmester, C. A. (1974–82): *National Library of Australia; guide to the collections*, 4 vols. Canberra.

Crowley, F. K. (1966): *South Australian History: a survey for research students.* Adelaide: Libraries Board of South Australia.

National Library (1959, 1960 and updates): *Union List of Newspapers in Australia.* Canberra: National Library.

——(1966): *Guide to Collections of Manuscripts relating to Australia.* Canberra: National Library.

——(1978): *Principal Collections in the National Library of Australia.* Canberra: National Library.

——(monthly and annually): *Australian National Bibliography.*

State Library of New South Wales (1967–8): *Catalogue of Manuscripts of Australia and the Pacific in the Mitchell Library, Sydney*, 2 vols. Sydney: State Library of New South Wales.

Part IV
What's What and Who's Who in Geography

Most students reading geography are given the opportunity to learn something of the discipline's past, its leading lights and influential texts. David Livingstone (chapter II.1) gives a broad view of the history of geographical thought, and in this section Alisdair Rogers provides an update on some of the dialogues which run through the discipline as a whole. Ron Johnston and Andrew Goudie make some suggestions as to the key texts in human and physical geography, while Brian Blouet presents biographical sketches of some of the subject's most prominent scholars writing today.

IV.1
Key Themes and Debates

Alisdair Rogers

Most geography degree courses feature a component which addresses the history and/or philosophy of the subject. Its existence draws attention to the fact that not only have the content and methods of geography changed over the years, but even in the present there is not total agreement as to what geography is or how it should be done. For some students this is a nuisance; for others it is the starting point for a stimulating intellectual journey. For many, the prospect of such a course is potentially interesting but they are put off by the frequently impenetrable and esoteric nature of much of the literature. It is to such students that this chapter is addressed. I shall try to set out the reasoning behind such courses, clarify some of the more important philosophical points and then review some of the contemporary debates in geography.

Why Study the History and Philosophy of Geography?

Most geography consists of statements about facts, whether they be distributions, histories or processes. Most geographical research is therefore aimed at producing new facts or relating hitherto unrelated facts. This may be termed technical geography, the ultimate aim of which may be greater understanding, prediction or control (Couclelis, 1982). In making such statements geographers use terms or concepts such as space, time, nature, class and region, which for the most part are taken for granted. They may also follow rules of procedure which are likewise adopted from their disciplinary forebears with little additional thought. But, as Einstein once observed, it is necessary to review such fundamental concepts in order that we may not be unconsciously ruled by them. It also helps to rethink the rules of enquiry, particularly if we agree with Feyerabend (1978) who argues that there have never been sets of rules for scientific discovery which can guarantee good results. A second level of enquiry, making statements about statements about facts, is therefore an essential part of any discipline. This critical or reflective enquiry may evaluate both whether or not concepts and rules actually work and whether we ought to be using them at all. In the past, for example, geographers have happily measured and written about the size and shape of people's heads as an indicator of their racial origins and therefore cultural characteristics. When they did so they were following scientific principles and working with commonly accepted ideas. We now regard such practice as bad and immoral science. Ethics, morals and political values may also enter our critical reckoning.

Couclelis points out that such critical geography cannot exist without the technical side (although one of the more worrying aspects of such critiques is that they often do assume they can exist independently of empirical enquiry). But technical geography, the stuff of the discipline, goes blind without critical inquiry. The history of science suggests that the great advances are generally made less by the uncovering of new facts and more by the elaboration of new terms and concepts to give them new meaning.

Geographers have not agreed and do not agree on how knowledge should be produced. Neither do they agree on what things can be said to exist, about which statements can be made. To clarify these points it is helpful to make a distinction between epistemology and ontology. Ontology is the area of thought which addresses the nature of being, or what exists and how it does so. If this seems far removed from geography, consider whether or not the following exist, i.e. are real, precede our identification of them and may possess independent causal powers: region, structure, class, Nature. The region, for example, has been regarded as a real thing, awaiting recognition, but also as an artefact of the geographer. Looking at it another way, can one say that a person is the same kind of thing as a tree, such that broadly similar types of statements can be made about them?

Epistemology is related to ontology, referring to ideas about knowledge itself, what it is and how it is produced. There are many diverse epistemologies, systems of thought which may be distinguished by the ways they answer a simple question: how do you judge the truth of a statement? Epistemological positions generally contain something that is insulated from doubt, against which statements can be judged. For example, logical positivism holds that the sense-datum of experience, the observable and measurable, is beyond questioning. Any proposition about the world is judged by whether or not such observed facts support or contradict it (although how this is actually done has been the subject of enormous philosophical debate). Marxism, by contrast, may assess the value of a statement according to whether it is internally coherent and logically compatible with the rigorous and abstract reasoning of historical materialism. Critics of Marxism often say that it protects itself from 'the facts' by sometimes ignoring them. This misses the point; Marxists demand that any statement is also consistent with a body of knowledge accumulated over decades. They, like many others, may be sceptical about whether facts do indeed exist independently of our theories of the world, and whether they can therefore be said to be fail-safe tests of propositions. Other such bedrocks of knowing include the idea that there are irreducible essences of knowing or deep structures governing social relations, absolutes which govern all thought and behaviour. Phenomenology and structuralism are so characterized, although they have had only passing relevance to the work of geographers (see Benton, 1977, for an introduction). Finally, in the human sciences at least a more modest requirement may be that any proposition is acceptable and intelligible to both analyst and analysed, i.e. the subject of enquiry.

There is obviously much more to epistemology than this, but the important point is that epistemological positions do not necessarily share the same absolutes against which statements can be judged. There are also many variations and disputes within epistemologies which make any attempts to regard them as clearly demarcated and separate somewhat difficult. Epistemologies also differ on questions of what exists, the procedures for knowing, whether or not such procedures are applicable to both natural and human sciences and finally what actually constitutes an explanation.

The Judgement of Paris

According to legend, Paris was asked by the three goddesses Hera, Athena and

Aphrodite to judge a dispute between them (plate IV.1.1). Spurning promises of greatness and prowess in war, he chose Aphrodite's promise of the most beautiful woman in the world. His subsequent abduction of Helen caused the Trojan War. About a decade ago it became common to view geography as being divided into three parts, called positivism, humanism (or hermeneutics) and structuralism (or sometimes Marxism) (see Johnston, 1983a; Gregory, 1978). Like Paris, we were supposed to judge between them and pick one (although some writers tried to show how we could have all three). While such an approach was not as disastrous as Paris's decision, it did have drawbacks. Much of what geographers do is not consistent with any one of these epistemologies. In fact the bulk of geographical writing can be described as an uncritical and incomplete positivism, leaving only small areas for the others. The impression that geography is fragmented into opposing camps is quickly dispelled by a perusal of any one of the major journals (which is a good exercise in any case). There is not much humanism in geography, if by humanism we mean the exotic philosophies of phenomenology, existentialism and idealism. Nor is there much structuralism, particularly if we accept that structuralism is not necessarily the same as Marxism. The big mistake of the Parisian approach was to over-simplify philosophical enquiry and then to try to find some geography that approximated to these simplifications. The result has been a generation of student's essays which take the form of 'the positivist view is . . . , the humanist view is . . .'.

There are different epistemological positions in geography, but they are difficult to pin down and categorize. It makes more sense to begin with the texts themselves and ask simple questions of them, e.g. how do they explain, what assumptions do they make, what things do they regard as existing, what do they leave out and do they indicate how they can be wrong? And the most important question to ask of any epistemological-type proposition is, what difference does it make? Does it use different facts, terms, concepts, processes of thought, does it arrive at original and distinct

Plate IV.1.1 Peter Paul Rubens, 'The Judgement of Paris'.

conclusions, does it depend on distinct moral, political and ethical positions etc.? After starting with geographical texts themselves then, if you wish, move on to consider the philosophies that might inform them. Texts can be both compared with each other, which is fairly easy, and judged by the epistemologies they purport to represent, which is fairly hard. To understand something of the philosophies that might inform geography see Keat and Urry (1982) (although bear in mind that they do have their own chosen position); and for physical geography, Haines-Young and Petch (1986) (they too have a clear preference). For an appreciation of some of the issues in the history of science try Chalmers (1982). Always bear in mind that most geographers are not perfect disciples of philosophy. Haines-Young and Petch, like Kennedy (1983) for example, decry the failure of geomorphologists to be good scientists rather than fair technicians. If you want to follow the debate on the history of knowledge, then Popper (1959), Kuhn (1970), Lakatos (1970) and Feyerabend (1978) are enjoyable if not directly accessible. Always try to keep geographical texts in mind, however; there is little point knowing all about Marxism if you do not know what difference it makes to geography.

Key Debates

There are plenty of good books which elaborate on epistemology but, as already indicated, viewing the discipline as partitioned into -isms is not necessarily the most constructive way of appreciating the subject's debates. Furthermore, geographical ideas are not the sole province of geographers. Geographers are engaged in a series of debates which extend beyond disciplinary boundaries and the flow of ideas is not always one way. Livingstone (chapter II.1) points out that it is probably futile to establish what geography is or should be, though that does not stop some people from trying (read a selection of presidential addresses in the *Transactions of the Institute of British Geographers* or the *Annals of the Association of American Geographers*). Nonetheless, geography does have a number of enduring dialogues. Livingstone has indicated their historical origins; here I address their contemporary forms.

Society and Nature

Livingstone (1985) has described geography as 'a fascinating experiment in holding together society and nature in a single explanatory framework' (p. 119). While such disciplines as economics and demography abandoned nature, geography (with anthropology) continued to relate human activity to the natural environment. This poses a critical question. If geography is to be a distinct discipline, then it must include within its subject matter both natural and social phenomena. But in doing so it cannot collapse one into the other entirely, making either the social a consequence of the natural or, the reverse, seeing nature as entirely a social product. Yet, keeping them conceptually separate raises awkward questions of causation and determination, which in turn have considerable moral, ethical and political consequences. The intellectual project of separating society from nature is a longstanding one (Glacken, 1967; Thomas, 1984). While geographers now generally agree that environmental determinism is neither scientifically sound nor morally acceptable, the boundary between society and nature remains blurred. Consider such contemporary issues as racism, sexism, green politics or sociobiology. All attest to the fact that the boundary is both unclear and contested. Are races naturally different, incompatible and ranked hierarchically? Is it natural for women to stay at home and raise children by virtue of their biology? Are there natural limits to the growth of the human population (the neo-Malthusian overpopulation views of the 1960s and 1970s were in some ways a revival of environmental determinism). Or does society dominate nature, meaning that societies can do whatever they want to?

Behind the current revival in interest about environmental issues there lie serious ethical and philosophical concerns about our proper relationship to the natural world. As a unique species, both inside and outside nature, humankind's conceptions of nature have always been contradictory (Smith, 1984). Nature is both universal (i.e. everything) and external (i.e. that which is not human), according to Smith. This contradiction runs through geographers' efforts to combine the study of human and physical processes, e.g. ecosystems, landscape and geographical information systems. Chorley (1973) points out that the ecosystem model is appropriate in so far as people can be considered to behave like any other life form (i.e. circulate energy, nutrients etc.), but is inappropriate in so far as people stand apart from nature (i.e. engage in class conflict, political power relations, reason etc.).

In fact it has proved exceedingly difficult to integrate human and physical phenomena, not least because of ontological problems. Humans, unlike trees and rocks, communicate meanings, have motivations, can anticipate future events etc., all of which mean that the same epistemological assumptions do not necessarily cover both kinds of phenomena. By and large, we are reasonably safe in assuming that they do. A great deal can be said about people as if they were the same as trees and rocks. Provided we do not make the mistake of thinking that people are actually the same kind of things as rocks, viewing human behaviour as similarly predictable and law governed, then few problems arise. After all, a lot of human behaviour appears predictable – as if it were the product of spatial laws, e.g. the rank–size distribution of settlements and the gravity model of migration. Some natural science analogies have been fairly useful in human geography, gases, molecules and plant communities among them. Then again, the world of meanings, emotions and class conflict is not one in which it makes sense to pursue such natural analogies. A debate ensues between those who believe that the methods and assumptions of the natural sciences are all that is needed in geography and those who argue that using such assumptions is unjustified philosophically and morally.

The form in which this society/nature problem has most visibly surfaced in geography itself is the question of whether or not physical and human geography are, or should be, kept together in a single discipline. This debate is often referred to as the 'unity of geography'. It is more than an intellectual worry; the survival of geography departments in the United States and the United Kingdom depends upon their ability to justify themselves to universities and governments. Smith (1987b) records the failure of Harvard's department to do so, leading to its closure. One way they justify themselves is by stressing the importance of an integrated subject, especially since it seems relevant to contemporary environmental concerns.

Not all geographers are happy with the marriage of human and physical. Those in favour of unity argue that since reality is unified so geography, as a synthesizing discipline, should reflect this natural unity (Mackinder, 1887; Hägerstrand, 1976). It provided geography's initial rationale as a bridge discipline, a role which is strengthened by the pressing need to tackle major environmental problems (Goudie, 1986;

Kates, 1987; Stoddart, 1987). Furthermore, human and physical geography, if split, would not be able to withstand being absorbed into other disciplines such as sociology or earth sciences. Finally, there are arguments that stress the pedagogic value of keeping them together, i.e. it is a good subject to teach (Pepper, 1987). Against this is the view that 'there is no natural necessity for the discipline of geography' (Johnston, 1986) and that all disciplinary boundaries are artificial creations of society (Taylor, 1986). Sayer (1979), drawing on insights from a realist epistemology, argues that reality is not unified but layered or stratified. Just as anthropologists interested in kinship do not need to know the biological facts of reproduction, so human geographers do not need to know about physical geography. Others point out that the combination of human and physical geography has had negative consequences in the past: environmental determinism, for example, seemed to justify racism and imperialism by conflating social differences with biological ones (Peet, 1985). Johnston (1983b) further claims that geographers are arrogant and unrealistic if they think that geography alone is the key to environmental problems. Even if insights about both physical and human phenomena are necessary in solving, say, land degradation, it does not necessarily follow that a single discipline must embrace both. Teams of specialists might be more effective. Finally, some physical geographers (for example Worsley, 1979) claim that they suffer by association with the less scientifically credible human geographers.

This debate is certain to continue because it taps into fundamental questions about society and nature. Among some geographers (e.g. Smith, 1987b), there is a feeling that attempts to define the discipline too narrowly or to place the survival of the discipline above the researcher's desire to ask whatever questions he or she wants to is authoritarian. Perhaps the debate needs to consider more fully what kind of unity is desired (Pepper, 1987). Pepper thinks the problem lies with a physical geography that is too pure, focusing too much on understanding physical processes as an end in itself, and divorced from real social concerns. A geography that was intrinsically relevant, problem oriented and practical would probably not need to discuss its unity; it would just happen.

Space

If the integration of society and nature has formed the major rationale for geography, then the idea of geography as the science or discipline of space is a close second. Anglo-American geographers' interest in space as an abstract concept dates from the post-war period, and the 1960s saw a number of classic books advancing a spatial science. It is fairly easy to see why; increased central state involvement in organizing the built environment demanded knowledge of optimum locations and efficient spatial relationships. At that time the philosophical debate about space revolved around the question of whether it was absolute or relative (Sack, 1980). We instinctively seem to regard space as an absolute dimension, empty, containing objects, a backdrop to social activity, something which exists independently of ourselves and which can therefore exert influence over our actions. This is the space of Euclid's geometry, Newton's physics and, in a different sense as a primary mode of knowing the world or organizing the sensations of outer reality, of Kant's philosophy. But ever since Einstein, if not before, it has been clear that space does not exist independently from either time or matter; in fact matter shapes space. This relative or relational concept of space assumes that it can be acted upon and is not independent of social activity and thought. Two important points stem from this insight. One is that there is no true, real or scientific space which is somehow more important than all other conceptions of space. The other is that, as Harvey (1973) observed, the conceptualization of space is not simply a philosophical matter, but bound up with human practice. To understand the contemporary debate on space we must grasp an apparently simple yet conceptually awkward idea – that space is produced. How space is produced depends upon the historical and material circumstances operating. 'Primitive' societies conceive of and produce space in quite different ways from technologically advanced or capitalist societies (Sack, 1980). Capitalism produces space in a different way from feudalism or socialism (Dodgshon, 1987). This is not an easy point to understand, partly perhaps because many of the insights into the production of space have come from non-English-speaking authors, whose ideas do not always translate well. Space needs to be understood as the complete assemblage of built environment, relations of distance, betweenness, proximity etc., as well as practices and representations of space. Harvey (1989) has tried to place all these spaces within a single framework, while the essays collected in Gregory and Urry (1985) go far in exploring what all this might mean.

It is a fair observation that geographers have not yet fully worked out all the implications of what has become a greatly expanded conceptualization of space. Some differences of opinion, however, are evident. After some initial contributions on the space of cognition (Downs and Stea, 1973), of experience (Tuan, 1977) and of being (Buttimer and Seamon, 1980), the debates have centred on Marxism, realism and what is generally termed critical theory (Gregory and Urry, 1985). An exchange between Smith (1986) and Browett (1984), for example, concerns Smith's contention that uneven spatial development is the 'hallmark' of capitalism. Smith argues that capitalism can only reproduce itself through producing space unevenly, generating inequalities of development, whereas Browett maintains that such unevenness is only inevitable, and not necessary. One of the themes of this exchange is that of spatial fetishism, the attribution of causal powers to space itself, suggested for example in the phrase 'the core dominates the periphery'. Geographers are usually anxious to avoid this misrepresentation of what are actually social relations as spatial relations, although it proves difficult to do in practice. A related debate concerns whether or not space (with time) is central to all social theory, or whether it is only relevant in certain contexts. Soja (1989) and Giddens (1985) argue persuasively that all social theory must be spatial and temporal, adding that until now social thought has been dominated by temporal concepts (progress, evolution, historical change etc.) at the expense of spatial awareness. They receive support from non-geographers in viewing the contemporary era as one in which space has asserted itself in our understanding. By contrast, Sayer (1985) and Saunders (1985), in different ways, suggest that we only need to consider space in a secondary role, for example in terms of its relevance to concrete situations rather than abstract theory. These debates can be difficult to follow, since it is not always clear what difference one or other position actually makes. What is undeniable, however, is that geographers are not alone in rethinking space, and it is one area which promises (or threatens) to dissolve the boundaries between it and other human sciences. The outstanding issues seem to be first how to develop a theory of the production of scale, i.e. why does social activity realize itself at international, national, urban or local levels;

second, how does the material production of space affect both practices with regard to it and representations of space, and vice versa; and last, how are social differences such as 'race', gender and class constituted spatially and how do these differences in turn affect the production of space.

Time and Change

Physical and human geographers have long discussed the nature of time and change. A basic issue has been whether or not to include a temporal dimension in geographical explanation. Within science it is generally acknowledged that there are experimental or theoretical sciences which are interested in discovering universal laws, i.e. statements which are independent of a particular spatial or temporal context, and historical sciences which are interested in singular events in their contexts. Ideally a good explanation should involve both kinds of consideration, what Kennedy (1979) terms the immanent and the configurational. This is what Darwin's theory of evolution managed to do, but in practice it has mostly eluded geographers. Within human sciences a parallel distinction exists between diachronic or longitudinal explanations (also called genetic), which seek to explain how a situation came about, and synchronic or functional explanations which attempt to explain a situation in terms of conditions existing at a point in time. Again, it is practically difficult to combine both. Indeed, many human geographers have sought to deny the relevance of a temporal perspective altogether (Hartshorne, 1959). Historical geographers have generally insisted on a temporal view, without which any explanation of a contemporary landscape is incomplete (Sauer, 1941). All too often, where human geography has considered time it has been in a simple static sense, either taking cross-sections through time or explaining something in terms of a series of stages (e.g. the demographic transition, Rostow's stages of growth).

It has therefore been physical geography that has been most sensitive to time and change, perhaps because of its close affiliation with geology and evolutionary biology, two of the historical sciences. At one level there has been, like space, a question of appropriate time-scale. Davis's cycle of erosion, for example, involved unwieldy lengths of time and was silent about the mechanisms which caused landscapes to change. The passage of time itself does not cause change. The shift to process studies in geomorphology (chapter II.2) was also a shift to shorter time spans. In addition, as Schumm and Lichty (1965) explained, the length of time chosen to study a physical system influenced whether variables could be considered dependent or independent. The question of time-scale also affects the basic geomorphological and biogeographical ideas of equilibrium and stability (Montgomery, 1989). It has proved difficult to conceptualize these system characteristics independently of time-scales (chapter II.3) and therefore to give them an immanent as well as a configurational meaning.

Closely related to the question of system behaviour over time is the problem of whether change is brought about through the constant operation of processes over a long period of time, or by sudden rare events. The distinction between uniformitarianism and catastrophism played a key role in geological debates in the early nineteenth century (Gould, 1988). The uniformitarian, by rejecting literal Biblical accounts of creation, came out on top, but in the 1960s and 1970s more attention was given to the role of big rare events (Dury, 1980). Such 'neo-catastrophism' originated in palaeontology, the study of the fossil record, and, as Dury argues, accorded with a general feeling of impending ecological doom current at the time. A belief in the power of mega-events, like sudden extinctions and the dambursts of glacial lakes, strengthened an environmentalist view of the power of nature over society. Some argue that geographers have overstressed the human influence on the environment and neglected the force of natural events or the ability of the earth to sustain itself (Lovelock, 1988). What remains to be explained is under what conditions and where in the landscape are low- and high-frequency events most efficacious (Schumm, 1979).

Not only is there a debate on the timing of change, but there is also a question of whether systems change through internal or external causes. Physical geographers are aware that sudden changes in the state of a system, for example the onset of turbulence in water, can occur as a result of gradual changes in that system's characteristics. Modelling such events mathematically proves difficult, and so researchers have increasingly turned to new branches of mathematics which are non-linear rather than linear, i.e. the solution of complex equations cannot be reached by simply adding up the solutions to simpler equations. A common type of such mathematics is catastrophe theory (Huggett, 1990). This concept has also been adopted by some human geographers (Wilson, 1981). The result of this mathematics has been a renewed sensibility to the possibilities of discontinuities in the evolution of spatial patterns, retail sites or industrial locations for example. The big problem is deciding not only how to model such situations correctly (all variables must be capable of being expressed mathematically), but also which situations are liable to catastrophic change.

Human geographers have also on occasion made time itself the subject of study. Like space, there are both absolute and relative conceptions of time. Thermodynamics gives time an absolute direction, while clocks give time a socially constructed meaning (Whitrow, 1989). There are many coexisting senses of time – biological, work related, psychological etc. – which shape our daily lives (Thrift, 1977). Just as with space, the question of how these various times relate to one another, and how to conceive time-scales, remains tricky (Harvey, 1989). Whatever the case, it is certain that time is not as simple as we intuitively think, so conditioned are we by thinking that clock-time, like Euclidean space, is the only true or real time.

Both physical and human geographers share a common basic problem of how to conceptualize continuity and change together. Ideas of equilibrium (in its various forms) offer one solution, as does the attempt to break up system behaviour into discrete time-scales with their own properties. In human geography the problem has recently resurfaced over the idea that the Western capitalist economies have undergone a 'sea change' (Harvey, 1989) and that we are living in New Times. A common way of thinking about this change is to divide capitalism into distinct periods, each with their own characteristics, e.g. Fordism and post-Fordism (see chapter II.9). Such models are often called transition or periodizing models. Although they provide useful simplifications, critics argue that they create false divisions, exaggerate change and assume that change is more pervasive than it actually is (Sayer, 1989; Thrift, 1990). It is worth recalling that historians' belief in another transition, from feudalism to capitalism, has been considerably altered by increasing knowledge of history; the time the transition took to happen appears to be getting longer and longer, and there is little evidence of any clean break. The post-Fordist idea is also open to the charge that it emphasizes temporal differences at the expense of spatial ones, but only does so by privileging certain core capitalist countries and omitting others, notably Japan. What the debate suggests, like the criticisms of Davis's cycle of erosion, is that stage or periodizing models should perhaps be regarded more as hypotheses than actual facts. If they are uncritically regarded as true, then they can channel geographers' enquiries into divisions which are too neat.

Space and Time

Geographers frequently theorize time and space separately from each other, but most would agree that a good geographical account should combine both. Putting them together has proved practically difficult. A common geographical strategy has been to formulate a spatial model or explanation and then see what might happen to it over time – Christaller's central place theory for example. Such explanations maintain a conceptual separation of space and time, however. Another technique, evident in Von Thünen's model of agricultural location, is to measure space in units of time and/or cost. More recently, various geographers have worked with the idea of building space and time constraints into models of human behaviour. The fact that a person cannot be in two places at one time, or that two people cannot be in the same place at the same time, are absolute constraints on human activity (Hägerstrand, 1970). Such 'time-geography' has had limited planning application in Sweden, but runs into great practical problems of combining masses of data for large numbers of people.

What all these approaches share, however, is a tendency to regard time and space as separate absolute dimensions, the co-ordinate system for social action or physical processes. Demonstrating how social action also shapes space and time has proved harder, although calls for a 'historical–geographical materialism' have at least stated that the problem exists (Soja, 1989). Harvey (1985a) provides a stimulating outline of how, in theory, space, time and money shape each other in the urban process, although the perspective proves difficult to sustain in the subsequent case study of nineteenth-century Paris. Giddens (1985) makes programmatic or conceptual statements of how space and time should be integrated into social theory, although once again actual attempts to make the conceptualizations work in practice (for example Pred, 1986) prove awkward and unconvincing. In many ways the practical difficulty of writing about spatial difference, temporal change and human action together represents an obstacle to their simultaneous conceptualization. As Soja (1989) has observed, geographers are obliged to use a linear or sequential form (the text) to describe what is actually simultaneous. Only maps represent a way of showing simultaneity, and it is perhaps no coincidence that there is a revived interest in their use among human geographers.

Individual and Society

Human geographers also have to address three closely related philosophical and methodological problems that are common throughout the social sciences (Agnew and Duncan, 1981). The first is whether macro-scale social situations are simply an aggregation of the actions and attitudes of individuals, or whether they have their own properties independent of individuals. There are two contrasting positions, methodological holism and methodological individualism. The former argues that there are entities, such as collective consciousness or class, which are ontologically real and which must be explained in their own specific terms. The latter position holds that any truthful statement about society must ultimately be reducible to a series of statements about individuals; the action of an army is simply the sum of the actions of its soldiers.

The second, related, problem is known either as the structure/agency debate or the structure/consciousness/ action debate. The essence of the problem can be found in Marx's celebrated statement that 'men make their own history, but they do not make it just as they please; they do not make it under circumstances chosen by themselves, but under circumstances directly encountered, given and transmitted from the past'. We are aware of ourselves as individuals capable of action, able to make a difference in the world, but we are also aware that the regularities of society appear to be rule governed. Giddens (1981) suggests that we are both knowledgeable and purposeful agents, but that social processes also seem to 'work behind our backs', in ways of which we are unaware. Within social sciences there have been two broad responses to this dilemma. One is to posit the existence of structures, either governing or determining social action, which lie either 'beneath' observable social acts or within the mind. Explanation consists of uncovering such structures and showing how they govern both consciousness and action. Examples include Levi-Strauss's anthropology and Chomsky's linguistics. The opposite response involves the proposition that action is explicable purely in terms of the thoughts, motivations and reasoning of individuals, e.g. Weber's sociology. So if we take a simple geographical question such as why migration occurs, we can identify two sorts of answers. One might be in terms of economic structures or the uneven development of capital that generates patterns of labour supply and demand, while the other might aggregate the reasons given by a sample of individuals.

The third variant of the basic society/-individual problem is perhaps more theological; are we free-willed or determined subjects? Within geography there have been various determinist positions: environmentalist, cultural, biological and spatial, all of which view our activities as being predetermined by things outside our control.

Together, these three dualisms pose a complex methodological, epistemological and ontological problem, one which is more than just academic. Ley (1980) for example points out the grave moral and ethical errors of explaining the actions of people as if they were determined by outside forces. In particular he is concerned that using analogies from the natural sciences, e.g. people as atomic particles in gravity models, can lead to the dehumanization of individuals. It we treat people like atoms or puppets in our enquiries, so the argument goes, then it is a short step to treating them like atoms or puppets in real life; in other words, it feeds a totalitarian conception of society. Most geographers would like to believe that we are capable of improving or changing the world. Yet, the more we act as if everything is determined, the more we contribute to a sense that the individual can change nothing. What geographers need, therefore, is a methodological and ontological position which views individual action as partly autonomous and partly determined. But actually finding a balance and demonstrating how we may change things is again practically difficult. It is one thing to make bold statements about ontological relations; it is quite another to show how they can inform our methodology and make a difference.

Geographers have been toying with this problem for a decade or so, looking to sociology and anthropology for clues. One solution seems to be found in a position generally known as structuration, an idea associated with a number of authors (Thrift, 1983). At the simplest level, structuration involves a series of ontological propositions that seek to claim a middle ground between structure and agency, holism and individualism, by demonstrating how they can be thought of as mutually shaping each other. In the form proposed by Giddens (1981), structuration also attempts to show how space and time are both conditioning of and conditioned by action. Giddens' attempts to resolve virtually all the problems of social theory, including the synchronic/diachronic dualism have already been mentioned. These theories are dense and offputting; they can appear to be no more than the right words. One way of grasping what they propose is to consider communication. We know that when we speak or write we use words and grammatical rules which are not of our making. Yet the grammar does not fully determine what we say or write. We can, if we choose, alter the rules, use new words. Provided our innovations are intelligible to others, they may then become part of the language. The language in turn has no real existence outside our use of it. If society is like language then we can also see perhaps how individual acts are dependent on the existence of society, that society would not exist but for these acts, and that our acts may change society.

So far so good; but the big question has been whether this makes a difference to what geographers do. Pred (1986) has tried to use these insights in a study of the effect of enclosure on Swedish peasants, though some reviews were sceptical. It may be that structuration provides a set of ideas about ontology which we can subscribe to, but which do not in themselves imply any novel way of conducting enquiry. Giddens himself suggested that the method and techniques of time-geography may provide the crucial translation of his general ideas into practical tools for analysis. But the practical problems of handling large amounts of detailed information on the paths individuals take through time and space make time-geography cumbersome. Geographers are attracted to these ideas, not least because social theorists like Giddens have declared that time and space matter. Their encounter with structuration illustrates the drawbacks of seizing upon intellectually attractive ideas without fully considering their practical implications.

Region, Place and Locality

The American geographer Hartshorne once posed the question of whether geographers 'seek to describe individual cases or formulate general laws' (1959). This question has also been asked by historians, anthropologists, psychologists and in a range of other disciplines. The world is full of things which may be regarded as either unique (Madrid, the Colorado River) or members of classes of things (cities, rivers). Explanation, Hartshorne argued, consists in abstracting either the unique aspects of phenomena or their general aspects (see Entrikin, 1981, for an excellent account). In the case of the former, a causal explanation would consist of showing how a series of unique phenomena were related to each other, e.g. the specific events leading up to the Russian Revolution or the stages in the growth of Chicago. By abstracting the general aspects of phenomena, explanation is achieved by showing how any one event or thing (the Russian Revolution, Chicago) is but a particular realization of the laws governing all similar events and things (revolutions, cities). A problem arises in that many do not regard the first of these routes of explanation as a real explanation at all;

'science is the deadly enemy of uniqueness,' said Bunge (1966).

The particular form in which this problem arose in geography was the region. Classical regional geography regarded each region as unique; Gilbert (1960) likened them to personalities. Scientific geography, by contrast, sought to show how each region or stream network or city was but one among a class of similar objects. Harvey (1969) observed that things were only unique when considered in absolute space. A relational or geometric transformation of a stream network into a graph network rendered it one of a class of like graph networks. In the 1960s, therefore, the unique (or idiographic) emphasis of regional geography was denigrated as unscientific. Geography, it was argued, should concern itself with the formulation of universal spatial laws. Furthermore, the idea that regions were real entities, pre-existing their identification by geographers, in which human and physical factors interacted to produce unique geographies gave way to a view of regions as simply one form of classifying the earth's surface: less a personality, and more a box.

The regional focus of geography never did disappear, however, which makes claims of its rediscovery in the 1970s and 1980s somewhat exaggerated. A number of developments converged to put the unique back into geography, although in a modified form. Among these were the renewed attention to ideas of a 'sense of place' as a key element in the geographical experience of the world (Tuan, 1977); a growing interest in art and literature as expressions of the unique identity of places and regions (Hart, 1982); a feeling that modern planning and architecture was destroying the valuable differences between places, creating a sense of 'placelessness' (Relph, 1976); and a recognition among political geographers that where one lives is a powerful indicator of how one votes. Furthermore, as economic historians had already realized, far from destroying regional distinctions, processes of economic development drew upon and recreated such distinctions. Geographers of the contemporary economic landscape also realized that homogenization was not happening and that new and unpredictable differences were being generated. By the late 1980s there was a sense that our general geographical models were failing to describe what was actually happening on the ground. At both national and global scales, the uneven development of capitalism was throwing up new places, regions and geographies (chapter II.9, II.12; Soja, 1985). The new sensibility involved more than the recognition that places were different and changing. It was recognized that the particular combinations of social, cultural, economic and political factors operating in places produced different responses to general or secular changes. One could speak of local cultures, class alliances, local politics etc. as distinctive responses to both past and present processes. Place mediated society and space, as revealed in local economic initiatives, urban boosterism and the efforts to package the heritage of places for consumption by tourism (Sack, 1988; Agnew and Duncan, 1989).

The rediscovery of place has its own vocabulary. A 'new regional geography' recognizes that regions are the real products of processes and not just a convenient method of classification. The former emphasis on the synthesis of social and natural has been dropped in favour of a focus on the specific combination of human processes. The term 'region' itself has become partly detached from its conventional subnational scale, to be replaced by a sense of all social life as regionalized and regionalizing, i.e. situated in, and reproducing, time–space contexts. Everything from a room to a nation-state can be considered a region. Sometimes the term 'locale' is invoked to describe this condition (Giddens, 1985). Place, far from being a rather inert and ahistoric form, may be thought of as a process (Pred, 1986). Pred describes place as a process of 'becoming', always in motion. Finally, a new term, 'locality', has been devised to capture both the differences between places and the fact that places are the context for collective acts of organization and the exercise of the rights of citizenship, as well as being the arena of everyday life (Cooke, 1989, 1990).

Much of the new vocabulary, however, is unclear. Urry (1987), for example, lists ten different definitions of local, while a long debate in the journal *Antipode* (1987–90) reveals a lack of consensus on the meaning of locality. It has become common to 'celebrate' the fact of difference or to state that place is important in advance of research which actually demonstrates how it is significant. Geographers have outlined a number of positions, but their substantiation remains partial. In the landmark volume *Localities*, summarizing the research of the Changing Urban and Regional Systems project, Cooke (1989) almost seems to gloss over what locality actually means. There is a danger of making a fetish out of place, as once both space and region were fetishized.

One way out of the clutter of meanings is to find a theory of the production of scale (Smith, 1987a). Geographers need to link their general theories about space with their ideas about the significance of place via some understanding of how various human activities are resolved at different scales (see also Barnes, 1989). A common analytical problem is the bounding or closure of parts of reality for investigation. These should not be arbitrary divisions but should correspond to some meaningful and real unit of investigation, e.g. the city conceptualized as the urban labour market (Harvey, 1985b). Furthermore, geographers require an idea of how these various units are nested in hierarchies. For this a theory of how global processes interact with local processes to produce scales of human activity is required, but such a theory proves elusive.

Despite the general feeling among the participants that the contemporary debate on locality has avoided the pitfalls of a simple opposition between the general and the unique mentioned by Livingstone in chapter II.1, by recognizing the need to combine them, many of the simple methodological problems of the old regional geography can be found in the new. These include the problem of selecting cases, of making generalizations from limited case studies, of comparing two or more case studies and of striking a balance between theoretical or abstract statements and empirical or concrete observations. For example, whereas in physical geography it may be possible to pick two very similar catchments and deforest one in a controlled experiment, such an option is not open to human geography. It is simply difficult to decide when two places are similar or are different. The localities debate has looked to realism, Marxism and post-modernism to answer these questions, but the result has been a degree of theoretical overkill. These problems can only be properly addressed by the logical nature of thought, the making of informed judgements, something which is not automatically the province of any particular epistemological position.

Chaos and Post-modernism

There is a widespread feeling that we are living in a period of fundamental and drastic change, in which an old order is giving way to a new one. The shift from Fordism to post-Fordism is one aspect; the emergence of non-class politics (of gender, sexuality and the environment) together with the partial dismantlement of the welfare state is another; the collapsing of boundaries between high and low culture (e.g. opera and soccer) and the lack of rules of cultural consumption is a third. The ever more rapid circulation of capital and the 'annihilation of space by time' as places become closer together in a more interdependent world economy, the de-alignment of old forms of political loyalties, and the fact that a great work of art can now fetch tens of millions of dollars although it no longer addresses our innermost feelings, these are all evidence of change. In the physical world, the litany of environmental problems attests to the possible instabilities and unpredictable behaviour of the planet itself. In the climate of change, two new modes of thought are emerging, chaos and post-modernism.

Although there is no necessary connection between the natural sciences' interest in chaos and the human sciences' encounter with post-modernism, they do share some features. Both are self-conscious efforts to break with old ways of thinking and are based upon a conviction that the world is more complicated than was once believed. Both question a simple idea of determination and of the essentially law-like nature of reality. They both imply a crossing or dissolution of disciplinary boundaries, and both pose radically new questions while also providing some answers to old ones.

Once it seemed that the universe could be described in terms of two kinds of systems (for a good account of chaos see Stewart, 1989). Some were deterministic and therefore predictable, tides for example. Because they were governed by relatively few variables which could easily be identified, it was possible to specify precise laws describing them. Others, governed by more variables, less easily identifiable and interacting in more complex ways, e.g. the weather, could only be approximated statistically. Since they appeared to behave randomly, the only statements one could make about them were probabilistic ones. What the insights now recognized as the theory or mathematics of chaos reveal is that this distinction is incorrect. Systems which appear random, like the switching of the earth's magnetic field, the changing size of an insect population or the onset of turbulence in a fluid, may in fact be deterministic and described by simple mathematical equations. It is not the number of variables which makes these systems appear unpredictable (indeed, chaotic behaviour can be produced by as few as three variables), but their way of combining or mixing. Provided one can discover the rules of mixing, it is theoretically possible to find the governing equations.

There are two problems, however. The first is that chaotic systems, as they are called, exhibit sensitive dependence on initial conditions. This means that tiny variations in the starting state may, though will not necessarily, produce large variations in the end state of a system. This imposes a practical difficulty of ever predicting the weather accurately for example. The other problem is that, although one might know the equations describing the overall behaviour of a system, one does not necessarily know precisely what it will do next. The consequence is that a fundamental belief, that explanation and predication are symmetrical (i.e. if you can do one then you can do the other), may have to be abandoned for certain systems. Thus, while chaos opens up the possibility of unlocking the secrets of hitherto unpredictable systems, it also creates a sense of finite predictability, introducing a level of modesty to the natural sciences which the social sciences are well accustomed to.

The implication of these ideas for geography are not yet clear. Many of the insights come from re-examining irregularities in data sets which were once discarded as either noise or measurement error. Meteorological data are possibly a good candidate for reappraisal, although they are rarely numerous enough to provide unambiguous evidence of chaotic behaviour. Climatologists may, for example, use the mathematics of chaos to simulate such events as El Niño or upper atmospheric shifts in wind direction which are partly regular. Species populations are another candidate for such modelling. In all such cases it is important to bear in mind that chaos is a property of mathematical models, and analysts have to judge whether real world systems do in fact share such properties. The identification of variables and system closure present thorny problems; in addition, even if chaos is found, it is another matter explaining where it comes from, i.e. locating the mechanisms.

Of perhaps more direct relevance is the use of a subset of chaos mathematics known as fractals (plate IV.1.2). A fractal is a space-filling curve which exhibits self-similarity (i.e. every part of the curve is a replica of the entire curve) and has the properties of both a line and an area; it is a shape between dimensions (Goodchild and Mark, 1987; Batty, 1989). Fractals appear to be irregular, but are in fact generated by a small number of equations. The unusual properties of fractals make them useful for simulating geographical forms like coastlines, terrains, cities or central place networks and for describing irregular shapes in terms of their position between one and two dimensions. As a branch of mathematics they provide new ways of analysing morphology, something which geographers have abandoned in favour of process studies. The fact that fractals can be quite simply generated on home computers has increased their applicability for simulation and modelling.

Although few geographical systems

Plate IV.1.2 A major development in recent years has been the study of fractals and chaos by mathematicians. This plate provides an example of fractal geometry in the form of a computer graphics representation of a detail from the Mandelbrot Set.

may actually be chaotic in the strict mathematical sense, the language of chaos provides a new set of metaphors for describing and simulating the real world, although not necessarily for explaining it. These metaphors bridge human and physical geography, and open up geography to other disciplines in which mathematics is used. The drawback is that there are too few geographers with the mathematical skills to make good use of them.

Post-modernism lacks the mathematical precision of chaos; indeed it has become in some ways a vogue term for anything that is new and different. Three recent geographical books, by Soja (1989), Harvey (1989) and Cooke (1990), do at least outline coherent arguments. All three recognize that there is a fundamental change happening in the world, one that affects economics, politics, culture and philosophy, not to mention geography. This change involves, paradoxically, both a negation of and a continuity with the past. This past is variously viewed as the post-seventeenth century Enlightenment or age of reason, Modernism (which may be said to begin at any time from the early nineteenth century onwards) or Fordism (starting in the early to mid-twentieth century). They recognize changes occurring simultaneously across the field of human experience, from architecture to philosophy, and wonder how they might be related and whether there is any single cause. Harvey argues that the increasing turn-over time of capital circulation has brought into being shifts in the organization of production and a new round of 'time–space compression' (the world is getting smaller everyday), which in turn is bound up with new ways of experiencing and representing space and time. The movies *Bladerunner* and *Wings of Desire* are used to illustrate his claims: in both there are two types of beings who exist in radically different speeds and spaces. Soja is more concerned to show how, within social theory, space is being reasserted. If the thinking of the Modern era was dominated by time, in the form of history, evolution, progress etc., then the late twentieth century has witnessed the emergence of an understanding of space as central to our thinking. He borrows a felicitous phrase from John Berger: 'prophesy now involves a geographical rather than historical projection; it is space not time that hides consequences from us' (Soja, 1989, p. 22). In a world dominated by the decision-making of global corporations and the rapid circulation of information and money, it is easy to imagine how actions in one place have immediate consequences in other places (see chapter I.2). This is what concerns Cooke, who tries to fix these post-modern changes more firmly in geography. He argues that we are witnessing the break-up of hierarchical and centralized forms of organization: the firm (flexible specialization), the state (the rise of regional and local claims and the devolution of responsibility for social services), the novel (the emerging voices of women, people of colour and other excluded peoples), and philosophy (the move away from grand theories which purport to explain everything). This period contains the possibilities of both a reactionary return to classical values, like Prince Charles's views on architecture or Mrs Thatcher's evocation of Victorian values, and a new decentralized localized control over daily life and work. His central concern is therefore with the rights of citizenship and belonging, and how the individual can make sense of the changing world and exercise some influence over it.

What does all this mean for geography? A number of developments seem apparent, including a new interest in cultural studies, a heightened role for, and theorization of, space, and a concern for how geography is actually written and presented. Soja's essay on Los

Angeles may be regarded as an experiment in geographical writing, using irony and juxtaposition rather than straight narrative to bring out the simultaneity of what he sees.

The implications can best be appreciated by considering one of the monumental texts of Modern geography, *Models in Geography*, edited by Chorley and Haggett in 1967. This collection of essays was regarded at the time as a landmark in the new geography (for a reappraisal see Macmillan, 1989). What the essays shared was a view of the world as more simple and ordered than it appears, and a sense that, by accumulating more knowledge and building better models of it, geographers would be in a position to predict, explain and change the world for the better. These models were to be built initially by analogy, from economics and biology for example. If one reverses all these assumptions then one has some idea of what post-modern geography might be all about. Reality is complex, there are no guaranteed ways of representing or modelling it, our explanations are partial and our insights are more likely to come from literary criticism or psychoanalysis. Human and physical geography part company. Our faith in rational planning and progress is weakened. Furthermore, all these things are viewed more desirably. The absence of single explanations and an inability to predict and control reality may be positive achievements if it is thought that such control is exercised principally by centralized and hierarchical powers. Openness, plurality and possibility are the watchwords of post-modernism.

Post-modernism, like chaos, may therefore involve an undoing of past certainties. They leave a void in our understanding which in turn represents a challenge to geographical endeavour. They impinge upon all the debates presented in this section. Chaos theory makes us sensitive to the possibility of sudden change in natural systems like the climate, which in turn makes us more aware of environmental ethics and society's relationship with nature. Our concepts of space and time are being radically altered, just as the world itself is undergoing dramatic shifts. These in turn pose questions of our place in the world, as individuals and societies. Although the debates have been treated separately, they are all in fact conjoined within the contemporary dialogues of geography.

References

Agnew, J. A. and Duncan, J. 1981: The transfer of ideas into Anglo-American human geography. *Progress in Human Geography*, 5, 42–57.

——and (eds) 1989: *The Power of Place: bringing together geographical and sociological imaginations.* Boston, MA: Unwin Hyman.

Barnes, T. 1989: Place, space and theories of economic value: contextualisation and essentialism in economic geography. *Transactions of the Institute of British Geography, New Series*, 14, 299–316.

Batty, M. 1989: Geography and the new geometry. *Geography Review*, 2, 7–10.

Benton, T. 1977: *Philosophical Foundations of the Three Sociologies.* London: Routledge & Kegan Paul.

Browett, J. 1984: On the necessity and inevitability of uneven spatial development under capitalism. *International Journal of Urban and Regional Research*, 8, 155–76.

Bunge, W. 1966: Locations are not unique. *Annals of the Association of American Geographers*, 56, 375–6.

Buttimer, A. and Seamon, D. (eds) 1980: *The Human Experience of Space and Place.* London: Croom Helm.

Carlstein, T. 1982: *Time, Resources, Society and Ecology*, Vol. I, *Preindustrial Societies.* London: Allen & Unwin.

Chalmers, A. F. 1982: *What is This Thing Called Science? An Assessment of the Nature and Status of Science and its Methods.* Milton Keynes: Open University Press.

Chorley, R. J. 1973: Geography as human ecology. In R. J. Chorley (ed.), *Directions in Geography*, London: Methuen, 155–70.

Cooke, P. 1989: *Localities: the changing face of urban Britain.* London: Unwin Hyman.

——1990: *Back to the Future: modernity, postmodernity and locality.* London: Unwin Hyman.

Couclelis, H. 1982: Philosophy in the construction of geographic reality. In P. Gould and G. Olsson (eds), *A Search for Common Ground*, London: Pion, 105–38.

Dodgshon, R. A. 1987: *The European Past: social evolution and spatial order*. London: Macmillan.

Downs, R. M. and Stea, D. (eds) 1973: *Image and Environment: cognitive mapping and spatial behaviour*. London: Edward Arnold.

Dury, G. 1980: Neocatastrophism? A further look. *Progress in Physical Geography*, 4, 391–413.

Entrikin, N. 1981: Philosophical issues in the scientific study of regions. In D. T. Herbert and R. J. Johnston (eds), *Geography and the Urban Environment*, vol. 4, Chichester: Wiley, 1–28.

Feyerabend, P. 1978: *Against Method: outline of an anarchistic theory of knowledge*. London: Verso.

Giddens, A. 1981: *A Contemporary Critique of Historical Materialism*. London: Macmillan.

——1985: Time, space and regionalisation. In D. Gregory and J. Urry (eds), *Social Relations and Spatial Structures*, London: Macmillan, 265–95.

Gilbert, E. W. 1960: The idea of the region. *Geography*, 45, 157–75.

Glacken, C. J. 1967: *Traces on the Rhodian Shore: nature and culture in Western thought from ancient times to the end of the eighteenth century*. Berkeley, CA: University of California Press.

Goodchild, M. and Mark, D. 1987: The fractal nature of geographic phenomena. *Annals of the Association of American Geographers*, 77, 265–78.

Goudie, A. S. 1986: The integration of human and physical geography. *Transactions of the Institute of British Geographers, New Series*, 11, 454–8.

Gould, S. J. 1988: *Time's Arrow, Time's Cycle*. Harmondsworth: Penguin.

Gregory, D. 1978: *Ideology, Science and Human Geography*. London: Hutchinson.

——and Urry, J. (eds) 1985: *Social Relations and Spatial Structures*. London: Macmillan.

Hägerstrand, T. 1970: What about people in regional science? *Papers of the Regional Science Association* 24, 7–21.

——1976: Geography and the study of the interaction between Nature and Society. *Geoforum*, 7, 329–34.

Haines-Young, R. and Petch, J. 1986: *Physical Geography: its nature and methods*. London: Harper & Row.

Hart, J. F. 1982: The highest form of the geographer's art. *Annals of the Association of American Geographers*, 72, 1–29.

Hartshorne, R. 1959: *Perspective on the Nature of Geography*, Monograph Series of the Association of American Geographers 1. Chicago, IL: Rand McNally.

Harvey, D. 1969: *Explanation in Geography*. London: Edward Arnold.

——1973: *Social Justice and the City*. London: Edward Arnold.

——1985a: *Consciousness and the Urban Experience, Studies in the History and Theory of Capitalist Urbanization I*. Oxford: Basil Blackwell.

——1985b: *The Urbanization of Capital, Studies in the History and Theory of Capitalist Urbanization II*. Oxford: Basil Blackwell.

——1989: *The Condition of Postmodernity: an enquiry into the origins of cultural change*. Oxford: Basil Blackwell.

Huggett, R. 1990: *Catastrophism: systems of earth history*. London: Edward Arnold.

Johnston, R. J. 1983a: *Philosophy and Human Geography: an introduction to contemporary approaches*. London: Edward Arnold.

——1983b: Resource analysis, resource management and the interaction of physical and human geography. *Progress in Physical Geography*, 7, 127–46.

——1986: Four fixations and the quest for unity in geography. *Transactions of the Institute of British Geographers, New Series*, 11, 449–53.

Kates, R. W. 1987: The human environment: the road not taken, the road still beckoning. *Annals of the Association of American Geographers*, 77, 525–34.

Keat, R. and Urry, J. 1982: *Social Theory as Science*, 2nd edn. London: Routledge & Kegan Paul.

Kennedy, B. 1979: A naughty world. *Transactions of the Institute of British Geographers, New Series*, 4, 550–8.

——1983: On outrageous hypotheses in geography. *Geography*, 68, 326–30.

Kuhn, T. S. 1970: *The Structure of Scientific Revolutions*, 2nd edn. Chicago, IL: Chicago University Press.

Lakatos, I. 1970: Falsification and the methodology of scientific research programmes. In I. Lakatos and A. Musgrave (eds), *Criticism and the Growth of Scientific Knowledge*, Cambridge: Cambridge University Press, 91–195.

Ley, D. 1980: Geography without man: a humanist critique. Research Paper 24, School of Geography, Oxford University.

Livingstone, D. N. 1985: Evolution, science and society: historical reflections on the geographical experiment. *Geoforum*, 16, 119–30.

Lovelock, J. A. 1988: *Gaia: a new look at life on earth*. Oxford: Oxford University Press.

Mackinder, H. J. 1887: On the scope and methods of geography. *Proceedings of the Royal Geographical Society*, 9, 141–60.

Macmillan, B. (ed.) 1989: *Remodelling Geography*. Oxford: Basil Blackwell.

Montgomery, K. 1989: Concepts of equilibrium and evolution in geomorphology: the model of branch systems. *Progress in Physical Geography*, 13, 47–66.

Peet, R. 1985: The social origins of environmental determinism. *Annals of the Association of American Geographers*, 75, 309–33.

Pepper, D. 1987: Physical and human integration: an educational perspective from British higher

education. *Progress in Human Geography*, 11, 379–404.
Popper, K. R. 1959: *The Logic of Scientific Discovery*. London: Hutchinson.
Pred. A. 1986: *Place, Practice and Structure: social and spatial transformation in southern Sweden: 1750–1850*. Cambridge: Polity Press.
Relph, E. 1976: *Place and Placelessness*. London: Pion.
Sack, R. D. 1980: *Conceptions of Space in Social Thought*. London: Macmillan.
——1988: The consumer's world: place as context. *Annals of the Association of American Geographers*, 78, 642–64.
Sauer, C. O. 1941: Foreword to historical geography. *Annals of the Association of American Geographers*, 31, 1–24.
Saunders, P. 1985: Space, the city and urban sociology. In D. Gregory and J. Urry (eds), *Social Relations and Spatial Structures*, London: Macmillan, 67–89.
Sayer, A. 1979: Epistemology and conceptions of people and nature in geography. *Geoforum*, 10, 19–44.
——1985: The difference that space makes. In D. Gregory and J. Urry (eds), *Social Relations and Spatial Structures*, London: Macmillan, 49–66.
——1989: Dualistic thinking and rhetoric in geography. *Area*, 21, 301–5.
Schumm, S. A. 1979: Geomorphic thresholds: the concept and its applications. *Transactions of the Institute of British Geographers, New Series*, 4, 482–515.
——and Lichty, R. W. 1965: Time, space and causality in geomorphology. *American Journal of Science*, 263, 100–19.
Smith, N. 1984: *Uneven Development: nature, capital and the production of space*. Oxford: Basil Blackwell.
——1986: On the necessity of uneven development. *International Journal of Urban and Regional Research*, 10, 87–104.
——1987a: Dangers of the empirical turn: some comments on the CURS initiative. *Antipode*, 19, 59–68.
——1987b: 'Academic war over the field of geography': the elimination of Geography at Harvard, 1947–1951. *Annals of the Association of American Geographers*, 77, 155–72.
Soja, E. W. 1985: Regions in context: spatiality, periodicity, and the historical geography of the regional question. *Environment and Planning D: Society and Space*, 3, 175–90.
——1989: *Postmodern Geographies: the reassertion of space in critical social theory*. London: Verso.
Stewart, I. 1989: *Does God Play Dice? The Mathematics of Chaos*. Oxford: Basil Blackwell.
Stoddart, D. R. 1987: To claim the high ground: geography for the end of the century. *Transactions of the Institute of British Geographers, New Series*, 12, 327–36.
Taylor, P. J. 1986: Locating the question of unity. *Transactions of the Institute of British Geographers, New Series*, 11, 443–8.
Thomas, K. 1984: *Man and the Natural World: changing attitudes in England 1500–1800*. Harmondsworth: Penguin.
Thrift, N. 1977: Time and theory in human geography: I and II. *Progress in Human Geography*, 1, 65–101, 413–57.
——1983: On the determination of social action in space and time. *Environment and Planning D: Society and Space*, 1, 23–57.
——1990: New times and spaces? The perils of transition models. *Environment and Planning D: Society and Space*, 7, 127–9.
Tuan, Y. -F. 1977: *Space and Place: the perspective of experience*. London: Edward Arnold.
Urry, J. 1987: Society, space and locality. *Environment and Planning D: Society and Space*, 5, 435–44.
Whitrow, G. J. 1989: *Time in History: views of time from prehistory to the present day*. Oxford: Oxford University Press.
Wilson, A. G. 1981: *Catastrophe Theory and Bifurcation: applications to urban and regional systems*. London: Croom Helm.
Worsley, P. 1979: Whither geomorphology? *Area*, 11, 97–101.

IV.2

What Human Geographers Write for Students to Read

Ron J. Johnston

Not so long ago it was common to talk about going to University to 'read for a degree', which provided a very good shorthand description of what students were expected to spend a great deal of their time doing. Today that phrase is rarely used. In any case, students are expected to do a wide range of things that were not on the curricula in many places even a few decades ago. Nevertheless, students are still expected to do a great deal of reading. Furthermore, the amount of material available to them is increasing rapidly; they need to be more selective in what they choose to read, and need more help in making their choices.

For any higher education course, the staff involved will almost certainly provide students with a basic reading list, comprising three main components: (a) the major text(s) for the course, which cover most if not all of the material and provide a framework for the lectures, tutorials, seminars, practicals and other course activities; (b) an indication of other important sources on major course topics that can be used as background reading; and (c) lists of specific items relevant to particular topics on the syllabus. Many reading lists indicate much more material than students can be expected to read, let alone digest, and so they must be selective; the most important items might be identified for them. In addition, however, students, especially those on the more specialized courses later in their degree programmes, will be expected to read more widely, and to explore the resources of the libraries available to them.

In approaching the task of reading for a degree, students need the answers to two questions. First, they need to understand why all of that material is on the bookshop and library shelves waiting for them: what is it trying to do? Second, before they start their degree and also prior to setting out on particular courses within it, they will want a general introduction to the literature of the subject: where can they get that? My goal is to try to answer each of these questions, by focusing on the nature of geographical literature.

Why, and What Do Geographers Write?

In order to evaluate something that we read, we need to appreciate why it was written – hence the question at the start of this section. So why do geographers write?

The simplest answer is that geographers write in order to stay in their jobs. Many would do so even if that were not the case, because it is an enjoyable experience in many ways and the result can bring a great deal of satisfaction as well as other, more tangible, rewards. But they have to anyway: why?

People are appointed to the staffs of universities, polytechnics and colleges to teach, to do research and to contribute to the administration of those activities. All will be required to teach, according to an agreed schedule. In universities, they will be contracted to do research; in colleges and polytechnics they will be expected to – and most do, because they will have done research previously, have enjoyed the many stimuli that it provides and will want to go on exploring the unknown. But what is the link between doing research and writing? Research involves exploring the unknown, in a variety of ways, and its results are new knowledge. (Some people distinguish between research, the discovery of new knowledge, and scholarship, the synthesis of existing knowledge in new ways, but the two are frequently used as synonyms: some academics prefer scholarship to research and vice versa, but overall the two activities are equally valued, for they are symbiotic.) Having discovered new knowledge, or produced a new synthesis, it is necessary to share it: to some, knowledge only exists when it is in a published form freely available to others (hence the oft-quoted phrase, 'unpublished research isn't research').

At this stage, as a brief aside, we should perhaps note that the pressures on academics to do research and to publish the results are increasing at the present time. There is a growing desire, especially among those who finance and manage the British university system, to ensure 'value for money' from the investment. This benefit is frequently measured in terms of research productivity – the volume of output relative to the financial input. For most disciplines, including geography, publication is the major output medium, so the evaluators assess both the volume and, as far as they can, the quality of publication. Increasingly, the financing of geography departments is based on such evaluations of their research output, and so the pressure is on to do research and to get it published.

Given the requirements and the pressures to write, what is the publication process? To discuss this, I am going to look separately at journal articles and at books.

The journals

Most British students will have come across some geography journals while still at school, notably *The Geographical Magazine* and *Geography Review*. These have slightly different purposes: the former aims to 'popularize' geography to a wide audience; the latter's major goal is to present up-to-date material to students, especially sixth-formers. These journals are essentially 'translation documents', therefore; they are not the vehicles for presenting research findings to other academic geographers but rather the means of taking new knowledge to wider readerships.

The majority of the journals in which geographers present their work are published for the relatively specialist audience of other professional geographers. Some are more specialized than others, however, which allows identification of three major types. The first comprises the *general journals*, which contain material covering the entire discipline, so that any one issue is likely to include a wide range of material, in both human and physical geography. Most of these journals are published by learned societies, such as the Institute of British Geographers and the Royal Geographical Society; the former publishes its *Transactions* quarterly and the latter produces the *Geographical Journal* three times a year. The comparable journals published in the United States are the *Annals of the Association of American Geographers* and the *Geographical Review* (the latter published by the American Geographical Society), and most other major national geographical societies produce similar journals – such as *Australian Geographical Studies*, the *Canadian Geographer, Irish Geography* and the *Scottish Geographical Magazine*. In all of these geographers publish both reports of completed research (new knowledge in the form of findings from original investigations) and pieces of scholarship (new syntheses in the form of reviews of major bodies of work); there are also book reviews and critiques of recently published articles, where scholars pass opinions on the work of others. The major learned societies also publish second journals to carry shorter articles, commentaries and reports – the Institute of British Geographers publishes *Area* quarterly, for example, and its American counterpart publishes the *Professional Geographer*.

The second type of journal caters for *specialist subdisciplinary* audiences, with all of the papers about the same field of study. Most of them are produced by commercial publishers and are relatively new, reflecting the great growth of geography and its division into many sub-fields in recent decades. The oldest is *Economic Geography*, launched at Clark University in 1926 and published quarterly since then; it was joined in the 1970s by the *Journal of Historical Geography* and then in the 1980s by *Political Geography Quarterly, Urban Geography* and the *Journal of Rural Studies* (which serves a wider constituency than geography alone but has a major geography component, and a geographer as the founder editor); *Geographical Analysis* was founded in 1969 to focus on theoretical, largely technical, material in the same year that *Antipode* was launched to provide a forum for more 'radical' material (hence the title, with its implication of 'turning things on their head'). In addition, there are a substantial number of other journals which are interdisciplinary in their coverage but which contain much geography and in several cases have geographers as editors: they include *Environment and Planning* (which has four parts: A is monthly and has the subtitle 'International journal of urban and regional research'; B, C and D are each quarterly and are subtitled, respectively, 'Planning and design', 'Government and policy' and 'Society and space'), *Regional Studies, Urban Studies* and the *International Journal of Urban and Regional Research*.

The third type is less coherent, but the journals share the basic goal of transmitting the results of geographical research to wider audiences of practising geographers. For example, *Geography* is published by the Geographical Association, which serves teachers of the discipline at all levels of the British educational system; it contains new material that can be used by teachers, plus discussions of pedagogical issues: the *Journal of Geography* performs the same task in the United States. There is also an important international journal concerned solely with pedagogical issues in universities, polytechnics and colleges – the *Journal of Geography in Higher Education*.

Of particular value to students in this third category, especially those wishing to find out about recent work in a field of the discipline, is the quarterly *Progress in Human Geography*. This is basically a journal of scholarship rather than research, concentrating on major review articles summarizing and debating new developments in the discipline. It also contains a series of annual 'Progress reports' reviewing the latest material in a range of fields, and a major book review section. For students, as well as for academic geographers, it provides a valuable way of keeping up to date with the burgeoning literature of a rapidly changing discipline.

In order to use the material in the journals, especially the journals of the first two types, students need to appreciate who the articles are written for and how they achieve publication. Most journal editors employ the same procedures. Articles are submitted to them and are immediately sent to two or three referees who are experts in the field. They are asked for their opinion on both the work contained in the article and its presentation and to recommend whether it should be accepted for publication, returned to the author(s) for revision or rejected as unsuitable. The editor then weighs the opinions and recommendations and decides whether or not to publish the article, and in what form. A substantial percentage of all submissions are rejected, for a variety of reasons, and few are accepted without some revision.

The journal literature is large; many journals are published and more are appearing each year. Fortunately, you can track down references fairly readily by using the innovative series of journals launched by geographer Keith Clayton in the 1960s – *Geographical Abstracts*. These not only list titles and authors, and cross-classify them, but also provide brief abstracts of the contents. They are invaluable sources when you are doing an initial search for material on a specific topic.

Journal articles, then, are written by professional geographers for their fellow professionals, and are evaluated by their peers to see whether they are suitable for publication. In most cases, the intended readership comprises people who know the field of study well and so appreciate the background within which the particular piece of research is set. Further, in many cases the originality of the piece, and thus the reason why it has been accepted for publication, is in the research methodology used, and so there may be more on the way in which the results were obtained than on their more general relevance. Thus students may find such articles too detailed for their use, unless they know the context in which they were produced – the history of the research field to which the author is making a new contribution. To provide such a guide to the context we have the other type of publication, to which this chapter now turns.

Books

Whereas journal articles are written for the dual purpose of advancing knowledge and advancing the author's career, with the two firmly intertwined, many books are produced for other reasons. To discover what they are, we need first to look at the types of books. I have identified six.

The first two are variants of the textbook, divided into the *instructional text* and the *synthesis text*. Books in the former category provide instruction, usually in techniques, and those on the use of statistical methods are the commonest produced for geographers. Many can be used as continuing works of reference, to be returned to whenever the owner is faced with a technical problem. The synthesis text, on the other hand, is a work of scholarship, which summarizes the knowledge in a field, providing both an overview of the detailed research literature and a guide to its use. Somewhat similar is the third type, the *basic reference work*, which for geographers should comprise an atlas (or more than one) and a dictionary: these are the working tools to which frequent recourse will be made for basic information.

The *research monograph* comprises the report of a major piece of work which cannot be handled within the confines of a short journal paper but needs the extended treatment that a volume of 100, 000 words or more allows. It is used in all disciplines, but much more so in some than in others because of the nature of the field. (In history, for example, the writing of books is much more common than the writing of journal articles, and major learned society journals, such as the *American History Review*, carry many more pages of book reviews than they do of articles.) Research monographs are less common in human geography, partly because of the nature of the discipline – not surprisingly, they are more common in historical geography than in most other sub-fields – and partly because geography lacks the wide, lay audience for its research output that characterizes much of history.

One other type of book which is less common in geography than in some other disciplines is the *essay* (or polemic) which is written to stimulate rather than to inform. Occasionally a scholar will want to goad her or his professional colleagues to consider some form of radical reorientation of their discipline, or part of it, and will find that this takes many more words than the editor of a learned society journal can allow – hence the production of a, usually relatively short, book.

Finally there is the *edited collection*, in which an editor brings together a set of chapters written by a range of authors. Within this type, we can identify three sub-types. The first involves the republication of 'classics', perhaps all by the one author, perhaps by several, which it is convenient to have together because of their importance to the development of a field, either for use as reference works by future researchers or for students' convenience; such collections are usually accompanied by editorial essays which set the re-published papers in context, indicating how and why they are important. Second, there is the set of new papers on a particular theme: these may be the papers presented at a conference or they may be specially commissioned by the editor(s) for the book; publication in this way rather than in journal articles is justified by the desirability of having them together in the one source rather than spread through the journal literature. (In some disciplines, notably in engineering, volumes of conference papers are the equivalent of journals, and the papers are refereed before acceptance for the proceedings; that is rare in human geography.) In a few cases the collection is put together to mark a special occasion, such as the retirement of a scholar whose students and/or colleagues publish a volume of essays to celebrate the scholar's career and indicate the importance of his or her work and leadership. The third type comprises collections of review material, somewhat like the papers in *Progress in Human Geography*, but built around a theme and intended to summarize all aspects of a part of the discipline.

Why are these various types of books produced and published? As with journal papers, it is in part because publication brings status and satisfaction to the authors, and potentially helps their career advancement. But that is not the whole story, because just as authors have to find editors who will accept their articles, so they have to find publishers who will accept their books – either their completed manuscripts or, more commonly, their proposals to write books. So what are the publishers' grounds for acceptance or rejection?

Nearly all publication of books by geographers is done by commercial companies; the learned societies do have limited publication lists and some university presses subsidize a few books, but by and large the decision whether or not to publish a book is taken by commercial outfits, applying the usual criteria of profitability. A book will only be published if it is believed that it will sell in substantial numbers to produce a profit for those who invest in it; there has to be a proven or potential market for it.

What is the market for geography books? Basically it is those who teach geography and those who study it: unlike some disciplines (such as history and literature) there is no wide general market, although some geographers believe that we should be able to develop one; and unlike some other disciplines (such as psychology) there is no large market of professionals practising as geographers after graduation. And there are not that many academic geographers, either, particularly when you consider that few books are written for all geographers, or even all human geographers, but just for certain specialists (economic geographers, say, or transport geographers). So, in assessing the market for a proposed book, most publishers are looking at the potential sales among students, in addition to the libraries that they use – hence the focus on the textbooks and reference works, plus some of the collected works. (This is not to say that publishers focus only on textbooks, of course, although for most these are the bread-and-butter of their existence; many do bring out books for small relatively specialized markets, but for these they must necessarily charge high prices because of the small print run.)

So how are geography texts produced? It involves publisher and author working together. In some cases, an author will approach a publisher with an idea for a book; in others the publisher will identify a 'market gap' where a book is desirable and will look for an author who can produce what is needed. In either case, the publisher will want to be convinced that the material is of good quality, which in many cases involves the use of referees, with regard to both the outline proposal and then the final manuscript. Production of a book involves substantial investment, and the publisher does not want the book to be badly received by the reviewers. And as large a market as possible will be sought, which for publishers of books in the English language means in particular the North American market. The large undergraduate market there means that a successful introductory text can be very profitable for the publisher (and for the author, who receives a royalty on every copy sold). American publishers invest a great deal in exactly tailoring books to the requirements of that market, whereas British publishers are more attached to the belief that a good book will probably create a small but viable market, given the greater range of introductory courses taught to British undergraduates and the lesser importance of the all-purpose textbook in most of them than in their American counterparts. (British introductory courses generally have many fewer students, because of the nature of the degree system, and the lectures are backed up by small-group tutorials which means that the students are less likely to have to rely on lectures and textbook alone than their American undergraduate counterparts.)

The commercial link between publishers and academics is a crucial one, but it should not be assumed that academia is ruled by mammon. Textbooks have a vital role to play in any discipline, irrespective of the mode of publication, because they are the foundation stones of its progress. Journal articles take the discipline forward, but as the movement proceeds so there is a continual need to 'take stock' and consolidate the present state of knowledge. This is what the textbooks do: they organize the sometimes anarchic situation of developments on the research frontiers by synthesizing what it is we know, and how it is we find things out. Look, for example, at the two editions of one of the classic texts produced in human geography in recent decades, Peter Haggett's *Locational Analysis in Human Geography* (the first edition was published in 1965; the second, jointly authored with Andrew Cliff and Allen Frey, appeared in 1977). It is organized in two parts (two volumes in the paperback version of the second edition): the first on models of locational structure – organizing what we know in the context of six elements of spatial structures – and the second on methods of locational analysis. Such books structure the discipline, ordering its past and present contributions to knowledge and also organizing its future, by suggesting what it is we need to find out next, and how we might go about it. Thus textbooks, both general ones and the more specialist volumes on particular aspects of geography, are crucial to students because they structure the discipline for them and provide a ready guide to its diverse journal literature; without them, lecture courses would be much more important, because they would provide the only available structure, and students would have to spend more effort recording what was said and less listening and being stimulated by new ideas.

Many textbooks are the outcome of lecture courses, most of which – even if they are concerned with the same general topic – have a range of unique features reflecting the emphasis and interests of the lecturer(s) concerned. This is in itself no bad thing, for there is no single 'right' way to approach a broad discipline such as geography and a range of competing books, each with its own particular slant on the discipline or subdiscipline, is far superior to any attempt to promote an orthodoxy through a single text. So don't think that any textbook you pick up is going to tell you the whole story, or that it is necessarily the right story: it is just one person's way of introducing you to a large literature. Is it a worthwhile story to read? If it is on a course reading list, somebody clearly believes that it is (although the recommender may be the author, of course!). One way of finding out is to read the review pages of the journals and see what other geographers thought about it.

So What Should You Read?

I have no intention of finishing this chapter by offering you a recommended list: it's more than my life's worth, and I might be accused of favouritism, or even self-advertisement (and using this book to promote my own and my friends' royalties)! What I can do is suggest how you approach the task of 'reading yourself in' to the discipline and its many components.

I should stress at the outset that human geography, like all other social sciences, has experienced a great deal of debate in recent decades about its philosophy and the attendant methodology. There is no orthodoxy, and in reading what geographers write you should realize that they do not all agree on the best way forward – as David Livingstone has made clear in his introductory chapter to part II of this book. So you need an overview of the major debates that have raged, and what they mean for definitions of geography. Probably the best introductory guide is provided in *The Dictionary of Human Geography* (edited by R. J. Johnston, D. Gregory and D. M. Smith and published by Basil Blackwell, Oxford and New York, 2nd edn. 1986). Its nearly 600 pages provide not only the definitions of terms you would expect to find in a dictionary (which is also done in Brian Goodall, *The Penguin Dictionary of Human Geography*, Penguin, London, 1987) but also a series of essays on human geography and its component fields: it is an encyclopaedia as well as a dictionary.

The material covered relatively briefly in that dictionary is treated in greater detail in a number of books that deal with aspects of the discipline's history. Arild Holt-Jensen has done this in *Geography: history and concepts* (Paul Chapman Publishers, London, 2nd edn 1988), for example, and T. W. Freeman produced *A History of Modern British Geography* in 1980 (published by Longman, London); you might also refer to two books by R. J. Johnston – *Geography and Geographers: Anglo-American human geography since 1945* (Edward Arnold, London, 4th edn 1991) and *Philosophy and Human Geography: an introduction to contemporary approaches* (Edward Arnold, London,

2nd edn 1986). Geography is practised somewhat differently in different countries, however, reflecting local cultures and histories, and the variations are discussed by the authors of a collection of essays *Geography since the Second World War: an international survey*, edited by R. J. Johnston and P. Claval (Croom Helm, London, 1984); a strongly North American view is provided by P. E. James and G. J. Martin's *All Possible Worlds* (Wiley, New York, 1981). The Association of American Geographers published a major review of the state of the discipline in the mid-1950s – *American Geography: inventory and prospect*, edited by P. E. James and C. F. Jones (Syracuse University Press, Syracuse, NY, 1954) – and a similar volume, not commissioned by the Association, was edited by G. L. Gaile and C. J. Willmott in 1989 (*Geography in America*: Merrill, Columbus, OH).

Whereas the books referred to so far provide overviews of the discipline's history, they lack detailed coverage of its substantive content and methods. Very few books are comprehensive in this regard, because of the lack of any orthodoxy within the discipline. Until about the 1950s there was some attempt at an orthodoxy, which was feasible because of the small number of practising geographers: in the United States, for example, Richard Hartshorne's *The Nature of Geography* (published by the Association of American Geographers in 1939) and its sequel *Perspective on the Nature of Geography* (Rand McNally, Chicago, IL, 1959) were very influential (but see some of the reminiscences published in the March 1979 issue of the *Annals of the Association of American Geographers* to celebrate its seventy-fifth anniversary and which discuss some of the alternatives to Hartshornian orthodoxy, and the special supplement published by that journal in 1989 to mark the fiftieth anniversary of Hartshorne's classic work); in Britain, *The Spirit and Purpose of Geography* by S. W. Wooldridge and W. G. East (Hutchinson, London, 1958) played a similar, though probably less influential, role.

In the mid-1960s a series of major books heralded a new attempt to devise a geographical orthodoxy. Two volumes edited by R. J. Chorley and P. Haggett led the way – *Frontiers in Geographical Teaching* (Methuen, London, 1965) and *Models in Geography* (Methuen, London, 1967). The latter, especially, was extremely influential: it is now somewhat dated, but still well worth consulting for the view it gives of what some saw as the beginnings of a major revolution in geography. (Others doubted that: see, for example, Michael Chisholm's *Human Geography: evolution or revolution?*, Penguin, London, 1975.) Peter Haggett's already-mentioned textbook carried the impetus forward, followed by similar American texts, notably those by R. F. Abler, J. S. Adams and P. R. Gould, *Spatial Organization: the geographer's view of the world* (Prentice Hall, Englewood Cliffs, NJ, 1971) and R. L. Morrill *The Spatial Organization of Society* (Wadsworth, Belmont, CA, 1970; it is useful to look at the successive editions of this book and see how it changes to reflect new emphases within the discipline). David Harvey produced a classic text on the philosophy of the new approach – *Explanation in Geography* (Edward Arnold, London, 1969) – and others produced important texts for particular parts of human geography, e.g. P. E. Lloyd and P. Dicken, *Location in Space* (Harper & Row, New York, 1974 and later editions).

The paradigm view of geography presented in those pioneering books did not go unchallenged, and during the subsequent decades alternative approaches were advanced. In 1973, for example, David Harvey published his *Social Justice and the City* (Edward Arnold, London), which heralded his acceptance of a Marxian approach to social science in general and geography in particular. It was followed in 1982 by another important book, *The Limits to Capital* (Basil Blackwell, Oxford), and then in 1985 by *The Urbanization of Capital* and *Consciousness and the Urban Experience*; David Harvey continued his remarkable series of seminal volumes with *The Condition of Postmodernity* (also published by Basil Blackwell, in 1989). Derek Gregory published a stimulating and widely read essay in 1978 – *Ideology, Science and Human Geography* (Hutchinson, London) – that opened up new lines of enquiry, which were illustrated in a major collection of essays that he edited with John Urry in 1985 – *Social Relations and Spatial Structures* (Macmillan, London). No book has been produced that embraces the wide diversity of views now current among geographers, in part because the different philosophies they represent are incommensurable; the closest is probably Peter Haggett's *Geography: a modern synthesis* (Harper & Row, New York, 1972 and later editions).

In recent years, a number of geographers have published personal – in some cases autobiographical – volumes promoting their views of the discipline. Thus you may wish to refer to Peter Gould's *The Geographer at Work* (Routledge, London, 1985), David Stoddart's *On Geography: and its history* (Basil Blackwell, Oxford, 1986), R. J. Johnston's *On Human Geography* (Basil Blackwell, Oxford, 1986), James Bird's *The Changing Worlds of Geography* (Oxford University Press, Oxford, 1989) and Peter Haggett's *The Geographer's Art* (Basil Blackwell, Oxford, 1990). Anne Buttimer's *The Practice of Geography* (Longman, London, 1983) contains a number of interesting autobiographical sketches, as does *Recollections of a Revolution* (Macmillan, London, 1984) by Mark Billinge, Derek Gregory and Ron Martin.

It is impossible in this brief review to give anything but a very incomplete picture of the wealth of material produced by geographers in the last twenty-five years. (A fuller listing is provided in the annotated bibliography edited by Chauncy Harris for the Association of American Geographers and published by the Association in 1985 – *A Geographical Bibliography for American Libraries.*) Some recent books do provide contemporary assessments of the state of the discipline. Some are collections of review essays covering most of the subdisciplines of human geography; edited by Michael Pacione, they have been published by Croom Helm (now Routledge) of London, beginning with *Progress in Urban Geography*, published in 1983. Probably of most value in providing a comprehensive view of the discipline, however, are the book edited by Gaile and Willmott referred to above and two collections produced to 'update' the classic volumes edited by Chorley and Haggett: *Horizons in Human Geography*, edited by Derek Gregory and Rex Walford (Macmillan, London, 1989) and the two-volume *New Models in Geography*, edited by Richard Peet and Nigel Thrift (Unwin Hyman, London, 1989). The debates over the nature of geography and how it should be practised are taken forward in a number of other edited contributions, including Doreen Massey and John Allen's *Geography Matters!* (Cambridge University Press, Cambridge, 1984); R. J. Johnston's *The Future of Geography* (Methuen, London, 1985); Bill Macmillan's *Remodelling Geography* (Basil Blackwell, Oxford, 1989); and Jennifer Wolch and Michael Dear's *The Power of Geography* (Unwin Hyman, Boston, MA, 1989).

Few books are written to induct students (usually graduate students) into academic life; one recent attempt is Martin Kenzer's *On Becoming a Professional Geographer* (Merrill, Columbus, OH, 1989).

In Summary

The literature of geography is large, and any introduction to it such as this can do no more than hint at its diversity and richness. As a student of geography, of course, you will almost certainly already have some perspective on the discipline and views of the relevance of its various parts to your own interests and orientations. But don't be blinkered in what you choose to read (and don't allow your tutors to blinker you either!); there is a wealth of insight there available to you on the shelves of the library and bookshop, and judicious selection, aided by up-to-date guides to the literature, will provide you with a whole range of new and profitable stimuli, whether you are looking for the overviews that textbooks provide or the developments on the research frontiers published in the journals. Go ahead and read for your degree.

IV.3
The Literature of Physical Geography

Andrew S. Goudie

The Journals

As with human geography, the range of journals with which one needs to be familiar as a physical geographer is huge, but it is in the journals rather than in the textbooks that the exciting new ideas and discoveries are first published. It is important to keep abreast of new discoveries as they occur and for this it is essential to browse through some of the weekly (e.g. *Nature*, *Science* and *New Scientist*) and monthly (e.g. *Scientific American*) general science journals. Likewise, *Progress in Physical Geography* carries not only regular up-to-date review articles, but also a series of annual progress reports on developments in a wide range of aspects of physical geography. One can also obtain a flavour of developments in the discipline by reading book reviews, for in some cases these may be quite extended critical analyses of new works. Among the journals that carry a good selection of lengthy book reviews are the *Geographical Journal*, *Professional Geographer*, the *Transactions of the Institute of British Geographers*, the *Annals of the Association of American Geographers*, the *Geographical Review* and *Progress in Physical Geography*.

There are relatively few journals which are dedicated to physical geography as a whole. Notable exceptions are *Physical Geography* and *Geografiska Annaler, Series A* (much of which is in English). Most journals seek to cover just one aspect of the subject. It is vital, however, to read beyond physical geography *sensu stricto* for journals in related fields carry articles of interest to, or written by, geographers. So, for example, geomorphologists will find much of value in the sedimentological journals (especially the *Journal of Sedimentary Petrology*, *Sedimentology* and *Sedimentary Geology*), besides that which is written in the specialist geomorphologist publications. Of these, the most notable are *Earth Surface Processes and Landforms*, *Geomorphology*, *Catena* and the *Zeitschrift für Geomorphologie* and its supplement volumes. Note that the last of these, in spite of its German title, carries many excellent articles in English.

Biogeographers also have to keep abreast of a huge and diverse literature. Thus in addition to reading the *Journal of Biogeography* it is necessary to read some of the specialist ecological journals, including *Ecology*, *Journal of Ecology* and *Ecological Monographs*. There is also a burgeoning conservation literature (e.g. *Biological Conservation*, *Ambio* and *Environmental Conservation*). Soil science impinges on both biogeography and geomorphology and a useful journal that bridges the three facets of the discipline is *Geoderma*.

The hydrological literature also appears in a wide range of journals, and notable amongst them are *Hydrological Processes*, *Journal of Hydrology*, *Water Resources Research* and the various publications of the International Association of Scientific Hydrology.

The climatological literature varies in

its comprehensibility and complexity, and some of the journals are not easy going for those with a weak mathematics or physics background. Among the more accessible journals are *Weather*, the *Journal of Climatology*, the *Monthly Weather Review* and the *Journal of Applied Meteorology*. However, an area in which there has been remarkable growth in recent years has been the publication of journals relating to past climatic and other environmental changes and these contain material of interest and value to all physical geographers (e.g. *Climatic Change*, *Quaternary Research*, *Journal of Quaternary Science*, *Quaternary Science Reviews* and *Global and Planetary Change*).

Finally, it is worth mentioning that there are various journals that deal with specific types of environment. Deserts are included in the *Journal of Arid Environments*, high latitude and high altitude areas in *Arctic and Alpine Research*, glaciated terrains in the *Journal of Glaciology* and *Annals of Glaciology*, and coastal areas in *Marine Geology*, *Coral Reefs* and the *Journal of Coastal Research*.

To keep up just with the journals listed above is a herculean task, without considering the thousands of other serials that we could have mentioned. However, *Geographical Abstracts: Physical Geography* and *Ecological Abstracts* make the task easier by providing abstracts of thousands of articles, together with an author and keyword index. Thus, if you need to write an essay or dissertation on, say, drumlins, look the word up in the keyword index for the last few years to get an immediate indication of what has been written on the subject. The keyword indexes also contain place-names, so that if you are planning fieldwork in, say, Iceland you can see what work has already been done and by whom.

Books

Physical geography is such a vast and ramifying field that it is almost impossible to cover it at more than a superficial level in any general text. This has not deterred authors from attempting to do so, and among the more recent attempts are those of A.N. and A.H. Strahler (*Modern Physical Geography*, Wiley, New York, 1983), A.S. Goudie (*The Nature of the Environment*, Basil Blackwell, Oxford, 1989) and D.J. Briggs and P. Smithson (*Fundamentals of Physical Geography*, Hutchinson, London, 1985). The terminology of physical geography is also daunting, and access is provided by J. B. Whittow, *The Penguin Dictionary of Physical Geography* (Penguin, London, 1984) and by *The Encyclopaedic Dictionary of Physical Geography* (edited by A. S. Goudie and others and published by Basil Blackwell, Oxford, 1985), which also contains bibliographic guidance with its definitions.

To obtain a general view of the history of the entire field of physical geography in recent decades one should read K. J. Gregory, *The Nature of Physical Geography* (Edward Arnold, London, 1985), while for a concise review of the history of geomorphology it is necessary to consult K. Tinkler, *A Short History of Geomorphology* (Croom Helm, London, 1985). However, if you have time and a true scholarly bent you should read the magisterial three-volume *History of the Study of Landforms*, written by R. J. Chorley, R. P. Beckinsale and A. J. Dunn and published by Methuen.

To learn how to *do* physical geography you should consult a valuable book on theory and methodology by J. Petch and R. H. Haines-Young, *Physical Geography: its Nature and Methods* (Harper & Row, London, 1986). Guidance on books dealing with specific laboratory and field techniques appears elsewhere

Table IV.3.1 Selected recent synthetic texts in physical geography

Subject	Author(s)	Title	Publisher	Date
Geomorphology				
General	Chorley, R. J., Schumm, S. A. and Sugden, D. E.	*Geomorphology*	Methuen	1985
	Selby, M. J.	*Earth's Changing Surface*	Oxford University Press	1985
Deserts	Thomas, D. S. G. (ed.)	*Arid Zone Geomorphology*	Belhaven Press	1989
Karst	Jennings, J. N.	*Karst Geomorphology*	Basil Blackwell	1985
	Ford, D. C. and Williams, P. W.	*Karst Geomorphology and Hydrology*	Unwin Hyman	1989
Periglacial	Clark, M. J. (ed.)	*Recent Advances in Periglacial Geomorphology*	Wiley	1988
Glacial	Drewry, D.	*Glacial Geologic Processes*	Edward Arnold	1986
Coastal	Carter, R. G. W.	*Coastal Environments*	Academic Press	1988
	Pethick, J.	*An Introduction to Coastal Geomorphology*	Edward Arnold	1984
Slopes	Selby, M. J.	*Hillslope Materials and Processes*	Oxford University Press	1982
Weathering	Yatsu, E.	*The Nature of Weathering*	Sozosha	1988
Tectonics	Vita-Finzi, C.	*Recent Earth Movements*	Academic Press	1986
Fluvial	Knighton, D.	*Fluvial Form and Process*	Edward Arnold	1984
Humid tropics	Douglas, I. and Spencer. T.	*Environmental Change and Tropical Geomorphology*	Allen & Unwin	1985
Climatic	Budel, J.	*Climatic Geomorphology*	Princeton University Press	1982
Applied	Cooke, R. U. and Doornkamp, J. C.	*Geomorphology in EnvironmentalManagement*	Oxford University Press (2nd edn)	1990
Climatology and meteorology				
General	Henderson-Sellers, A. and Robinson, P. J.	*A Contemporary Climatology*	Longman	1986
Meso-scale	Atkinson, B. W.	*Meso-scale Atmospheric Circulations*	Academic Press	1981
Local scale	Oke, T. R.	*Boundary Layer Climates*	Methuen	1978
Mountain	Barry, R. G.	*Mountain Weather and Climate*	Methuen	1981
Tropical	Riehl, H.	*Climate and Weather in the Tropics*	Academic Press	1979
Hazards	Perry, A.	*Environmental Hazards in the British Isles*	Allen & Unwin	1981
Palaeo	Goudie, A.	*Environmental Change* (3rd edn)	Oxford University Press	1992
Applied	Hobbs, J. E.	*Applied Climatology*	Dawson	1980
Hydrology				
General	Shaw, E. M.	*Hydrology in Practice*	Van Nostrand Reinhold	1983
Groundwater	Freeze, R. and Cherry, J.	*Groundwater*	Prentice Hall	1979
Management	Dunne, T. and Leopold, L. B.	*Water in Environmental Planning*	Freeman	1978

Table IV.3.1 Continued

Subject	Author(s)	Title	Publisher	Date
Floods	Ward, R.	*Floods: A Geographical Perspective*	Macmillan	1978
Lakes	Lerman, A. (ed.)	*Lakes, Chemistry, Geology, Physics*	Springer	1978
Hillslope hydrology	Kirkby, M. J. (ed.)	*Hillslope Hydrology*	Wiley	1978
Applied	Shaw, E. M.	*Hydrology in Practice*	Van Nostrand	1983
Biogeography				
General	Simmons, I. G.	*Biogeography: Natural and Cultural*	Arnold	1979
Ecology	Begon, M., Harper, J. and Townsend, C.	*Ecology*	Blackwell Scientific	1988
Zoogeography	Illies, J.	*Introduction to Zoogeography*	Macmillan	1974
Soils	Fenwick, I. and Knapp, B.	*Soils; Process and Response*	Duckworth	1982
Rainforest	Mabberley, D.	*Tropical Rain Forest Ecology*	Blackie	1983
Savanna	Cole, M. M.	*The Savannas*	Academic Press	1986
Mediterranean	Castri, F. di and Mooney, H. A. (eds)	*Mediterranean Type Ecosystems Origin and Structure*	Chapman and Hall	1973
Grasslands	Coupland, R. T. (ed.)	*Grassland Ecosystems of the World*	Cambridge University Press	1979
Deserts	Goodall, D. W., Perry, R. A. and Howes, K. M. W. (eds)	*Arid Land Ecosystems*	Cambridge University Press	1979
Tundra	Bliss, L. C., Heal, D. W. and Moore, J. J. (eds)	*Tundra Ecosystems in a Comparative Perspective*	Cambridge University Press	1981
Palaeoenvironments				
Pleistocene	Lowe, J. J. and Walker, M. J. C.	*Reconstructing Quaternary Environments*	Longman	1984
Holocene	Roberts, N.	*The Holocene*	Basil Blackwell	1989
Human impact				
General	Goudie, A. S.	*The Human Impact* (3rd edn)	Basil Blackwell	1990
Historical	Simmons, I.	*Changing the Face of the Earth*	Basil Blackwell	1989
Urban	Douglas, I.	*The Urban Environment*	Edward Arnold	1983

in this volume (see chapters III.8 and III.9) and will not be repeated here.

When it comes to a consideration of the availability of textbooks in physical geography we can do no more than list a selection of some of the more general and recent synthetic texts which will help you to get into a particular field. These are given in table IV.3.1. Physical geography, as this list indicates, is well provided with texts, many of which contain useful case studies.

IV.4
Biographical Dictionary
Brian W. Blouet

The purpose of this chapter is to provide short biographical sketches of some of the geographers a student is likely to encounter when reading the current geographical literature published in English. Most of the scholars listed are associated with universities in Britain or North America.

Although over sixty individuals are listed any such selection is bound to reflect the author's preferences and judgement. Some important geographers have undoubtedly been omitted. Nevertheless, the selection does at least take account of one form of 'objective' evaluation, namely the citation analyses given in tables IV.4.1 and IV.4.2. Although there are various means of calculation, such analyses are simply counts of the number of times an author is cited by others in a range of publications (including journals). It is a quantitative method, and therefore includes no qualitative assessment. Table IV.4.3 adds an indication of the most frequently cited books until the end of 1984.

Academic geography in Britain and the United States can be traced to the last quarter of the nineteenth century. In Britain the modern subject is usually seen as starting with the work of Halford Mackinder at Oxford, beginning in 1887, and leading to the founding of the School of Geography in 1899. Separate departments of geography began to appear in British universities around the end of the First World War and in the 1920s. Among the early departments offering honours degrees in geography were Aberystwyth, Cambridge, Liverpool, the London School of Economics, Manchester and Sheffield.

In the United States a leading early figure was W. M. Davis (1850–1934) who established geography at Harvard and helped to create the Association of American Geographers in 1904. Geography did not take root at Harvard or other Ivy League universities. The subject became firmly established in state universities and land grant colleges (institutions of higher education established by the Morrill Act 1862; the Act gave federal lands to help finance new universities), particularly in the mid-west and the west. Frequently, the type of geography taught was physical with practical applications: soils, conservation, biogeography and climatology were well represented in the late nineteenth century. Cultural geography is associated with the approach developed by Carl O. Sauer, at the University of California at Berkeley, from 1923.

An excellent guide to the history of geography is Preston James and Geoffrey Martin, *All Possible Worlds* (Wiley, New York and Chichester, 1981). Still very useful are T. W. Freeman, *A Hundred Years of Geography* (Duckworth, London, 1971), and Robert E. Dickinson, *The Makers of Modern Geography* (Praeger, New York, 1969). For detailed descriptions of the emergence of academic geography in Britain see *Institute of British Geographers Transactions, New Series*, 8, 1 (1983), a special issue on 'The Institute of British Geographers 1933–1983' compiled by D. R. Stoddart, and Robert W. Steel, *The Institute of*

Table IV.4.1 Citation analyses of human geographers: human geographers receiving 100 or more citations in the Social Science Citation Index, 1971–1975

Rank	Author		Citations
1	Berry, B. J. L.	(1934)	890
2	Wilson, A. G.	(1939)	366
3	Harvey, D. W.	(1935)	276
4	Haggett, P.	(1933)	255
5	Morrill, R. L.	(1934)	242
6	Hägerstrand, T.	(1916)	235
7	Brown, L. A.	(1935)	224
8	Gould, P. R.	(1932)	220
9	Butzer, K. W.	(1934)	217
10	Sauer, C. O.	(1889–1975)	182
11	Pred, A. R.	(1936)	176
12	King, L. J.	(1934)	173
13	Lowenthal, D.	(1923)	169
14	Olsson, G. P.	(1935)	166
15	Brookfield, H. C.	(1926)	151
16=	Chisholm, M. D. I.	(1931)	149
16=	Johnston, R. J.	(1941)	149
18	Christaller, W.	(1893–1969)	146
19	Hall, P. G.	(1932)	139
20	Dacey, M. F.	(1932)	134
21	Bobek, H.	(1903)	131
22	Cox, K. R.	(1939)	129
23	Mabogunje, A. L.	(1931)	123
24	Curry, L.	(1922)	116
25	Wolpert, J.	(1932)	115
26	White, G. F.	(1911)	114
27	Ullman, E. L.	(1912–76)	113
28	Bunge, W. W.	(1928)	111
29	Tobler, W. R.	(1930)	104
30=	Harris, C. D.	(1914)	101
30=	Simmons, J. W.	(1936)	101
30=	Taaffe, E. J.	(1921)	101

Years of birth and death are shown in parentheses.
Source: Based on J. W. R. Whitehand, *Transactions of the Institute of British Geographers, New Series*, 10, 1985, table 1

British Geographers: the first fifty years (Institute of British Geographers, London, 1984). Both Institute of British Geographers publications trace events well before 1933. For the United States see Brian W. Blouet (ed.), *The Origins of Academic Geography in the United States* (Shoe String Press, Hamden, CT, 1981).

On individual geographers the leading reference source is the biobibliographic series established by the late professor Walter Freeman (T. W. Freeman and Philippe Pinchemel (eds), *Geographers: biobibliographical studies*, vols 1–12, Mansell, London, 1977–). The editor of this series is now Geoffrey Martin, Southern Connecticut State University. The biobibliographic studies concentrate on geographers of the past. The entries below mainly focus on scholars who are publishing at present.

Ronald F. Abler (b. 1939) was educated at the University of Minnesota (PhD 1968) and has taught at Penn State

Table IV.4.2 Citation analyses of human geographers: human geographers receiving 100 or more citations in the Social Science Citation Index and Science Citation Index, 1981–1985

Rank	Author	Citations	Rank	Author	Citations
1	Berry, B. J. L.	808	28	Christaller, W.	177
2	Johnston, R. J.	679	29	Keeble, D. E.	173
3	Harvey, D. W.	666	30	Brookfield, H. C.	172
4	Wilson, A. G.	518	30	Tobler, W.	172
5	Massey, D.	440	32	Scott, A. J.	171
6	Cliff, A. D.	358	33	Lowenthal, D.	170
7	Hall, P.	333	34	Downs, R. M.	164
8	Pred, A. R.	331	35	Batty, J. M.	159
9	Butzer, K. W.	309	36	Goddard, J. B.	156
10	Gould, P.	302	36	Burton, J.	156
11	Hägerstrand, T.	273	38	Buttimer, A.	153
12	Tuan, Y. F.	270	39	Openshaw, S.	152
13	Brown, L. A.	262	39	Clark, G. L.	152
14	Taylor, P. J.	258	41	Mabogunje, A. L.	149
15	Smith, D. M.	253	42	Chisholm, M. D. I.	141
16	Dear, M.	236	43	Wrigley, N.	134
17	Haggett, P.	231	44	O'Riordan, T.	127
18	Cox, K. R.	226	45	Short, J. R.	125
19	Bennett, R. J.	214	46	Knox, P. L.	122
20	Sauer, C. O.	208	47	Williams, H. C. W. L.	117
20	White, G. F.	208	48	Olsson, G.	112
22	Morrill, R. L.	203	48	Rushton, G.	112
23	Ley, D.	193	50	Herbert, D. T.	111
24	Clark, W. A. V.	192	51	Soja, E. W.	107
25	Gregory, D.	188	52	Dicken, P.	106
25	Golledge, R. G.	188	53	Curry, L.	104
27	Wolpert, J.	187	54	King, L. J.	101
			54	Harris, C. D.	101

Source: Based on N. Wrigley and S. A. Matthews, *Area*, 19, 1987, table 1

University since 1967. Dr Abler served as Director of the Geography and Regional Science Program at the National Science Foundation (1984–8) and was appointed Executive Director of the Association of American Geographers in 1989. He has made large contributions to the study of urban geography but in recent years has taken a lead in energizing American academic geography (see his presidential address to the Association of American Geographers, 'What shall we say? To whom shall we speak', *Annals of the Association of American Geographers*, 77, 4, December 1987).

Alan R. H. Baker was educated at the University of London and now teaches at Cambridge where he is Head of Department and a Fellow of Emmanuel, John Harvard's old college. Dr Baker is editor of the *Journal of Historical Geography* and of the Cambridge Studies in Historical Geography series. In this series he has edited some excellent volumes including Robert Sack, *Human Territoriality* (1986), Denis Cosgrove and Stephen Daniels, *The Iconography of Landscape* (1988) and J. M. Powell's stimulating *An Historical Geography of Modern Australia: the restive fringe* (1988).

Roger G. Barry was born in Sheffield (1935) and educated at the universities of Liverpool, McGill and Southampton

Table IV.4.3 A list of the most cited books written by geographers (based on citations in Social Science Citation Index and Science Citation Index until the end of 1984)

Rank	Citations	Author	Date	Title
1	485	I. B. Leopold, M. G. Wolman and J. P. Miller	1964	Fluvial processes in geomorphology
2	419	D. Harvey	1973	Social justice and the city
3	378	P. Haggett	1965	Locational analysis in human geography
4	336	H. H. Lamb	1972	Climate: past, present and future
5	306	A. G. Wilson	1970	Entropy in urban and regional modelling
6	242	D. Harvey	1969	Explanation in geography
7	208	A. D. Cliff and J. K. Ord	1973	Spatial autocorrelation
8	191	T. Hägerstrand	1967	Innovation diffusion
9	167	L. J. King	1969	Statistical analysis in geography
10	161	M. Chisholm	1962	Rural settlement and land use
11	148	I. J. Gottmann	1961	Megalopolis
12	139	A. G. Wilson	1974	Urban and regional models in geography and planning
13	137	P. R. Gould and R. White	1974	Mental maps
14	129	R. F. Abler, J. Adams and P. R. Gould	1971	Spatial organization
15	128	P. Hall, H. Gracey, R. Drewett and R. Thomas	1973	Containment of urban England
16	126	M. A. Carson and M. J. Kirkby	1972	Hillslope form and process
17	105	W. Bunge	1962	Theoretical geography
18	98	D. M. Smith	1971	Industrial location
19	96	R. J. Johnston	1971	Urban residential patterns
20	96	R. J. Chorley and P. Haggett (eds)	1967	Models in geography
21	93	Y.-F. Tuan	1974	Topophilia
22	92	B. J. L. Berry and D. F. Marble (eds)	1968	Spatial analysis
23	90	K. J. Gregory and D. Walling	1973	Drainage basin form and process

Source: N. Wrigley and S. A. Matthews, *Area*, 18, 1986, table 4; 19, 1987, p. 279

(PhD 1965). He is a member of the Department of Geography and the Institute of Arctic and Alpine Research at the University of Colorado, where he studies long-term climatic variation. Major work includes *Mountain Weather and Climate* (Routledge, London, 1981) and with R. J. Chorley, *Atmosphere, Weather and Climate* (Methuen, London, 5th edn 1987).

Robert J. Bennett (b. 1948) was educated at Cambridge (PhD 1974) and appointed Professor of Geography at the London School of Economics in 1985. In his research Dr Bennett analyses the impact of public policies on geographic patterns (see *Geography of Public Finance*, Methuen, London, 1983).

Brian J. L. Berry (b. 1934) was brought up in Gainsborough, Lincolnshire, the town in which Halford Mackinder spent his early years. After undergraduate work at University College London, Berry moved to the United States in 1955 and gained higher degrees at the University of Washington, Seattle (PhD 1958). From Seattle Dr Berry went to the University of Chicago (1957–76) as an assistant professor, rising to hold the Irving B. Harris Chair of Urban Geography. From Chicago Dr Berry moved to Harvard (1976–81) and then Carnegie-Mellon (1981–6). At present, he is Founders Professor of Political Economy at the University of Texas at Dallas. Professor Berry is the author of over 300 publications. Major work includes *Geography of Market Centers and Retail Distribution* (Prentice Hall, Englewood Cliffs, NJ, 1967) (still an important work in central place studies), *Comparative Urbanization* (St Martin's Press, New York, 1982) and *Economic Geography: resource use, locational choices, and regional specialization in the global economy* (Prentice Hall, Englewood Cliffs, NJ, 1987).

Gerald H. Blake (b. 1936) took a degree at Oxford before completing a PhD at Southampton in 1964. Since that time, he has taught at the University of Durham. The Department of Geography at Durham emerged as a major centre for Middle Eastern studies under the leadership of Professor W. B. Fisher (1916–1984). Dr Blake has contributed to the Durham tradition of Middle Eastern and Islamic Studies. Major works include Blake, J. Dewdney and J. Mitchell, *Cambridge Atlas of the Middle East and North Africa* (Cambridge University Press, Cambridge, 1987), Blake and R. N. Schofield (eds), *Boundaries and State Territory in the Middle East and North Africa* (Menas, Wisbech, Cambs, 1987) and Blake, Peter Beaumont and J. M. Wagstaff, *The Middle East: a geographical study* (Wiley, Chichester, 2nd edn 1988). Dr Blake is the Principal of Collingwood College, a residence which houses nearly 400 Durham students.

John R. Borchert (b. 1918) holds a PhD (1949) from the University of Wisconsin. He is an authority on metropolitan areas and urban geography. His 'American metropolitan evolution', *Geographical Review*, 57 (1967), remains an excellent introduction to the development of American urban places. More recently, his *America's Northern Heartland* (University of Minnesota Press, Minneapolis, MN, 1989) has been seen as an exceptional analysis of life and landscapes in the upper mid-west.

Lawrence A. Brown (b. 1935) was trained at Northwestern University in the mid-1960s when theoretical approaches to geography dominated the programme. Subsequently he has become a leading member of the Ohio State School of Theoretical Geography. Dr Brown's *Innovation Diffusion: a new perspective* (Routledge, London, 1981) is a leading book in diffusion studies which is used in several disciplines in addition to geography.

Ian Burton (b. 1935) was educated at the University of Birmingham (MA 1957), Oberlin College, and the University of Chicago (PhD 1962). He has taught at many universities including Indiana, Queen's (Ontario) and East Anglia and is at present a professor at the University of Toronto. Dr Burton is a leading authority on environmental hazards and is involved in constructing public policies to meet environmental problems. Major works include *Living with Risk: environmental risk management in Canada* (University of Toronto Press, Toronto, 1982) and Robert W. Kates and Ian Burton (eds), *Geography, Resources, and Environment*, 2 vols (University of Chicago Press, Chicago, IL, 1986). This last work is a selection from the writings of Gilbert F. White. Dr Burton is a Fellow of the Royal Society of Canada (1983).

Robin Butlin was educated at the University of Liverpool and has taught at University College, Dublin, Nebraska and Queen Mary College; he is now Professor of Geography at Loughborough University of Technology. Dr Butlin's contributions in historical geography include, with Robert Dodgshon, *Historical Geography of England and Wales* (Academic Press, London, 1978) and *Transformation of Rural England c. 1500–1800* (Oxford University Press, Oxford, 1982). Professor Butlin is an authority on field systems in Britain. A. R. H. Baker and Butlin, *Studies of Field Systems in the British Isles* (Cambridge University Press, Cambridge, 1973), is still the fundamental work on the topic.

Anne Buttimer was born in County Cork, Ireland, in 1938. She holds degrees from the National University of Ireland and the University of Washington, Seattle (PhD 1965). Dr Buttimer has been associated with many universities including the Catholic University of Louvain, Glasgow, Clark, Lund and Ottawa. She has published widely on the philosophy of geography and urban geography. Recent studies have concentrated on the context of creativity in which scholars operate: *The Practice of Geography* (Longman, London and New York, 1983) and *Creativity and Context* (CWK Gleerup, Lund, 1983). *Life Experience as Catalyst: adventures in Dialogue 1977–1985* (University of Lund, Lund, 1986) lists the interviews made by the Dialogue project started by Buttimer and T. Hägerstrand in 1977.

Karl W. Butzer (b. 1934) spent his early years in Mülheim, Germany, but took degrees at McGill in Canada prior to gaining a DSC at the University of Bonn in 1957. His *Environment and Archaeology* (Aldine, Hawthorne, NY, 1964) made an important contribution to the study of prehistoric environments. His work in the Nile Valley resulted in *Early Hydraulic Civilization in Egypt* (University of Chicago Press, Chicago, IL, 1976). Many of Butzer's ideas on cultural ecology in prehistory are drawn together in *Archaeology as Human Ecology* (Cambridge University Press, Cambridge, 1982). In recent years Dr Butzer has published on the origin, development and diffusion of agriculture in Iberia and the New World (see 'Cattle and sheep from old to new Spain', *Annals of the Association of American Geographers*, 78 (1988), 29–56). Dr Butzer is Dickson Professor in the Department of Geography, University of Texas at Austin. Previously he has held appointments at the University of Wisconsin-Madison and the University of Chicago.

Michael Chisholm (b. 1931) was educated at Cambridge and has held appointments at the universities of Oxford, London, Ibadan, Bristol and Cambridge. An early book *Rural Settlement and Land Use; an essay in location* (Hutchinson, London, 3rd edn 1979) made an important contribution to the analytical approach which characterized the geography of the 1960s. The excellent volume is still in print, in a revised edition, and provides an excellent introduction to central place theory, Von Thünen type models and the influence of accessibility on land-use patterns. Later work includes *Modern World Development: a geographical perspective* (Hutchinson, London, 1982).

Richard J. Chorley was born at Minehead, Somerset, in 1927. He was educated at Oxford prior to going to the United States, as a Fulbright scholar, and working at the universities of Columbia and Brown. In 1958 he joined the Department of Geography at the University of Cambridge, becoming professor in 1974. Together with Peter Haggett, Dick Chorley became a leading influence in the methodological revolution that marked British geography in the 1960s. Professor Chorley has a wide range of interests and has published in many fields including physical geography, methodology, hydrology, systems analysis, geomorphology and the history of the study of landforms. Recent work includes, with Stanley A. Schumm and David E. Sugden, *Geomorphology* (Methuen, London and New York, 1985).

Andrew Cliff (b. 1943) earned degrees at the universities of London, Northwestern and Bristol. He lectures at the University of Cambridge and has made contributions in diffusion studies, particularly in relation to the spread of diseases. Major works include Cliff and J. K. Ord, *Spatial Processes: models and application* (Pion, London, 1981), Cliff, P. Haggett and J. K. Ord, *Spatial Aspects of Influenza Epidemics* (Pion, London, 1987), and Cliff and P. Haggett, *Atlas of Disease Distributions* (Basil Blackwell, Oxford, 1988).

Hugh D. Clout (b. 1944) was educated at University College London (PhD 1979) in the department where he now teaches. He has researched and published extensively on France and Western Europe. Major works include *Land of France, 1815–1914* (Allen & Unwin, London, 1983), *Massif Central* (Oxford University Press, Oxford, 1973), *Regional Development in Western Europe* (Fulton, London, 3rd edn 1987) and *Regional Variations in the European Community* (Cambridge University Press, Cambridge, 1986).

John P. Cole was born in Sydney, Australia, and educated at the University of Nottingham, where he currently teaches. He is a prolific writer who has authored a dozen books including *Geography of World Affairs* (Butterworth, London, 6th edn 1983), *The Development Gap* (Wiley, Chichester, 1981), *Geography of the Soviet Union* (Butterworth, London, 1984) and *Development and Underdevelopment: a profile of the Third World* (Methuen, London, 1987).

Ronald U. Cooke (b. 1941) was educated at University College London, where he is now a Professor and Head of Department. Professor Cooke is a desert geomorphologist, with field experience in the Atacama, the south west United States and in the Middle East. He is particularly noted for his studies of pediments, stone pavement, salt weathering and arroyos. In addition, Professor Cooke has developed a major interest in applied geomorphology and is the author of, for example, with J. C. Doornkamp, *Geomorphology in Environmental Management* (Oxford University Press, Oxford, 2nd edn 1990).

Henry Clifford Darby (b. 1909). Three academic geographers have been knighted: Sir Halford Mackinder, Sir Dudley Stamp and Sir Clifford Darby. Mackinder's knighthood came in 1920 towards the end of his parliamentary career. Stamp was knighted for his land utilization survey work in 1965, and Darby was honoured in 1988. Professor Darby's reputation stands on the *Domesday Geography of England* (Cambridge University Press, Cambridge, 7 vols, 1952–77). Other outstanding work includes *A New Historical Geography of England* (Cambridge University Press, Cambridge, 1973) and studies of the medieval fenland. Dr Darby has held chairs at Liverpool, University College London and Cambridge. He has been a visiting professor at the universities of Chicago, Harvard and Washington in the United States, besides delivering the Carl Sauer lecture at Berkeley (1985) and the Ralph Brown lecture (1987) at the University of Minnesota. The connection with the late Ralph Brown (1898–1948) is interesting for, if you were to pick the scholars who gave historical geography form on the two sides of the North Atlantic, Brown and Darby would head the lists. Brown's classic *Historical Geography of the United States* appeared in 1948 and is still in print.

Harm De Blij was born in the Netherlands (1935) and educated in South

Africa (BSc, Witwatersrand, 1955) prior to graduate work at Northwestern University (PhD 1959). Dr De Blij (pronounced duh-Blay) is author of over twenty books, including several leading North American textbooks. He has written research monographs on Africa, political geography and viticulture. While on leave from the University of Miami, he was editor of *National Geographic Research* (1984–90) at the Society's headquarters in Washington, DC. Recent works include *Systematic Political Geography* (Wiley, New York, 4th edn 1989) and *Geography Regions and Concepts* (Wiley, New York, 6th edn 1991). Professor De Blij is a most effective teacher and advocate for geography on television and makes frequent appearances on *Good Morning America*. He holds a teaching appointment at Georgetown University in Washington, DC.

George S. Demko (b. 1933) was educated at Penn State University (PhD 1964). Dr Demko taught for many years at The Ohio State University before heading the Geography and Regional Science Program at the National Science Foundation in Washington, DC. He subsequently served as The Geographer at the US Department of State. At present, he is Director of the Rockefeller Center for the Social Sciences at Dartmouth College. Dr Demko is well known for his research on Europe and the Soviet Union and in recent years he has been a leader among those trying to underscore the value of geography in American affairs (see Dr Demko's presidential address to the Association of American Geographers, 'Geography beyond the ivory tower', *Annals*, 78, 4, December 1988).

Peter Dicken (b. 1938) was educated at the University of Manchester and is now reader in geography at that university. He is well known for his contributions in the field of economic development and for work on shifts in the distribution of economic activity at the global scale. Major works include a book jointly authored with Peter E. Lloyd, *Modern Western Society, A Geographical Perspective on Work, Home and Well-Being* (Harper & Row, London, 1981), and *Global Shift: industrial change in a turbulent world* (Harper & Row, London, 1986).

Robin Doughty was born in Britain (1941) and educated at the University of Reading (BA 1966) before gaining the PhD at Berkeley (1971). Dr Doughty is a leading authority on the manner in which cultures interact with the environment and the impact that humans have on plant and animal species. Major works include *Wildlife and Man in Texas: environmental change and conservation* (Texas A & M University Press, College Station, TX, 1983), *At Home in Early Texas: early views of the land* (Texas A & M University Press, College Station, TX, 1987) and *The Whooping Crane: struggle for existence* (University of Texas Press, Austin, TX, 1989). Dr Doughty teaches at the University of Texas at Austin.

John Burgess Goddard (b. 1943) holds the Henry Daysh Chair of Regional Development Studies at the University of Newcastle-upon-Tyne. In the 1930s the basic heavy industries of the northeast began a decline which the Second World War slowed but did not halt. The closing of Palmer's shipyard, on the Tyne at Jarrow, highlighted the problems of communities dependent upon one industry. At Newcastle the late Professor Daysh initiated studies on the structure of northern industries. These studies found their way into planning concerning Second World War regional reconstruction. Professor John House contributed in the same field at Newcastle before moving to Oxford and the Halford Mackinder chair in 1974. Professor Goddard works in the Daysh–House tradition and is active in research concerning northern economic development. Major works include *Leisure, Recreation, and Tourism* (Pergamon, Oxford, 1981), with A.G. Champion (eds), *Urban and Regional Transformation of Britain* (Methuen, London, 1983), and with Ash Amin, *Technological Change, Industrial Restructuring and Regional Development* (Unwin Hyman, London, 1986).

Reginald G. Golledge (b. 1937) was educated in Australia prior to moving to the United States to gain a PhD (1966) at the University of Iowa. In the mid-1960s, Iowa had a group of graduate students interested in theoretical and behavioural geography. The group, which looked to Harold McCarty (1901–87) for leadership, included Doug Amedeo, Bob Stoddard, John Hudson and Gerry Rushton. Golledge now teaches at the University of California, Santa Barbara. Major works include Douglas Amedeo and Golledge, *An Introduction to Scientific Reasoning in Geography* (Wiley, New York and Chichester, 1975), Golledge and Gerard Rushton (eds), *Spatial Choice and Spatial Behavior: geographical essays on the analysis of preferences and perceptions* (Ohio State University Press, Columbus OH, 1976), Golledge and Robert J. Stimson, *Analytical Behavioral Geography* (Croom Helm, London, 1987), and Golledge and Harry Timmermans (eds), *Behavioral Modelling in Geography and Planning* (Routledge, London, 1988).

Jean Gottmann (b. 1915) was educated at the Sorbonne. Professor Gottmann has held appointments at the Sorbonne, Princeton, Johns Hopkins, the University of Paris and Oxford. Few scholars have held as many visiting positions at universities around the world. Jean Gottmann is best known for *Megalopolis*, a book first published in 1961 (Twentieth Century Fund, New York) but still widely studied today. *The Coming of the Transactional City* (University of Maryland Institute for Urban Studies, College Park, MD, 1983) examined the role of the city in modern economies and pointed out that, although many city functions and population may have moved to the suburbs, major key transactions are still performed at the central nodes of urban areas.

Peter R. Gould (b. 1932) was born in Britain but educated in the United States, receiving a BA at Colgate (1956) and higher degrees from Northwestern University (PhD 1960). He taught at the University of Syracuse before moving to Penn State University in 1964, where he still teaches. Dr Gould's best known book *Mental Maps* was authored with Rodney White and published by Allen and Unwin in 1985 (2nd edn 1986). *The Geographer at Work* (Routledge, New York, 1985) is another excellent book.

Derek Gregory (b. 1951) was educated at the University of Cambridge. He is best known for his writings relating geography to social theory in general and the ideas of Anthony Giddens, Professor of Sociology at Cambridge. These begin with *Ideology, Science and Human Geography* (Hutchinson, London, 1978) and continue in a volume edited with J. Urry, *Social Relations and Spatial Structures* (Macmillan, London, 1985), and inform his own research on the historical geography of industrialization in Britain; see *Regional Transformation and Industrial Revolution: the geography of the Yorkshire woollen industry* (Macmillan, London, 1982).

Kenneth Gregory (b. 1938) was educated at London University (PhD 1962). From 1962 until 1976 he taught at the University of Exeter and then became Professor of Geography at Southampton. Professor Gregory has broad interests in physical geography and hydrology. Major works include *Background to Paleohydrology* (Wiley, Chichester, 1983), *Nature of Physical Geography* (Edward Arnold,

London, 1985), editor, *Energetics of Physical Environment* (Wiley, Chichester, 1987), and with D. E. Walling (eds) *Human Activity and Environmental Processes* (Wiley, Chichester, 1987).

David Grigg was educated at the University of Cambridge before taking a teaching post at the University of Sheffield where he is now a professor. Dr Grigg's early work was concerned with the agricultural revolution in his native Lincolnshire, but he now works on agricultural problems on a broad scale. Major work includes *Dynamics of Agricultural Change* (Hutchinson, London, 1982), *Introduction to Agricultural Geography* (Hutchinson, London, 1984), *World Food Problem, 1950–80* (Basil Blackwell, Oxford, 1985) and *English Agriculture: an historical perspective* (Basil Blackwell, Oxford, 1989).

Torsten Hägerstrand (b. 1916) was educated at the University of Lund, Sweden (PhD 1953), and taught at that institution for many years. Today, Dr Hägerstrand is an emeritus professor but travels widely to deliver lectures (see, for example, Hägerstrand, 'Some unexplored problems in the modelling of culture transfer and transformation', in Peter J. Hugill and D. Bruce Dickson (eds), *The Transfer and Transformation of Ideas and Material Culture*, Texas A & M University Press, College Station, TX, 1988). His work was a key component in the methodological shift that took place in geography during the 1950s and 1960s. In his PhD thesis Hägerstrand introduced new ideas on the process of diffusion. The thesis was published in English as *Innovation Diffusion as a Spatial Process*, translated from the Swedish by Allan Pred (University of Chicago Press, Chicago, IL, 1967). His international reputation rests on his innovative contributions to diffusion studies. In more recent years, along with other Swedish geographers, he has studied at a micro-scale how humans utilize space and time. In addition, Professor Hägerstrand has made contributions to the history of geography (see Hägerstrand and A. Buttimer, *Geographers of Norden: reflections on career experiences* (Chartwell-Bratt, Bromley, 1988).

Peter Haggett (b. 1933). Of all the names associated with the methodological revolution which marked British geography in the 1960s none is better known than that of Professor Haggett. Such was Haggett's eminence that he was appointed to a Chair of Geography at Bristol in 1966, when still in his early thirties. The titles of some of the books written by Professor Haggett, often in conjunction with R. J. Chorley, indicate the scope of his contributions: *Locational Analysis in Human Geography* (Edward Arnold, London, 1965; enlarged and revised 1977), with R. J. Chorley (eds), *Models in Geography* (Methuen, London, 1967) and, with R.J. Chorley, *Network Analysis in Geography* (Edward Arnold, London, 1969). These works helped to give the subject new directions and their influence is still felt. *Geography: a modern synthesis* (Harper & Row, London, 3rd edn 1983) is used as a text in North American universities offering introductory courses that combine human and physical geography. His own semi-autobiographical account of the developments in geography is presented in *The Geographer's Art* (Basil Blackwell, Oxford, 1990).

Peter G. Hall (b. 1932) was educated at Cambridge (PhD 1959) and taught at Birkbeck College, the London School of Economics and the University of Reading before taking his current appointment as Professor of City and Regional Planning and Geography at the University of California at Berkeley. Professor Hall has written a large number of books including *Europe 2000* (Duckworth, London, 1977), *Great Planning Disasters*

(Weidenfeld & Nicolson, London, 1980), *The World Cities* (Weidenfeld & Nicolson, London, 3rd edn 1984), *Cities of Tomorrow* (Basil Blackwell, Oxford, 1988) and *London 2001* (Unwin Hyman, London, 1989). Current research focuses on the location of high-tech industries and the impact of economic long waves on the structure of the industrial sector.

R. Colebrook Harris was born in Vancouver in 1936. After attending the University of British Columbia (BA 1958) he went to Montpellier and then Wisconsin (PhD 1964). Cole Harris has published prolifically. His major contribution to date has been to interpret the experience of European culture groups in the context of North American environments. Major works include, with John Warkentin, *Canada Before Confederation* (Oxford University Press, New York, 1974) and *The Seigneurial System in Early Canada* (McGill-Queen's University Press, Montreal, 1984). Dr Harris edited the *Historical Atlas of Canada*, vol. 1 (University of Toronto Press, Toronto, 1987). He is a Fellow of the Royal Society, Canada.

David W. Harvey (b. 1935) was educated at the University of Cambridge (PhD 1961) and has taught at the universities of Bristol, Penn State, John Hopkins and Oxford, where he is the Halford Mackinder Professor of Geography. He began as a historical geographer studying his native Kent. His international reputation rests on his ability to deal with methodological arguments. *Explanation in Geography* (Edward Arnold, London, 1969) is an incisive commentary on the methodological revolution of the 1960s. *Social Justice and the City* (Johns Hopkins University Press, Baltimore, MD, 1973; new edition, Basil Blackwell, Oxford, 1988) examines the functions and structures of urban areas and how they influence the economic life of city dwellers. These themes are further developed in *Limits to Capital* (Basil Blackwell, Oxford, 1982) and the companion volumes *Consciousness and the Urban Experience* and *The Urbanization of Capital* (Basil Blackwell, Oxford, 1985). His latest work, *The Condition of Postmodernity* (Basil Blackwell, Oxford, 1989) enquires into the economic origins of contemporary cultural change and has been widely acclaimed outside geography itself.

David T. Herbert (b. 1935) was educated at the University of Wales prior to completing a PhD at Birmingham in 1964. He has held a number of visiting appointments in North America and currently is professor at the University College of Swansea. He is the author of numerous books on urban geography including, with David M. Smith, *Social Problems and the City: new perspectives* (Oxford University Press, Oxford, 1989) and, with Colin J. Thomas, *Urban Geography: a first approach* (Wiley, Chichester, 1982).

Ronald J. Johnston (b. 1941) took degrees at the University of Manchester before gaining a PhD at Monash University, Australia (1966). After lecturing at Monash, Dr Johnston moved to the University of Canterbury, New Zealand, in 1967 and then returned to a chair of geography at Sheffield in 1974. He is a prolific writer on urban geography, political geography, the history of geography, Australia, New Zealand, North America and the United Kingdom. Major works include *Dictionary of Human Geography* (Basil Blackwell, Oxford, 2nd edn 1986) and *City and Society: an outline of urban geography* (Hutchinson, London, 1984). *Geography and Geographers: Anglo-American human geography since 1945* (Edward Arnold, London, 3rd edn 1987) is a valuable overview of trends since the Second World War.

Terry Jordan was born in Dallas (1938) and was educated at the Southern Methodist University (BA 1960) and the universities of Texas (MA 1961) and Wisconsin (PhD 1965). He taught at Arizona State and North Texas State before returning to the University of Texas at Austin as the Walter Prescott Webb Professor of Geography. Major works include *American Log Buildings: an Old World heritage* (University of North Carolina Press, Chapel Hill, NC, 1981), *Trails of Texas: southern roots of western cattle ranching* (University of Nebraska Press, Lincoln, NE, 1981) and *Texas: a geography* (Westview, Boulder, CO, 1983).

David E. Keeble (b. 1939) was educated at the University of Cambridge (PhD 1966) where he now lectures. He is an authority on industrial geography and regional development. Major works include, with Egbert Weaver (eds), *New Firms and Regional Development in Europe* (Croom Helm, London, 1986) and, with Philippe Aydalot, *High Technology Industry and Innovative Environments* (Routledge, London, 1988).

Leslie J. King (b. 1934) was brought up and educated in New Zealand prior to gaining a PhD at Iowa (1960). From 1964 to 1970 Dr King taught at The Ohio State University which, together with Iowa and Northwestern, was regarded as a leading department in theoretical approaches to geography. In 1970 King moved to McMaster, as Professor of Geography, and subsequently served there as Dean of Graduate Studies and then Vice-President for Academic Affairs. Major works include *Central Place Theory* (Sage, London, 1984) and, with Reginald G. Golledge, *Cities Space and Behavior: the elements of urban geography* (Prentice Hall, Englewood Cliffs, NJ, 1978). Dr King is a Fellow of the Royal Society, Canada.

Paul Knox (b. 1947) was educated at the University of Sheffield (PhD 1972). He has taught at a number of universities including Dundee, Oklahoma, Stirling and Wisconsin-Madison. At present he is Director of the Center for Urban and Regional Studies at Virginia Polytechnic Institute. Professor Knox's specialty is socio-economic geography. Major works include, with John Agnew, *The Geography of the World-Economy* (Edward Arnold, London, and Routledge, New York, 1989), *Urban Social Geography* (Longman, London, and Wiley, New York, 2nd edn 1987) and *The Geography of Western Europe: a socio-economic survey* (Croom Helm, London, 1984).

Luna B. Leopold was born in 1915 in Albuquerque, New Mexico. Educated at the universities of Wisconsin, California at Los Angeles and Harvard (PhD 1950), Dr Leopold has made contributions to the fields of geology, meteorology, hydrology and the conservation of resources. His work in fluvial geomorphology had a great influence upon physical geographers and Leopold, M. Gordon Wolman and John P. Miller, *Fluvial Processes in Geomorphology*, first published in 1964 by Freeman (San Francisco, CA), remains a classic in the literature. Dr Leopold is the son of Aldo Leopold (1886–1948), the ecologist and authority on environmental conservation. Luna edited a selection from Aldo Leopold's journals under the title *Round River* (Oxford University Press, New York, 1972; original publication 1953).

David Lowenthal (b. 1923) holds degrees from Harvard, Berkeley and Wisconsin (PhD 1953). He is known for contributions in several fields of geography including conservation, the Caribbean, environmental perception, historical geography and the history of geography as an academic discipline. Dr Lowenthal served as research associate and then secretary of the American Geographical Society (1956–72) before becoming Professor of Geography at University College London in 1972. His books include *George Perkins Marsh: versatile Vermonter* (Columbia University Press, New York, 1958) (an important contribution to the history of the conservation movement), *West Indian Society* (Oxford University Press, Oxford, 1972), with Martyn Bowden, *Geographies of the Mind* (Oxford University Press, New York, 1976) and *The Past is a Foreign Country* (Cambridge University Press, Cambridge, 1985).

Akin L. Mabogunje (b. 1931) was raised in Nigeria and educated at the University of London where he was awarded the BA and the PhD. He has made contributions to the study of African urbanization and the analysis of economic development in the Third World. Major works include *The City of Ibadan* (Cambridge University Press, Cambridge, 1967) and *Development Process: a spatial perspective* (Hutchinson, London, 1980). Professor Mabogunje has taught for many years at the University of Ibadan besides holding visiting appointments at institutions in Britain, Sweden and the Americas.

Doreen B. Massey was educated at Oxford and the University of Pennsylvania. At present she is Professor of Geography at the Open University, a department with a strong emphasis on applied regional studies. Professor Massey's publications reflect this emphasis. Major work includes *Spatial Division of Labour: social structures and geography of production* (Macmillan, London, 1984). She has played a leading role in raising the profile of women in British geography.

Donald W. Meinig (b. 1924) entered the School of Foreign Service at Georgetown University in Washington, DC, after service in the Second World War, with the intention of joining the US Foreign Service. The geographical study he was exposed to at Georgetown convinced him that he wanted to be a historical geographer and he has risen to be the leader in that field in the United States. Dr Meinig's reputation is based upon an authoritative group of books. The volumes include *The Great Columbian Plain: a historical geography, 1805–1910* (University of Washington Press, Seattle, WA, 1968), *Imperial Texas* (University of Texas Press, Austin, TX, 1969) and *The Shaping of America: a geographical perspective on 500 years of history*, vol. 1, *Atlantic America 1492–1800* (Yale University Press, New Haven, CT, and London, 1986). Eventually, *The Shaping of America* will constitute a three-volume survey of the historical geography of the region from initial European settlement to the twentieth century. Dr Meinig has taught at the University of Syracuse since 1959, where he is the Maxwell Professor of Geography. Some of his ideas on historical geography, including an assessment of the late Ralph Brown's work, can be found in his article 'The historical geography imperative', *Annals of the Association of American Geography*, 79, 1 (March 1989).

Janet Henshall Momsen, who holds degrees from Oxford, McGill and London, is an authority on Brazil, the Caribbean, gender roles in economic development and the impact of tourism on traditional societies. A pathfinding book, jointly compiled with Janet Townsend (Durham University), *Geography of Gender in the Third World* (Hutchinson, London, 1987), opens up the study of gender roles in traditional societies. Dr Momsen lectured at the University of Newcastle-upon-Tyne before moving to the University of California at Davis.

Janice Monk was born in Australia (1937) and educated at the University of Sydney (BA 1958) before graduate work at the University of Illinois (PhD 1972). She is executive director of the Southwest Institute for Research on Women at the University of Arizona. A recent book with Vera Norwood is *The*

Desert is No Lady: southwestern landscapes in women's writing and art (Yale University Press, New Haven, CT, and London, 1987).

Richard L. Morrill (b. 1934) was educated at Dartmouth College and the University of Washington, Seattle (PhD 1959). After teaching briefly at Northwestern University, Dr Morrill returned to the University of Washington in 1961 and has been a member of the Department of Geography in Seattle ever since. His publications are concerned with migration, diffusion, spatial organization, political geography and poverty. Major works include *Spatial Organization of Society* (Chadsworth, Belmont, CA, 1975) and *Spatial Diffusion* (Sage, London, 1988).

Stanley Openshaw holds a doctoral degree from the University of Newcastle-upon-Tyne where he now teaches. He is a leading authority on the problems of nuclear energy. His widely quoted studies include *Doomsday: Britain after nuclear attack* (Basil Blackwell, Oxford, 1983), *Nuclear Power: siting and safety* (Routledge, London, 1986) and *Britain's Nuclear Waste: safety and siting* (Belhaven Press, Lymington, 1988).

Risa Palm (b. 1942) was educated at the University of Minnesota (PhD 1972). She has studied human responses to earthquake hazards in California and found that there is a tendency to discount dangers. Her presidential address to the Association of American Geographers, 'Coming home', dealt with earthquakes and the need for geographers to study real rather than abstract problems (*Annals of the American Association of Geographers*, 72, 4, December 1986). Dr Palm is Dean of the Graduate School, University of Colorado.

John H. C. Patten (b. 1945) was educated at Cambridge (PhD 1972) and then lectured in geography at Oxford. Andrew Clark and Dr Patten were the founding editors of the *Journal of Historical Geography*. Major works include *English Towns 1500–1700* (Dawson, Folkestone, 1978) and, with Paul Coones, *The Penguin Guide to the Landscape of England and Wales* (Penguin, Harmondsworth, 1986). Dr Patten has been a Member of Parliament since 1979 and at present represents Oxford West and Abingdon. The group of geographers who have been elected to the British Parliament is small. It includes John Patten, Sir Halford Mackinder and Robert Hicks, MP for Cornwall South-East. Patten is, and Sir Halford was, a Privy Councillor.

Allan R. Pred (b. 1936) was educated at Antioch College, Penn State and the University of Chicago (PhD 1962). Early in his career Pred became interested in Sweden and Swedish geographers. In 1967 he published a translation of Hägerstrand's *Innovation Diffusion as a Spatial Process* and has continued to be an important link between Swedish and American geographers. Professor Pred studies the manner in which regions and localities transform spatially in response to changing economic and social circumstances. His books include *Urban Growth and City Systems in the United States, 1840–1860* (Harvard University Press, Cambridge, MA, 1980), *Place, Practice and Structure: social and spatial transformation in southern Sweden 1750–1850* (Barnes & Noble, New York, 1986), *Lost Words and Lost Worlds: modernity and everyday language in late-nineteenth century Stockholm* (Cambridge University Press, Cambridge, 1990) and *Making Histories and Constructing Human Geographies: essays on the local transformation of practice, power relations and consciousness* (Basil Blackwell, Oxford, 1990). As the titles reveal, Professor Pred looks to social theory to help evaluate the nature of spatial change at local and regional

Plate IV.4.1 Carl Otwin Sauer, who was greatly concerned with landscape and with the human modification of the surface of the earth.

levels. He teaches at the University of California at Berkeley.

Carl Ortwin Sauer (1889–1975) (plate IV.4.1) received his PhD in geography at the University of Chicago in 1915. He then taught at the University of Michigan, where he undertook pioneering land-use studies. In 1923 Sauer was appointed to the University of California at Berkeley where he developed a school of cultural geography that examined the way human groups imprinted their cultures on the land to create cultural landscapes. The best introduction to the work of Sauer was assembled by his colleague, John Leighly, *Land and Life: selections from the writing of Carl Ortwin Sauer* (University of California Press, Berkeley and Los Angeles, CA, 1963). Sauer's brand of cultural geography is still widely taught. Several of his students went on to direct numerous PhD degrees. Leslie Hewes at Nebraska, Andrew Hill Clark at Wisconsin, Fred Kniffen at Louisiana State University and James Parsons at Berkeley supervised well over a hundred doctoral dissertations between them.

Stanley A. Schumm (b. 1927) was educated at the University of Columbia (PhD 1954). He is recognized as a leading fluvial geomorphologist and his work is extensively used by physical geographers. Major works include *The Fluvial System* (Wiley, New York, 1977), *River Morphology* (Van Nostrand Reinhold, New York, 1982) and *Experimental Fluvial Geomorphology* (Wiley, New York, 1987). Professor Schumm teaches at Colorado State University in Fort Collins, thus working in an area where the rivers of the Rockies flow out onto the flat lands of the Great Plains.

Allen J. Scott was born in Liverpool (1938) and educated at Oxford prior to gaining a PhD at Northwestern University (1965). Dr Scott has taught at the universities of Toronto and Paris. Currently, he is a member of the Department of Geography, University of California at Los Angeles. Major work includes *The Urban Land Nexus and the State* (Pion, London, 1980), with Michael Storper (eds), *Production, Work, Territory: geographical anatomy of industrial capitalism* (Unwin Hyman, London, 1986), and *Metropolis: From the Division of Labor to Urban Form* (University of California Press, Berkeley, CA, 1988). Dr Scott has made important contributions to the literature on the relationships between industrial location, labour and urbanization.

David M. Smith (b. 1936) was educated at the University of Nottingham (PhD 1961). He has taught at several universities including Manchester, Southern Illinois (Carbondale), Florida, New England (New South Wales), Natal and Witwatersrand. Dr Smith's interests include industrial geography, past and present (see *Industrial Location: an economic geographic analysis*, Wiley, Chichester, 1981), social well-being (see *Geography, Inequality and Society*, Cambridge University Press, Cambridge, 1988; *North and South: Britain's growing inequality*, Penguin, Harmondsworth, 1989) and South Africa (*Apartheid in South Africa*, Cambridge University Press, Cambridge, 1987). He has been Professor of Geography at Queen Mary and Westfield College, London, since 1973).

David R. Stoddart (b.1937) was educated at Cambridge (PhD 1964) and taught there for many years before going to Berkeley to chair the department. He is not the first man from the North of England to have charge of geography at Berkeley. George Davidson, born in Nottingham, was appointed the first Professor of Geography in 1890. Professor Stoddart's research interests are in coastal geomorphology, coral reefs and the history of geography. Major works include *Biogeography and Ecology of the Seychelles Islands* (W. Junk, Dordrecht, 1984) and *On Geography: and its history* (Basil Blackwell, Oxford, 1986).

Peter J. Taylor was educated at the University of Liverpool and is now Reader in Political Geography at the University of Newcastle-upon-Tyne. Dr Taylor has been a leading figure in the resurgence of political geography in recent years. He has edited the *Political Geography Quarterly* from the beginning. Work includes *Political Geography: world economy, nation-state and locality* (Longman, London, 1985), *Geopolitics Revisited* (Department of Geography, Newcastle-upon-Tyne, 1988) and Taylor and J. N. House (eds), *Political Geography: recent advances and future directions* (Croom Helm, London, 1984).

David Thomas (b. 1931) was educated at Aberystwyth, University of Wales, and has taught at the University of London, St David's University College, Lampeter, and the University of Birmingham, where he is Professor of Geography. Research interests include agricultural geography, Wales and planning. Major works include *Wales: a new study* (David & Charles, Newton Abbot, 1977) and Prys Morgan and Thomas, *Wales: the shaping of a nation* (C. Davies, Swansea, 1981).

Yi-Fu Tuan was born in China (1930), educated at Oxford, and developed his professional career in the United States. Dr Tuan has taught at the universities of Indiana, New Mexico, Toronto, Minnesota and Wisconsin besides holding many visiting positions. His early publications were in physical geography. His most influential contributions are concerned with human perceptions and relationships with environments. Major works include *Topophilia* (Prentice Hall, Englewood Cliffs, NJ, 1974), *Landscapes of Fear* (Pantheon, New York, 1980) and *The Good Life* (University of Wisconsin Press, Madison, WI, 1986).

David Ward (b. 1938) was brought up in Manchester which gave him an early understanding of the qualities of large nineteenth-century industrial cities. At the University of Leeds he developed an interest in urban geography and heard Asa Briggs lecture on Victorian cities. After completing his BA (1959) and MA (1961) he went to the University of Wisconsin-Madison where he worked with the great historical geographer Andrew Clark (1911–1975), gaining a PhD in 1963. The mix of European and North American training allowed Dr Ward to produce new insights on the development of North American cities. Major works include *Cities and Immigrants: geography of change in nineteenth-century America* (Oxford University Press, New York, 1971) and *Poverty, Ethnicity and the American City, 1840–1925: changing conceptions of the slum and ghetto* (Cambridge University Press, New York, 1989).

David Watts (b. 1935) holds degrees from

the University of London, the University of California at Berkeley and McGill (PhD 1963). He is the author of a textbook on biogeography and has served as editor of the *Journal of Biogeography*. His brand of biogeography examines the interaction of humans and plants in the environment employing an approach that is in the Sauer–Berkeley tradition. Dr Watts's *West Indies: patterns of development, culture and environmental change since 1492* (1988), in the Cambridge Studies in Historical Geography series, is an important contribution to human impact research in a Caribbean setting.

Gilbert F. White (b. 1911) was educated at the University of Chicago (PhD 1942) and taught at that university until 1969 before moving to the University of Colorado (Boulder) where he is now Gustavson Professor Emeritus. White is a leading authority on water resources, natural hazards and hazards created by humans. He developed the idea that natural hazards are often made worse by human behaviour, e.g. flood insurance may promote disasters by encouraging building on flood plains. He has also been concerned with the environmental effects of nuclear disasters (Julius London and White (eds), *The Environmental Effects of Nuclear War*, Westview, Boulder, CO, 1984). Dr White's intellectual influence on the study of resource use is immense and several of his students, e.g. Burton, have developed his ideas.

Alan G. Wilson (b. 1939) took a degree in mathematics at Cambridge before working with the Ministry of Transport to develop models that estimated traffic flows with more precision than previously used gravity models. He is now Professor of Geography, University of Leeds. Major works include *Catastrophe Theory and Bifurcation* (Croom Helm, London, 1981) and *Geography and the Environment: system analytical methods* (Wiley, Chichester, 1981).

Sidney W. Wooldridge (1900–63) is one of the few geographers to have been elected a Fellow of the Royal Society. Wooldridge's major contribution to geomorphology was to establish an erosion chronology for southeast England, identifying erosion surfaces from a peneplain (700–900 feet) down to present-day sea level. This classic, Davisian, approach is well illustrated in Wooldridge and D. L. Linton, *Structure, Surface and Drainage in South-east England* (George Philip, London, 1939). Wooldridge made contributions in many fields of geography. Wooldridge and W. G. East, *Spirit and Purpose of Geography* (Hutchinson, London, 1958) remains an excellent guide to the nature of British geography prior to the methodological revolution of the 1960s.

E.A. Wrigley (b. 1931) was educated at the University of Cambridge and is now a Fellow of All Souls College, Oxford. Dr Wrigley has opened new horizons in the study of demographic and economic forces as societies change from pre-industrial to modern. Major works include, with R. S. Schofield, *The Population and History of England 1541–1871: a reconstruction* (Cambridge University Press, Cambridge, 2nd edn 1989), *People, Cities and Wealth: the transformation of traditional society* (Basil Blackwell, Oxford, 1987) and *Continuity, Chance and Change: the character of the Industrial Revolution in England* (Cambridge University Press, Cambridge, 1988).

Maurice Yeates (b. 1938) was educated at Reading (BA 1960) and Chicago (PhD 1963). He has taught at Queen's University, Ontario, since 1965, serving as Head of Geography and Dean of Graduate Studies. Major works include, with Barry Garner, *North American City* (Harper & Row, London, 3rd edn 1980), a standard work on urban geography. Dr Yeates was elected a Fellow of the Royal Society of Canada in 1980.

Part V
A Geographical Directory

In his introductory essay Nigel Thrift referred to the information revolution which has transformed our experience of the contemporary world (chapter I.2). Today's student of geography has available an extensive and ever-expanding range of geographical information sources. Students undertaking their own research are often unaware of the variety of sources or how to go about gaining access to them. This section gives an introduction to some of these sources, from on-line databases to the world's great libraries. Ron Abler and Shane Winser also provide advice on obtaining funding for overseas travel and exploration.

V.1

World Libraries and Museums

Helen Wallis

The great library founded by the Greeks in the Museum at Alexandria in about 283–2 BC contained many geographical and cartographic works. The two leading geographers of the classical world, Eratosthenes (c. 275–194 BC), who was to become the library's director, and Claudius Ptolemy (c. AD 90–168) both worked and studied there. Yet in the following centuries geographical collections were to suffer many vicissitudes. Geography became a neglected subject in the universities of western Europe. Up to the sixteenth century the loss of maps was greater than for any other class of historical materials.

The chequered history of geography and map making explains why geographical and cartographic materials have to be sought in many different types of institution, and why the publication of map catalogues dates mainly from the post-war period. The great increase in map production during the Second World War and the consignment of cartographic material to libraries afterwards gave impetus to map rooms as specialist units within general libraries and encouraged the growth of professional map librarianship. On the whole, many map collections in Europe remained unexploited up to the end of the Second World War.

Interest in the history of geography and cartography had developed earlier. The International Geographical Union (IGU), founded in 1871, appointed a Commission for the Reproduction of Early Maps in 1908. Its successor, renamed the Commission on Early Maps in 1949, brought out its first (and only) work in 1964, *Monumenta cartographica vetustioris aevi A.D. 1200–1500* (N. Israel, Amsterdam, 1964). In that year the commission was dissolved and became a working party.

The International Cartographic Association (ICA) was founded in 1959 in association with the IGU and has taken over part of the IGU's role. The ICA's Standing Commission on the History of Cartography, founded in 1976, is concerned with many aspects of historical documentation. In 1969 the International Federation of Library Associations (IFLA) established a Geography and Map Libraries Section, which took up as its first project the compilation of a *World Directory of Map Collections*, published in 1976.

National initiatives complemented the international developments. In Great Britain, for example, the British Cartographic Society was founded in 1963. More recently the International Map Collectors Society (IMCOS) was established in 1980 as a forum for collectors, curators, map dealers and map lovers in general. The Charles Close Society, founded in 1981, is concerned with the study of Ordnance Survey maps. The quarterly periodical *The Map Collector* (Tring, Herts) regularly includes illustrated accounts of major map collections.

In recent years the 'Eurocentric' tendency in cartographic studies has become a matter of comment. Work on the history of map making in non-European communities is now progressing, and an established literature is available documenting the great map making and geographical traditions of China, Japan and India. Many artefacts, which are now regarded as maps, are preserved in museum collections throughout the world. The six-volume *History of Cartography*, edited by J. B. Harley and David Woodward, to be published by the University of Chicago Press (volume 1 was published in 1987), will cover all societies. The international journal for the history of cartography, *Imago Mundi* (c/o King's College London), published annually, also deals with the full range of map making.

General References – Libraries and Museums

The World of Learning 1990, 40th edn, 1990. London: Europa Publications (ISBN 0-946653-54-2; ISSN 0084-2117). Published annually. The main text is arranged in alphabetical order by country and includes academies, learned societies, libraries and archives, museums and art galleries.

World Guide to Libraries, Internationales BibliotheksHandbuch, 7th edn, 1986, Handbook of International Documentation and Information vol. 8. München: K. G.Saur (ISBN 3-598-20531-7; ISSN 0000-0221).

World Guide to Special Libraries, 1st edn, 1983, Handbook of International Documentation and Information vol. 17. München: K. G. Saur (ISBN 3-598-20528-7; ISSN 0724-8717).

Darnay, B. T. and DeMaggio, J. A. (eds) 1990: *Directory of Special Libraries and Information Centers*, 13th edn, 2 vols. Detroit, MI: Gale (ISBN 0-8103-4810-1; ISSN 0731-633x). Mainly covers the United States and Canada, but has an international section (in vol. 1, pt 2).

Walford, A. J., 1980: *Walford's Guide to Reference Material*, vol. 1, *Science and Technology*, 5th edn. London: The Library Association (ISBN 0-85365-978-8). Class 52/529 Astronomy and surveying, pp. 63–83 (includes cartography). Class 55 Earth sciences, pp. 131–75.

—— 1982: *Walford's Guide to Reference Material*, vol. 2, *Social and Historical Sciences, Philosophy and Religion*, 4th edn. London: The Library Association (ISBN 0-85365-564-2). Class 9 Geography, biography, history, pp. 365–701. Class 908 Area studies, pp. 365–433. Class 91 Geography, exploration, travel, pp. 434–47. Class 912 Atlases and maps, pp. 448–69. References to the catalogues of libraries and other institutions are included.

Museums of the World, 3rd revised edn, 1981, Handbook of International Documentation and Information vol. 16. München: K. G. Saur (ISBN 3-598-10118-x).

Woodhead, P. and Stansfield, G. 1989: *Key-guide to Information Sources in Museum Studies*. London and New York: Mansell (ISBN 0-7201-2025-x).

Cartographic and Geographical Directories

Wolter, J. A. (General Editor), Grim, R. E. and Carrington, D. K. 1986: *World Directory of Map Collections*, 2nd edn, IFLA Publications 31. München: K. G. Saur (ISBN 3-598-20374-8 and ISBN 0-86291-296-2). Prepared under the auspices of the Geography and Map Libraries Section of the Division of Special Libraries of the International Federation of Library Associations (IFLA). The Directory is arranged in alphabetical order by country. It is compiled from returns of questionnaires sent out to the leading map libraries of the world. The text is in English except for institutional names and addresses. A third edition is in preparation.

Perkins, C. R. and Parry, R. B. 1990: *Information Sources in Cartography*. London: Bowker-Saur (ISBN 0-408-02458-5). This covers both contemporary and historical cartography.

Bibliographic Guide to Maps and Atlases, 1979– (annual). Boston, MA: G. K. Hall (ISSN 0360-5889). The Bibliographic Guide lists selected publications catalogued during the past year by the Research Libraries of the New York Public Library and the Library of Congress.

Ehlers, E. (ed.) 1988: *Orbis Geographicus 1988/92, World Directory of Geography*, 6th edn. Stuttgart: Franz Steiner (ISBN 3-515-04326-8).

Great Britain

General guides to libraries and museums

Hudson, K. and Nicholls, A. 1987: *The Cambridge Guide to the Museums of Britain and Ireland.* Cambridge: Cambridge University Press (ISBN 0-521-32272-3).

Libraries in the United Kingdom and the Republic of Ireland. London: The Library Association. Published annually.

Adkins, R. T. (ed.) 1988: *Guide to Government Department and Other Libraries 1988,* London: The British Library, Science Reference and Information Service (ISBN 0-7123-0746-x). Published every two years.

Note: Six libraries in the British Isles (known as the Copyright Libraries) benefit from the Copyright Act, requiring legal deposit of books, atlases and maps published in Great Britain.

Map libraries and geographical institutions

The British Library (plate V.1.1) The principal copyright library of the United Kingdom. The library departments of the British Museum were incorporated into the newly created British Library in 1973. The map collections have grown steadily since the British Museum's foundation in 1753 and are now spread in several parts of the Library. The principal repository of printed maps is the Map Library, which has been a separate unit since 1867. It houses King George III's Topographical and Maritime Collections, probably the finest eighteenth-century collection in the world. Many maps are also held in the Manuscript Collections (formerly the Department of Manuscripts).

Pamphlet: *The Map Collections of the British Library,* July 1989.

Catalogues: Catalogue of Printed Maps . . . Photolithographic edition to 1964, published 1967, and ten-year supplement to 1974.

Catalogue of cartographic materials in the British Library, 1975–1988, 3 vols, 1989. London: K. G. Saur (ISBN 0-86291-765-4). Since 1975 accessions have been entered on the automated 'Cartographic Materials File', including, after 1986, manuscript items. This file is available on microfiche and on line.

Address: British Library, Map Library, Great Russell Street, London WC1B 3DG.

India Office Library and Records The map collections form part of the India Office Records which, with the India Office Library, are now part of the British Library.

Pamphlet: *Map Collections in the India Office Records,* July 1988.

Address: India Office Library and Records, Orbit House, 197 Blackfriars Road, London SE1 8NG.

Note: In the 1990s the British Library will be moving from the British Museum to a new building near St Pancras Station. The India Office Library and Records will also be located there.

National Library of Scotland A copyright library. The Map Room of the National Library of Scotland was created in 1958. It holds the largest and most comprehensive map collection in the United Kingdom north of Oxford. The Map Library moved to a new building in 1988.

Address: The Map Library, National Library of Scotland, Causewayside Buildings, 33 Salisbury Place, Edinburgh EH9 1SL.

National Library of Wales A copyright library, with worldwide coverage of national atlases and sheet maps, and a large collection of manuscript estate maps relating to property in Wales.

Address: Department of Pictures and Maps, National Library of Wales, Aberystwyth, Dyfed SY23 3BU.

Bodleian Library A copyright library, with a large worldwide collection of maps and atlases. The Gough collection

Plate V.1.1 The reading room of the British Library, London.

of British local topography, and the Todhunter–Allen Collection are notable historical collections.

Address: Map Section, Bodleian Library, Oxford, Oxfordshire OX1 3BG.

University of Cambridge A copyright library. The Map Room holds atlases, thematic mapping and topographic mapping worldwide.

Address: Map Room, University Library, West Road, Cambridge, Cambridgeshire CB3 9DR.

University of London *A Guide to Geography and Map Collections*, 1985, is published for the Geography Subcommittee of the Library Resource Co-ordinating Committee (ISBN 7187-0712-5). The Geography and Geology Librarian and Map Curator is in charge of the general and special map collections of the University of London Library, Senate House, Malet Street, London WC1E 7HU.

Trinity College Library, Dublin A copyright library, with a large collection of geographical books and maps relating to Ireland.

Public Record Office Receives records from central government departments and courts of law. Holds maps of many parts of the world from the sixteenth century onwards. Some material is at Chancery Lane. Published catalogues of maps and plans cover (a) the British Isles, (b) America and the West Indies and (c) Africa.

Address: Map Department, Public Record

Office, Ruskin Avenue, Richmond, Surrey TW9 4DU.

Public Record Office of Northern Ireland Printed and manuscript maps of Northern Ireland and the Republic of Ireland, Ordnance Survey maps of Northern Ireland and maps of landed estates.

Address: Public Record Office of Northern Ireland, 66 Balmoral Avenue, Belfast, County Antrim, Northern Ireland BT9 6NY.

Ministry of Defence Mapping and Charting Establishment, R.E., Map Research and Library Group, Ministry of Defence Map Library, Tolworth, Surbiton, Surrey. Holds a large world-wide collection of topographical and aeronautical mapping.

Ordnance Survey Books and Periodicals Library. Map Record Library. Technical Information and Support Services. Central Register for Air Photographs for England (future location of the Register to be determined).

Address: Ordnance Survey, Romsey Road, Maybush, Southampton SO9 4DH.

The Royal Geographical Society The Royal Geographical Society, founded in 1830, has been responsible for the major part of the nineteenth- and twentieth-century British exploration. Its collections comprise books, maps and atlases, archives, museum items, paintings, prints and photographs. The Map Room is funded by government grant and is open to the public.

Cameron, I. 1980: *To the Farthest Ends of the Earth, The History of the Royal Geographical Society*. London: Macdonald and Jane's (ISBN 0-354-04478-8).

Address: Royal Geographical Society, Map Room, 1 Kensington Gore, London SW7 2AR.

Directories

Watt, I (compiler) 1986: *A Directory of U.K. Map Collections*, 2nd edn. British Cartographic Society, Map Curator Group Publication 3 (ISBN 0-904482-08-1).

Royal Commission on Historical Manuscripts, 1987: *Record Repositories in Great Britain. A Geographical Directory*, 8th edn. London: HMSO (ISBN 0-11-440210-8).

Gibson, J. and Peskett, P. 1982: *Record Offices: how to find them*, 2nd edn. Plymouth: Federation of Family History Societies (ISBN 0-907099-16-5).

Visvalingen, M. and Kirby, G.H. 1987: *Directory of Research and Development Based on Ordnance Survey Small Scale Digital Data*. Produced by the Cartographic Information Systems Research Group (CISRG) on behalf of the Ordnance Survey, Hull: University of Hull. It aims to publicize and promote research based on Ordnance Survey small-scale digital data. The directory includes academic institutions, government and public services, and the private sector, including individuals.

Barker, M. J. C. (ed.) 1986: *Directory for the Environment Organisations in Britain and Ireland 1986–7*, 2nd edn. London: Routledge & Kegan Paul (ISBN 0-7102-0961-4). A directory of organizations concerned with the environment, which gives aims, activities and publications. Library resources are included where appropriate.

Rest of the World

Australia

Rauchle, N. M. and Alonson, P. A. G. (eds) 1977: *Directory of Map Collections in Australia*, 2nd edn. Canberra: National Library of Australia (ISBN 0-642-99098-0).

Canada

Langelier, G. 1985: *National Map Collection, Collection nationale de cartes et plans*, General Guide Series 1983. Ottawa: National Archives of Canada (ISBN 0-662-53825-0). A guide to the map division of the National Archives of Canada.

Dubreuil, L. 1986: *Directory of Canadian Map*

Collection, Repertoire des collections Canadiennes de cartes, 5th edn. Ottawa: Association of Canadian Map Libraries. Appendix 2 gives a 'List of Deposit Agreements' with map agencies.

See also the United States for a joint directory.

France

Briend, A. M. and Gabay, D. 1980–1: *Repertoire des Cartotheques de France.* Extrait d'*Intergeo Bulletin*, Paris: Organe trimestriel des instituts et centres de recherches de géographie, Laboratoire d'information et de documentation en géographie; 1980, no. 60; 1981 no. 61.

Federal Republic of Germany

Zogner, L. (ed.) 1983: *Verzeichnis der Kartensammlugen in der Bundesrepublik Deutschland einschliesslich Berlin (West).* Wiesbaden: Otto Harrassowitz (ISBN 3-447-02193-4).

Netherlands

van Slobbe, A. (ed.) 1980: *Gids voor Kaartenverzamelingen in Nederland.* Amersfoort: Nederlandse Vereniging voor Kartografie (ISBN 90-6469-543-1).

New Zealand

Directory of New Zealand Map Collections, 1989. Christchurch: New Zealand Map Society (available from Mr W. H. Cutts, Treasurer, New Zealand Map Society, University of Canterbury).

United States

Carrington, D. K. and Stephenson, R. W. (eds) 1985: *Map Collections in the United States and Canada. A Directory*, 4th edn. New York: Special Libraries Association (ISBN 0-87111-306-6). Compiled from the returns to questionnaires mailed to more than 3000 libraries. The library holdings listed divide into six general categories: academic, public, geoscience, state and federal, historical societies, and private libraries. The first MAGERT (Map and Geography Round Table) publication.

(See also V.3, this volume, p. 299.)

V.2
International Data Sources

Christopher Talbot

Introduction

Statistical information is a basic requirement for the understanding of geographical phenomena. For human geography statistics are a primary source of information, as they are also, to a lesser extent, in certain aspects of physical geography. There is a bewildering array of statistical information, and the geographer needs to be aware of how to find and to select the information needed for a particular geographical enquiry. Statistics are usually compiled as a result of two processes:

1 as the end-product of a census or survey where the main purpose is to collect the statistics;
2 as a by-product of a statutory function such as customs requirements or value-added tax (VAT).

Problems in the Use of Statistics

Statistics by their very nature can be manipulated, misinterpreted or abused. It is not unknown for deliberately misleading figures to be published for political or propaganda purposes, or to be quoted out of context. When using statistics the geographer should check the authority of the publishing or compiling agency, to see that the source of the information is reliable or unbiased. Statistics should never be quoted without citing their source.

Although deliberately inaccurate data are relatively rare, unintentional errors can appear if the data are compiled from several sources. It is important to establish the means used to collect the tabulated data and, if necessary, the size and method of selection if a sample has been used. The purpose of compilation should also be noted. The best safeguard against errors arising from the careless use of statistics is to check carefully any introductory notes, footnotes and other annotations provided. Most statistical works include information on the methods of compilation, scope and area covered, and any anomalies such as estimates or a variable database.

Another common source of difficulty with statistical data is compatibility. Units of measurement are not always consistent. This problem may arise when international statistical sources are assembled from the returns of a variety of agencies who have different methods of gathering information. In such cases it is necessary to convert the statistics into comparable units. Tables such as the *Geographical Conversion Tables* (edited by D. H. K. Amiran and A. P. Schick, International Geographical Union, Zurich, 1961) are available for this purpose. Other problems of compatibility arise when a series of figures include some that are based on slightly varying criteria. This should be noted in a footnote or similar reference.

The most recent statistics on a particular topic may be difficult to obtain. Generally speaking the most readily available and convenient statistics are found in comprehensive compilations which are only digests of data previously published elsewhere, possibly from sources that are obscure or difficult to obtain. Secondary sources such as the *New Geographical Digest* (George Philip, London, 1986) are handy for quick reference but are not so reliable for very recent information. Information contained in secondary sources is often less detailed than that contained in the original source.

More statistical data may have been collected than actually appear in print. The reasons for this could be that some of the information is confidential, or because publication of the information cannot be economically justified. It is quite possible that such information may be made available on request. In these circumstances it is worth tracing the potential producers of these statistics and making enquiries. Local statistics are an example of information that is very difficult to obtain from regular published sources but may be obtained from local government offices (see chapter V.5), chambers of commerce, local trade associations and societies. On a national scale there are various publications which list the central agency for collecting statistics for each country.

Selected Guides to, and General Sources of, Statistical Information

Brewer, J. G. 1973. *The Literature of Geography: a guide to its organisation and use.* London: Clive Bingley. One of the very few guides devoted to the subject of geography; out of date, but includes a useful introduction to statistical sources.

The New Geographical Digest, 1st edn, 1986. London: George Philip.

Reader's Digest Almanac and Yearbook, 1966– (annual). New York: Random House. Inexpensive quick reference source.

Whitaker's Almanack, 1868– (annual). London: J. Whitaker. Reliable up-to-date information on Britain and the Commonwealth.

World Almanack and Book of Facts, 1868– (annual). New York: Newspaper Enterprise Association. Contains statistics on social, industrial, religious, educational and other subjects including political organizations, societies and historical events. Up to date and generally reliable; sources for many of the statistics are given.

National government statistical publications

The most prolific generators of statistics are national governments. Most countries have a centralized system for collecting and dissemination: Germany's Statistisches Bundesamt is one such statistical agency; Netherlands is another country with a Central Bureau of Statistics. Most of the agencies have developed a house style for their publications which makes them very easy to identify. By contrast the United Kingdom has a decentralized system where each principal government department has its own statistical division which collects, analyses and publishes data, with the Central Statistical Office (CSO) having a co-ordination role.

Statistical Guides to National Official Publications

Harvey, J. M. 1970: *Statistics: African sources for market research.* Beckenham: Research Ltd; London: Mansell.

Jeffries, J. 1981. *A guide to the Official Statistics Publications of the European Communities*, 2nd edn. London: Mansell.

National Statistical Offices of the World, 1988. London: Statistics and Market Intelligence Library, Department of Trade and Industry. Very useful up-to-date listing of names and addresses of national statistical offices.

The Stateman's Yearbook, 1964– (annual). Basingstoke: Macmillan. Gives a brief description of individual countries together with basic facts and figures; also has names and addresses of official statistical agencies.

Slattery, M. 1986: *Official Statistics.* London: Tavistock Publications.

Statistical Yearbook, 1960–. Council for Mutual Economic Assistance (CMEA), Moscow: Statistika (English edition 1978).

Provides statistics for the USSR and Eastern Europe, including population, transport, industry, agriculture and forestry.

Westfall, G. 1986. *Bibliography of Official Statistical Yearbooks and Bulletins.* Alexandria, VA: Chadwyck-Healey.

International organizations

International organizations such as the United Nations, the Organization of Petroleum Exporting Countries (OPEC) and the Organization for Economic Co-operation and Development (OECD) act mostly as co-ordinating bodies, collecting data from each national source. They have great problems ensuring that all the information is properly comparable and reliable.

The Statistical Office of the European Communities has had a tremendous task in trying to harmonize the results of all the different statistical systems of its members. However, the published output of the international agencies is easy to use and provides just one source for a variety of figures. Listed below are some of the publications available from these organizations.

Publications of the Organization for Economic Cooperation and Development (OECD) The OECD was established in 1961 and now has twenty-four member countries, including European nations, the United States, Japan, Turkey and Australia. It is a prolific publisher of statistics, and many of its publications are available in different format (print, micro-fiche and machine readable). Some of the statistical publications of particular interest to geographers are as follows.

Energy Statistics, 1961– (annual). Paris: Organization for Economic Co-operation and Development.

Main Economic Indicators, 1965– (monthly). Paris: Organization for Economic Co-operation and Development.

Quarterly Oil and Gas Statistics, 1976–. Paris: Organization for Economic Co-operation and Development.

Also the *Annual Oil and Gas Statistics* which is available on-line from I. P. Sharp Associates.

The State of the Environment, 1985. Paris: Organization for Economic Co-operation and Development.

Also a companion volume *OECD Environmental Data Compendium*, 1985.

Publications of the United Nations and affiliated organizations The United Nations publishes a number of yearbooks, specialized studies, bulletins and other compilations of interest to the geographer. The annual *Catalogue of United Nations Publications* and the *United Nations Documentation in Microfiche, UNDOC: Current Index* provide up-to-date coverage of United Nations documentation.

The United Nations Statistical Office makes the following machine-readable products available:

United Nations Demographic Database
Commodity Production Statistics
Energy Statistics Database
General Industrial Statistics/Industrial Statistics Database
Maritime Transport Statistics

Some United Nations databases are commercially available on-line from commercial vendors. For further information on over 600 databases see the *Directory of United Nations Databases and Information Systems*, 1985.

The following section covers some of the United Nations statistical publications of particular interest to geographers.

African Statistical Yearbook, 1976–. Rome: United Nations. Gives demographic, social, economic and agricultural production data by country.

Agricultural Review for Europe, 1958– (annual). New York: United Nations. Provides information on European dairy, meat, grain and produce industries. Formerly titled *Review of the Agricultural Situation in Europe.*

Annual Bulletin of General Energy Statistics for Europe, 1968– (annual). Geneva: Economic Commission for Europe. Provides basic data on the energy situation as a whole, and more detailed information on energy production and consumption.

Compendium of Human Settlement Statistics, 1974– (irregular). New York: United Nations. Based on the national housing censuses and other surveys; data are presented for more than 150 countries and areas.

Compendium of Social Statistics, 1963– (irregular). New York: United Nations. Contains basic statistical indicators of social changes and trends including population, health care, food consumption, housing and education.

Demographic Yearbook, 1949– (annual). New York: United Nations. A central source for demographic data from approximately 220 countries. Topics include rates of increase in population, birth and death rates, urban population and international migration.

Directory of Environment Statistics, 1983. New York: United Nations. Contains environmental data and facts collected by 150 member countries. Also contains a listing of environmental statistics publications published by other international organizations.

Energy Statistics Yearbook, 1952– (annual). New York: United Nations. Formerly titled *World Energy Supplies/Statistics*, this publication presents a historical view of world and regional production of commercial energy sources, and a ten-year series for individual countries. Updated production figures are included in the *Monthly Bulletin of Statistics.*

Population and Vital Statistics Report, 1949– (quarterly). New York: United Nations. Provides the latest census data for over 200 countries, plus worldwide demographic statistics on birth and mortality.

Production Yearbook, 1947– (annual). Rome: United Nations Food and Agricultural Organization. Contains data on all important aspects of food and agriculture, including production, supplies, means of production and employment.

The State of Food and Agriculture, 1947– (annual). Rome: United Nations Food and Agricultural Organization. A narrative and statistical review of the year for food, agriculture and related areas including forestry and fisheries. The emphasis is on developing countries.

Statistical Abstract of the Arab World, 1977. Beirut: Economic Commission for Western Asia.

Statistical Yearbook for Asia and the Pacific, 1973– (annual). Bangkok: United Nations, Economic and Social Commission for Asia and the Pacific (ESCAP). Supplemented by *Quarterly Bulletin of Statistics for Asia and the Pacific* and *Statistical Indicators for Asia and the Pacific.*

Statistical Yearbook for Latin America and the Caribbean, 1975– (annual). New York: United Nations.

United Nations Statistical Yearbook, 1949– (annual). New York: United Nations. Worldwide coverage on statistics; categories include population, manpower, agriculture, forestry, fishing, mining, transport, energy, housing, health and education.

World Agricultural Statistics, 1986. Rome: United Nations Food and Agricultural Organization. Provides indicators relating to agriculture, fisheries and food for more than 200 countries. Data are gathered for selected years beginning in 1965.

V.3

Major National Data Sources in the United States

Chauncy D. Harris

The United States offers a fertile and rewarding field for the use of national data in geographical studies. The challenge lies in the richness of the sources of quantitative information, in the size and importance of the country, and in its great regional diversity. Since the amount of data available is so vast it is well to start with the basic general references and then to proceed to more detailed and specialized sources. Published data in convenient form, available in many libraries, can provide the necessary statistical basis for studies ranging from simple student papers completed in a weekend, say on the general regional distribution of ethnic groups, to detailed research investigations involving many researchers over an extended period of time, say on the suburban trends in all the major metropolitan centres.

The place to start is the basic assemblage of national data in the *Statistical Abstract of the United States*, which has been published annually since 1879. The 109th edition for 1989 in 956 pages presents key national data on population; geography and environment (area, weather records, land utilization by types, federal lands, elevations, rivers, water bodies, water withdrawals and consumptive use, air and water pollution, and solid waste); vital statistics, health and sanitation; education; law enforcement; parks, recreation and travel; elections; labour force and employment; income; finance; communications; energy; transportation; agriculture; forests and fisheries; mining and mineral wealth; construction and housing; manufactures, domestic trade and service; foreign trade and aid; and other topics. Both the table of contents and the index aid in locating specific types of information. For many of the data figures are given by some regional breakdown (such as major regions or states). Each table indicates its source. Appendix 1 provides a guide to major sources of statistics by topics. Also the *Bureau of the Census Catalog* is helpful.

The range of the current national figures of the *Statistical Abstract* is extended by special volumes that provide much greater detail in areal patterns and a more extended historical depth.

In a country as large and diverse as the United States national totals often mask striking regional contrasts. Two volumes, both supplements to the *Statistical Abstract*, are treasure troves for geographers in providing detailed areal data by an ever smaller spatial mesh. *State and Metropolitan Area Data Book*, published every few years (1982 and 1986 for example), provides data by increasingly small areal breakdowns: for four major regions (Northeast, Midwest, South, and West), for nine geographic divisions, and for fifty states and the District of Columbia. Data on some 1873 different variables are thus reported, from change in population 1980–5 to area of cropland, built-up area or number of black elected officials. A subject index provides an alternative and convenient entry into these diverse data. Another large section of this volume is devoted to data for 230 metropolitan areas, listed alphabetically by state. These data include 298 variables with totals for each metropolitan area and for each of the constituent counties. In a separate table eighty-six variables are presented for each metropolitan area and its central cities or areas within the metropolitan area but outside central cities.

A companion volume, *County and City Data Book*, has been issued every few years since 1949. The 1988 volume provides data on states, counties, cities and other places. For the fifty states and the more than 3000 counties data on 203 items are reported, for cities of 25,000 or more inhabitants (in 1980) data on 134 items, and for places of 2500 or more inhabitants, four items on population and income.

Any geographer worth his salt can find in these sources abundant recent socio-economic data for a wide range of size of areal units from the country as a whole, through regions, states, counties, metropolitan areas and down to quite small urban places. These can provide a statistical base for a broad variety of geographical studies, either of areal distributions of specific phenomena over the country as a whole, large regions or local areas, or of a statistical profile of a specific area including a wide range of types of data. Sometimes the data can be read directly but in other cases a student may wish to combine the data of several units. So much for areal detail, we now turn to historical evolution.

Data for studies in the field of historical geography may be found in *Historical Statistics of the United States: colonial times to 1970*, also a supplement to the *Statistical Abstract*, published in two large volumes in 1976, on the Bicentenary of the founding of the country. Here are presented in handy summary form data from the earliest available records through 1970. The series cover the same sort of data as in the other summary statistical volumes. Altogether data are presented on about 12,500 different series. Of course not all of them are of geographic interest but the key socio-economic characteristics of the population, society and economy are well represented. Detailed notes on the sources are valuable as leads for further investigation.

Four other sources of data are valuable for historical geography of the United States. John L. Andriot's *Population Abstract of the United States* (1973) conveniently assembles available data by fifty states, 3064 counties and 2217 incorporated places (with 1980 populations of 10,000 or more) for each census back to the earliest census for each unit, and also the 1970 and 1980 populations for townships (minor civil divisions). An index volume contains some 70,000 entries. A companion volume *Township Atlas of the United States* (1987) provides maps of states, counties and townships, with an index to the 46,900 townships. It should be remembered that during the first century of censuses of the United States the area of settlement was still expanding toward the west. The *Guide to Cartographic Records in the National Archives* (1971) is a key source for older maps and aerial photographs. One can profitably also examine the data sources listed and annotated in *Historical Geography of the United States: a guide to information sources* by Ronald E. Grim (1982).

Censuses conducted by the United States Bureau of the Census provide a wealth of information. Data from these are available in published volumes, on microfiche and on computer tape files. First and foremost is the *Census of Population*, which has been taken every ten years since 1790. The latest of these was taken two hundred years later on 1 April 1990. It contains elaborate data on socio-economic and demographic characteristics of the population. Data on years between the censuses can be found in *Current Population Reports*.

The associated *Census of Housing* has been taken every ten years since 1940.

A series of specialized economic censuses are also conducted by the Bureau of the Census. Currently these are taken every five years for the years ending in 2 and 7 (i.e. 1982 and 1987). The *Census of Agriculture* formed part of the decennial *Census of Population* from 1840 to 1950 (with special censuses in 1925, 1935 and 1945). Since then it has been taken at five-year intervals. *Censuses of Business* have been taken since 1929, currently divided into three economic censuses: *Wholesale Trade, Retail Trade* and *Service Industries.* A *Census of Manufactures* has been taken since 1810 at ten-year intervals in the nineteenth century and at shorter intervals in the twentieth century. A *Census of Transportation* has been conducted since 1963. A *Census of Mineral Industries* was taken as part of decennial censuses from 1840 to 1950. Since then it has been on a five-year basis.

Four specialized annual volumes provide continuing current economic data. Two are compiled by the Bureau of the Census: *Annual Survey of Manufactures* and *County Business Patterns.* The *Annual Survey of Manufactures* provides annual figures on employment, value added, cost of materials, value of shipments etc. by industry groups for the country as a whole and for states, major metropolitan areas, industrial counties and selected cities. Much less detailed than the censuses of manufactures, it is nevertheless of value in providing more frequent and therefore more current information. The *County Business Patterns* publishes data on employment and payrolls, number of establishments and size of establishments by detailed industry for states and counties. *Agricultural Statistics,* compiled by the Department of Agriculture and published as a separate volume since 1936, provides data on agricultural production by individual crops and types of livestock, finances and conservation for the country as a whole and generally also by states for several recent years. It also notes the many specialized series issued by the Department of Agriculture. The *Minerals Yearbook,* compiled by the Bureau of Mines, separately published since 1932, includes both systematic reviews of individual metallic and non-metallic mineral commodities and regional reviews for each of the fifty states.

For the geographer the map is a key data source. A good starting point for national maps, charts, aerial photographs, satellite images, cartographic literature and geographic information systems is the *Scholars' Guide to Washington, D.C. for Cartography and Remote Sensing Imagery* by Ralph E. Ehrenberg (1987). It inventories collections with 50,000 atlases, 9.2 million maps, 17 million aerial photographs and remote sensing images, key national bibliographies and map reference centres. The United States Geological Survey is the principal mapping agency. It issues a *Map Sales Catalog,* revised periodically, which describes the principal series of topographical, geological, water resource and other maps. Then on a state-by-state basis the Geological Survey issues two guides. The *Index to Topographic and other Map Coverage* provides sectional maps depicting names, locations and coverage of quadrangles. Each index is valid indefinitely. A separate large folded map of each state is also useful and is tied into the index. A *Catalog of Topographic and other Published Maps,* also on a state-by-state basis, reports on the published and available quadrangles in alphabetical order, a separate list for each scale. It is revised every few years. It also notes the commercial dealers for topographic maps and the map depository libraries in each state. The National Cartographic Information Center of the Geological Survey, in Reston, Virginia, is a convenient source of information on maps and mapping of the United States.

Within the compass of a brief introductory review it is possible to mention only a few of the more important specialized data sources of interest in physical geography. The National Oceanic and Atmospheric Administration (NOAA) maintains a National Climatic Center in Asheville, North Carolina, where the basic climatic data for the country are stored, and from which detailed data can be secured. The Environmental Data Service issues *Selective Guide to Climatic Data Sources* and *Climatological Data*, both in national summary and by states. NOAA has published or republished *Climatic Atlas of the United States* and *The Climates of the States* in fifty-one parts, also commercially reprinted in two large volumes. It includes extensive tables based on the first-order weather reporting stations. The Soil Conservation Service publishes the county soil maps, which, incidentally, are often the best source for detailed areal patterns of soil, physical geography and agriculture for small areas in the United States. The Soil Conservation Service publishes from time to time a *List of Published Soil Surveys.* The Geological Survey publishes both retrospective and current lists of its publications.

The national database from governmental and other sources is so extensive that one needs a guide to the range and types of other data collected and published. Detailed current information is to be found in *ASI. American Statistics Index: a comprehensive guide and index to statistical publications of the U.S. government*, published annually since 1973. For earlier statistics one can usefully consult John L. Andriot's *Guide to U.S. Government Statistics* (1973). Non-governmental statistics are covered by *SRI. Statistical Reference Index: a selective guide to American statistical publications from private organizations and state government sources*, published monthly with quarterly and annual cumulations, from 1980. Also of value are two listings of a broad range of types of social and economic data, Joan M. Harvey's *Statistics America: sources for social, economic, and market research* (1980) and Steven Wasserman's *Statistics Sources 1990: a subject guide to data on industrial, business, social, educational, financial and other topics for the United States and internationally* (1989).

The *Monthly Catalog of United States Government Publications* prepared by the Superintendent of Documents, a comprehensive listing arranged by individual government departments, bureaus, centres, agencies, services and institutes, is invaluable in tracing down the rich, diverse and sometimes elusive government documents and in ferreting out specialized statistical data.

References

Unless otherwise noted all entries are published in Washington, DC, by the Government Printing Office. The Library of Congress call number is provided in square brackets; in many major American research libraries this call number helps identify and locate the document.

ASI. American Statistics Index: a comprehensive guide and index to the statistical publications of the U.S. government, 1973– (annual, with monthly supplements). Washington, DC: Congressional Information Service. Cumulative indexes 1980–4, 1985–8 [Z7554.U5A46].

Andriot, J. L. 1973. *Guide to U.S. Government Statistics*, 4th edn. McLean, VA: Documents Index, 431 pp. (1st edn 1956) [Z7554.U5G8].

—— 1983: *Population Abstract of the United States*, revised edn, vol. 1 *Tables*, 895 pp.; vol. 2, *Indexes and Rank Tables*, 406pp. McLean, VA: Andriot Associates (earlier edn 1980) [HA202.A686 1983].

—— 1987: *Township Atlas of the United States*. McLean, VA: Andriot Associates, 969 pp. (earlier edn, 1979, 1184 pp.) [G1201.F7A5].

Ehrenberg, R. E. 1987: *Scholars' Guide to Washington, D.C. for Cartography and Remote Sensing Imagery*, Scholars' Guide to Washington, D.C., no. 12. Washington, DC: Smithsonian Institution Press, 385 pp. [GA193.U5E37 1987].

Grim, R. E. 1982: *Historical Geography of the United States: a guide to information sources*, Geography and travel information series vol. 5. Detroit, MI: Gale Research, 291 pp. [Z1247.G74].

Harvey, J. M. 1980: *Statistics America: sources for social, economic and market research (North, Central and South America)*, 2nd edn. Beckenham, Kent: CBD Research, 385 pp. [Z7554.A5H37].

SRI. Statistical Reference Index: a selective guide to American statistical publications from private organizations and state governmental sources, vol. 1, no. 1–, January 1980–. Washington, DC: Congressional Information Service. Monthly with quarterly cumulative index, and annual cumulations. Frequency varies, bimonthly 1984– [Z7554.U5S73].

US Bureau of Mines. *Minerals Yearbook*, 1932–3– (annual). Predecessor volumes from 1880: 1880–1923 in U.S. Geological Survey. *Annual Report of the Director*; 1924–31 as *Mineral Resources of the United States* [TN23.U642].

US Bureau of the Census. *Annual Survey of Manufactures*, 1949–50 (annual) [HD9724.A211].

—— *Census Catalog and Guide*, 1946– (annual). Supplement: Monthly product announcement. 1946–84 as *Bureau of Census Catalog* [Z7554.U5U32].

—— *Census of Agriculture*. 1840, 1850, 1860, 1870, 1880, 1890, 1900, 1910, 1920, 1930, 1940, 1950 with the *Census of Population*; also 1925, 1935, 1945, 1954, 1959, 1964, 1969, 1974, 1978, 1982, 1987 [HD1769.C46].

—— *Census of Business*, 1929, 1933, 1935, 1939, 1948, 1958, 1963 and 1967, then divided into:

Census of Wholesale Trade, 1972, 1977, 1982, 1987 [HF5421.C4].

Census of Retail Trade, 1972, 1977, 1982, 1987 [HF5429.3.C4].

Census of Service Industries, 1972, 1977, 1982, 1987 [HD9981.4.C46].

—— *Census of Housing*. Taken in conjunction with decennial *Census of Population*, 1940, 1950, 1960, 1970, 1980, 1990 [HD7293.A4883].

—— *Census of Manufactures*. Every ten years 1810–1900 (except 1830), every five years 1904–19, biennial 1921–39, then 1947, 1954, 1958, 1963, 1967, 1972, 1977, 1982, 1987 [HD9724.C422].

—— *Census of Mineral Industries*. As part of each decennial *Census* 1840–1950, then in 1954, 1958, 1963, 1967, 1972, 1977, 1982, 1987 [HD9506.U62C45].

—— *Census of Population*, 1790–. Every ten years. 1790, 1800, 1810, 1820, 1830, 1840, 1850, 1860, 1870, 1880, 1890, 1900, 1910, 1920, 1930, 1940, 1950, 1960, 1970, 1980, 1990 [HA201].

—— *Census of Transportation*. 1963, 1967, 1972, 1977, 1982, 1987 [HE203.C44].

—— *County and City Data Book*. 1949–, 1952–. Irregular. Issued for 1949, 1952, 1956, 1962, 1972, 1977, 1983 and 1988. A *Statistical Abstract* supplement [HA202.A36].

—— *County Business Patterns*, 1946–. Annual. Separate volume for each state [HC101.A184].

—— *Current Population Reports*, 1047–. Several series, such as P-20, *Population characteristics* [HA195.A53], and P-25, *Population estimates and projections* [HA195.A533].

—— *Historical Statistics of the United States: colonial times to 1970*, Bicentennial edition, 1975, 2 vols, 1200 pp., plus appendices. A *Statistical Abstract* supplement. Supersedes two previous editions: *Historical Statistics of the United States, 1789–1945* (1949) and *Historical Statistics of the United States: colonial times to 1957* (1960) [HA202.B87].

—— *State and Metropolitan Area Databook* 1979–, 1980–. Irregular. Issued for 1979, 1982, 1986. A *Statistical Abstract* supplement [HA202.S84].

—— *Statistical Abstract of the United States*, 1878–, 1879, vol. 1– (annual) [HA202].

US Department of Agriculture. *Agricultural Statistics*, 1936– (annual) [HD1751.A43].

US Environmental Data Service. *Climatological Data: national summary*, Asheville, NC, 1950–, vol. 1– (monthly and annual) [QC983.A2C63]. Also reports for each state [QC983.A5].

—— 1979: *Selective Guide to Climatic Data Sources*, National Climatic Center, Asheville, NC. Washington, DC: Environmental Data and Information Service, 142 pp.

US Environmental Science Service. *Climatic Atlas of the United States*, 1968, 80 pp. Reprinted by US National Oceanic and Atmospheric Administration, Asheville, NC: Climatic Data Center, 1983, 80 pp. Also reprinted with new material as *The Climates of the States*, 2nd edn, 2 vols, Detroit MI: Gale Research, 1980; earlier edition, as *Weather Atlas of the United States*, 1975 [G1201.C8U55].

US Geological Survey. *Catalog of Topographic and Other Published Maps*. Issued directly by the Geological Survey. Separate catalog for each state. Frequently revised.

—— *Index to Topographic and other Map Coverage*. Issued directly by the Geological Survey. Separate index for each state. Each index is considered valid indefinitely.

—— *Map Sales Catalog*. Issued directly by the Geological Survey. Frequently revised.

—— *Publications of the Geological Survey*, 1879–1961, 1964, 457 pp.; 1962–70, 1972. Annual issues supplement these cumulative catalogues [Z6034.U49U53].

US National Archives. 1971: *Guide to Cartographic Records in the National Archives*, 444 pp. [Z6028.U575].

US National Oceanic and Atmospheric Administration. *Climates of the states, climatology [climatography] of the United States*, no. 60, parts 1–49, 51–52 (no. 50 never issued), 1976–1978. 2 vols [QC983.A46]. Reprinted with revisions as *The Climates of the States*, 2nd edn, 2 vols, Detroit, MI: Gale Research, 1980, 1182 pp.

US Soil Conservation Service. *List of Published Soil Surveys*. Frequently revised.

US Superintendent of Documents. *Monthly Catalog of United States Government Publications*, 1895– (monthly). Decennial cumulative indexes. Various supplements [Z1223.A18].

Wasserman, S. 1989: *Statistics Sources 1990: a guide to data on industrial, business, social, educational, financial and other topics for the United States and internationally*, 13th edn, 2 vols. Detroit, MI: Gale Research (1st edn ed. by Paul Wasserman, 1962) [HA42.S82].

V.4
Major National Data Sources in the United Kingdom

Christopher Talbot

In the United Kingdom the Government Statistical Service comprises the statistical divisions of all the major departments plus the two big collecting agencies, the Business Statistics Office and the Office of Population Censuses and Surveys (OPCS), and the Central Statistical Office (CSO), which co-ordinates the whole system. The following publications detail and describe the statistics that are available in the United Kingdom.

Rhind, D. (ed.) 1983: *A Census User's Handbook*. London: Methuen. Provides an overview of the censuses, both past and present, geographical applications including mapping and statistical analysis techniques.

Central Statistical Office 1986: *CSO Guide to Official Statistics*. London: HMSO. The so-called 'purple book' includes a list of sources, and government department contact points, alphabetical keyword index and a bibliography. The guide covers all official and some important non-official sources of statistics for the United Kingdom, Isle of Man and the Channel Islands.

Dewdney, J.C. 1981: *The British Census*, Concepts and Techniques in Modern Geography (Catmog) Series 29. Norwich: Geobooks.

Government Statistical Service 1989: *Government Statistics. A Brief Guide to Sources*. London: CSO.

Maunder, W. F. (ed.) 1978: *Reviews of United Kingdom Statistical Sources*, 5th edn, Royal Statistical Society and Social Science Research Council. Oxford: Pergamon. Latest series of eight guides to UK statistical sources; out of date, but an excellent series that includes:

Coppock, J. T. and Gebbit, J. 1978: Vol. VIII, *Land Use and Town and Country Planning*;

Munby, D.L. and Watson, A. H. 1978: Vol. VII, *Road Passenger Transport and Road Goods Transport*.

Statistical News, Central Statistical Office. London: HMSO. Provides a monthly update on the availability of the latest official statistical publications; there is also a quarterly journal.

Townsend, A.R., Blakemore, M.J. and Nelson, R. 1987: The Nomis Data Base: availability and uses for geographers. *Area*, 19, 1, 43–50.

In addition to the official sources of UK statistics there is a large number of organizations involved in compiling and disseminating information. They include trade associations, private companies, professional bodies, local authorities and academic institutions. For further information on these organizations refer to the *Sources of Unofficial Statistics*, compiled by D. Mort and L. Siddall, Warwick Statistics Service, University of Warwick, and published by Gower, Aldershot, 1985. The following publications are the basic minimum list of statistical data for the United Kingdom available in central public libraries and academic libraries.

Annual Abstract of Statistics. Central Statistical Office. HMSO.

British Business. Department of Trade and Industry. Weekly. HMSO.

Economic Trends. Central Statistical Office. Monthly. HMSO.

Employment Gazette. Department of Employment. Monthly. HMSO.

Monthly Digest of Statistics. Central Statistical Office. HMSO.

Regional Trends. Central Statistical Office. Annual. HMSO.
Social Trends. Central Statistical Office. Annual. HMSO.
Scottish Abstract of Statistics. Scottish Office. HMSO.
Digest of Statistics. Welsh Office.
Northern Ireland Annual Abstract of Statistics.

In addition to the above publications, government departments and agencies produce statistical information of interest to geographers as follows.

Agriculture, fisheries and food

Agricultural Statistics for the United Kingdom. Ministry of Agriculture, Fisheries and Food (MAFF).
Farm Classification in England and Wales. MAFF. HMSO.
Household Food Consumption and Expenditure. Annual Report of the National Food Survey Committee. MAFF. HMSO.

Energy

CEGB Statistical Yearbook (CEGB is now privatized as National Power, PowerGen and other companies).
Development of the Oil and Gas Resources of the United Kingdom. A Report to Parliament by the Secretary of State for Energy (Brown Book). Annual. HMSO.
Digest of United Kingdom Energy Statistics. Department of Energy. Annual. HMSO.
Energy Trends. Department of Energy. Monthly.
Handbook of Energy Supply Statistics. Electricity Council. Annual.
National Coal Board Statistical Tables. Annual.

The annual reports of the following public corporations provide further statistical information.

British Gas Corporation (now privatized)
British National Oil Corporation
Electricity Council
National Coal Board (now British Coal)
North of Scotland Hydro-Electric Board
Northern Ireland Electricity Service
South of Scotland Electricity Board

Human welfare, medical and social geography

Cancer Statistics (Series MB1). OPCS. Annual. HMSO.
Communicable Disease Statistics (Series MB2). OPCS. Annual.
On the State of the Public Health. Annual Report of the Chief Medical Officer of the Department of Health and Social Security.
Social Security Statistics. Department of Health and Social Security. Annual. HMSO.

Housing

English House Condition Survey. Department of the Environment. HMSO.
Housing and Construction Statistics. Department of the Environment, Scottish Development Department and Welsh Office. Annual. HMSO.
Local Housing Statistics for England and Wales. Department of the Environment and Welsh Office. Quarterly. HMSO.

Industrial Production

Business Monitor: Production (P) Series. Business Statistics Office HMSO.
Business Monitor: PA 100 Analysis of UK Manufacturing (local) Units by Employment Size. Business Statistics Office. HMSO.

Crime

Crime and Criminal Statistics: England and Wales. Home Office. HMSO.

Tourism, Entertainment and Leisure

Digest of Tourist Statistics. British Tourist Authority. Annual.
Overseas Travel and Tourism (Business Monitor MQ6). Business Statistics Office.
Overseas Travel and Tourism (Business Monitor MA6). Business Statistics Office. Annual. HMSO.

Transport and Environment

Basic Road Statistics. British Road Federation. Annual.
British Shipping Statistics. General Council of British Shipping.
CAA. Annual Statistics/Monthly Statistics. Civil Aviation Authority.
Digest of Environmental Pollution and Water Statistics. Department of the Environment. Annual. HMSO.
Port Statistics. Department of Transport and British Ports Association. Annual.
National Travel Survey. Department of Transport. Irregular. HMSO.
Transport Statistics of Great Britain. Department of Transport, Welsh Office and Scottish Development Department. Annual. HMSO.

Population, households and vital statistics

Census of Population. London: Office of Population Censuses and Surveys.

Statistics of population and households are collected in the periodic censuses conducted by the OPCS and the Registrars General for Scotland and Northern Ireland. The most recent census was taken in 1981, and the results are published by the HMSO. Some cover the United Kingdom or Great Britain, and some are limited to regions and counties. Coverage of the census reports for the 1981 Census is as follows.

County Reports.
County Leaflets.
Regional Reports (Scotland).
Population Estimates (Series PP1). OPCS. Annual.
Population Projections: National Figures (Series PP2). OPCS. Annual.
Population Projections: Sub National Figures (PP3). OPCS. Annual.
Annual Estimates of the Population of Scotland. Registrar General of Births, Deaths and Marriages for Scotland. HMSO.
Annual Reports of the Register General for Northern Ireland. HMSO.

V.5

Local Authorities as an Information Source in the United Kingdom

John Hope

Introduction

Registering births and running cemeteries; constructing roads and caring for the elderly; strategic planning and stray dogs; art galleries and allotments – with involvement in so many facets of everyday life the Councils which administer local government are a rich potential source of information for the student of geography.

Local authorities, however, are complex organizations which, while well suited to delivery of the services which are their primary concern, may often be less well attuned to the needs of people carrying out academic enquiry or research. The aim of this chapter is to give advice on how to maximize your chances of getting the best out of the system by obtaining access to the information which is of real value to you.

What are the Local Authorities?

The bulk of local government services are provided by what are known as 'principal local authorities'. Together these authorities provide a full cover of the whole country. The major conurbations in England are administered by groups of single tier local authorities. Elsewhere in England, and in Scotland and Wales, responsibility for the provision of services is shared between two tiers of local authority. Table V.5.1 shows the general arrangement within Britain.

Other authorities

In addition to the principal authorities there are a number of other local authorities which may be in operation in some areas.

Parish and Town Councils (Community Councils in Wales) generally only exist for villages and small towns in rural areas. They have a very limited range of powers which overlap with certain of the district council powers.

Joint Authorities are set up by groups of principal authorities where services need to be integrated across a wider area. Good examples are the Thames Valley Police Authority, covering Berkshire, Buckinghamshire and Oxfordshire, and the Fire and Civil Defence Authorities serving the metropolitan areas.

Table V.5.1 Principal local authorities in Britain

England and Wales			Scotland
Conurbations		Elsewhere	
London			
London Boroughs (and the City of London)	District Councils	County Councils	Region and Island Councils
		District Councils	District Councils

Who Does What?

Where a single tier system is in operation all the 'principal authority' functions for an area will be the responsibility of the appropriate London Borough or Metropolitan District. Where the two tier system operates the situation is more complicated, as shown in table V.5.2.

Even table V.5.2 is a simplification because, for some functions, authorities are allowed to enter into agreements to employ other authorities to act for them on an agency basis. The table is therefore best taken as a guide to where to look first for a particular function. Each authority will also have 'central service' functions necessary to run its own operations. These include financial and legal services and a committee secretariat. Personnel and property management functions may also be separately identifiable.

Table V.5.2 Division of main functions between counties and districts in England and Wales[a]

County Councils	District Councils
Consumer protection	Allotments
Education	Building regulations
Fire service	Cemeteries and crematoria
Libraries	Coastal protection
Police	Environmental health
Registration of births etc.	Housing
Smallholdings	Markets and fairs
Social services	Rate and community charge collection
Traffic/transport/highways	Registration of electors
Certain functions are shared	
Gypsy	sites
Provision	Management
Refuse	(not Wales)
Disposal	Collection
Town and	country planning
Structure plans	Local plans
National parks	Development control
Control of mineral extraction	Advertisement control
Some powers may be exercised by either authority	
Airports	
Arts support	
Recreation	
Refuse collection and disposal (Wales)	
Civil emergencies	
Footpath creation etc.	
Land drainage	
Building preservation	
Conservation area designation	

[a]The division of functions in Scotland is broadly similar but not identical. A much fuller background on the organization and functions of local government may be found in *Local Government in Britain* by Tony Byrne, published by Pelican, 1986.

Council business is usually carried out by way of committees of elected members. A number of standing committees meet at regular intervals, each responsible for a major area of the Council's work. These committees will report significant items of business to meetings of the whole Council. Meetings of the whole Council must be open to attendance by the public. Committee meetings are often open too.

Each Council employs a body of paid workers who carry out the day to day operations of the authority. These workers are grouped into departments, usually based around the different func-

Table V.5.3 Structure of committees and departments (Oxford City Council)

Committees	Departments
Direct Labour Organization	Chief Executive's
Employment and Economic Development	Competitive Services
Estates	Engineering and Recreation
Finance	Environmental Health
Health and Environmental Control	Estates
Highways and Traffic	Housing
Housing	Personnel
Personnel	Planning
Planning	Secretary and Solicitor's
Public Affairs	Treasurer's
Recreation and Amenities	

tions of the council and/or the different professional disciplines. Departments are headed by chief officers or directors whose job it is both to advise the council and committees on the services for which they are responsible and to run their departments. Table V.5.3 shows the committee and departmental structure for a typical District Council.

What Information is Available?

The information held by local authorities is used for planning services, controlling resources, communicating intentions and a host of other purposes. In this section we seek to illustrate the kind of information that may be available, principally by reference to the town and country planning function. This has been chosen because it is exercised at both county and district levels; it has a very wide frontier of interest; and it has much common ground with geography.

The breadth of planning involvement is shown by the range of issues dealt with in a typical structure or local plan, as illustrated in table V.5.4. The precise range of issues will vary according to the nature of the area and the local priority attached to each issue. However, those that are of importance will each need to be considered in terms of broadly the same range of questions:

What is happening now?
How did the present situation arise?
Where are things heading?
Is a change of course desirable?
What tools can best be used to do this?
Are they having the intended effect?

Table V.5.5 shows the kind of information that is needed if these questions are to be answered on the topic of housing. A similar table could be drawn up for most of the other topics in table V.5.4.

How then are these questions to be answered? Table V.5.6 illustrates the wide variety of sources which may be used. (The whole range is vast. Only a few representative examples are shown.) Items towards the top of table V.5.6 may be more accessible in libraries than at the Council offices. Further down, however, the Council may be the only place where the information can be obtained.

The existence of information does not guarantee its availability to members of the public. Some will be confidential because it deals with personal information or for some other reason. For example it would not be possible to inspect a 'Children at Risk' register in a Social Services Department. However, information in such cases may be avail-

Table V.5.4 Issues in plan preparation

Population
Housing
Employment
Industry and commerce
Shopping
Recreation and leisure
Education
Tourism
Conservation/townscape/landscape
Transport and communication
Development intentions
Resources

Table V.5.5 Planning for housing – information requirements

Supply	
Existing stock	How many dwellings
	In what condition
	Levels of occupancy
Recent trends	House completion rates
	Vacancy rates
Future commitments	Outstanding permissions
	Availability of land
Demand	
Population change	Natural change
	Migration
Other demands	Household formation rates
	Replacement of poor stock
	Higher space standards

Table V.5.6 Planning department data sources

Sources	Examples
Nationally based	
Published (usually give data only for fairly large areas)	Population Census – County Reports – Topic tables General Household Survey National Travel Survey Family Expenditure Survey Employment Gazette Ordnance Survey maps and plans
Unpublished (a finer area breakdown or more specialized versions of some of the above)	Population Census – Small area statistics – Special statistics on work and migration Census of Employment
Locally based	
Local surveys (may include both raw data and analyses)	Housing conditions Shopping catchment Traffic origin and destination Public attitude to proposals Land use Historic buildings Photographs (including aerial)
Local authority operational records	Register of Planning Applications Register of Electors Committee/council reports/minutes Annual reports Periodic returns to central government
Informal sources	Newspapers (including community broadsheets) 'The grapevine' Local knowledge and experience

able in summary form, often as part of a regular monitoring report to a committee. Conversely some information held by Councils *must* be available for public inspection, for instance the statutory registers of planning applications, electors etc.

Availability of information may also depend on the method of storage. Large data sets such as Small Area Census information, traffic survey data etc. will frequently be held on computer. Summary tables may be available in paper form or on microfilm or fiche but may not reflect the full range of data or be analysed or presented in the way which best suits your purposes.

How do I go About Getting the Information?

Local government officers these days are generally working under considerable pressure. Your request may be competing with a vocal complaint from a resident or tomorrow's committee deadline. If you are to get the information you need it is essential to target your enquiry carefully and to limit your questions to those the Council is best placed to answer.

Tables V.5.1 and V.5.2 will help you to decide which local authority to begin with. You will then have to select a department to start at. Table V.5.3 may help a bit here but of course each Council is different and even departments with the same functions may be labelled differently from authority to authority. A departmental label clearly related to your area of interest will be an obvious place to start. If you can find nothing as obvious as this try to establish whether the Council has a research department or section and start there. Otherwise go to one of the central service departments such as the Chief Executive's or the committee secretariat. These departments may not be able to respond to your enquiry directly but are best placed to put you in touch with whoever can. If you have to move between departments in this way, always try to get direction to a named person.

If you go direct to a relevant department or have been given no name, ask first whether there is a departmental research officer or section. People with this specific role are always to be preferred as contacts as they are more likely to be geared up to answering information queries. Again, try to find out the name of the person you should talk to.

Finally a basic list of dos and don'ts:

DO Be prepared to make an appointment. The person you need to see may not be available if you turn up unannounced. Better to wait to see the right person than to see a less appropriate person instantly.

DO Be prepared to make more than one visit. It may not always be possible to produce information at short notice. It may be sensible to limit your first visit to explaining your project, defining your needs and establishing what is available.

DO Be as clear as possible on what your needs are. You should have talked this through with your tutor or lecturer well before visiting the Council. Your ability to be entirely clear may be limited by a lack of knowledge as to what the Council has to offer but at least define the knowledge gaps which you would ideally like to fill.

DO Remember that some information is subject to limitations of confidentiality and copyright.

DON'T Go to the Council offices for information which is readily available elsewhere.

DON'T Expect the Council officers to do your job for you. Access to the relevant information may often be the most you can reasonably expect. Extraction and analysis of data will generally be left to you.

DON'T Ask for the same information more than once. When working as a group delegate one person to the task of visiting the Council and then share the information collected.

V.6
On-line Data Sources

Christopher Talbot

The general availability of computers and in particular microcomputers has meant that statistical information is increasingly being made available through on-line services and by subscription to companies providing statistical information on floppy discs and CD-ROM. There is a great potential in the development of computerized statistical information which has only recently begun to be developed in the United Kingdom. The advantage of machine-readable statistical databases is that the data can be manipulated by software packages such as LOTUS 1-2-3 and SMART. It is also possible to combine different sets of data, so long as they are in a compatible format such as the standard ASCII format. Developments in computer graphics techniques and programmes allow the flexible presentation of statistics; for example the ability to combine statistical analysis with computer-generated maps is a prime feature of geographical information systems. The information held in machine-readable format is more readily updated than the printed version. Databases will also provide long series data quickly, thus avoiding a long search through printed volumes.

The disadvantage with on-line databases is finding out what is available. A number of organizations provide the data, but it is difficult to find out who they are and what information they provide. In most cases special training is required to make full use of statistical databases, particularly with the manipulation of information, and there is the expense of subscribing to a commercial database. Examples of on-line statistical databases for geographers are included in the general list of bibliographic and numeric databases listed below.

As an alternative to on-line databases, more data are being made available for microcomputers on discs and CD-ROM. During 1989 the British Central Statistical Office has made all the data sets available on 5.25 and 3.5 inch discs as ASCII text which can be read by LOTUS (1-2-3 and Symphony) and SMART. The first data set to be made available on disc is the 1989 edition of the *Economic Trends Annual Supplement*.

Information on discs and CD-ROM is a growing market. Already there is a variety of data available. For example George Philip Services have produced 'World Patterns', a floppy disc for the BBC B or Master microcomputer, which holds over forty of the world series of statistics and allows the user to manipulate data and present data in a variety of ways, including lists, graphs and map forms.

On CD-ROM, Microinfo have available the Central Intelligence Agency (CIA) 'World Factbook'. This comprises an annual world almanac of every country, including geographic, economic and demographic data. Also available on CD-ROM is the Domesday project, consisting of a UK database map based on the Ordnance Survey 1:625,000 showing locations of towns and villages, together with text and photographs of any selected area. Chadwyck-Healey Ltd have produced a digital map of the whole world on CD-ROM which claims to be the only detailed and accurate map of the world for the personal computer user. The system is called MundoCart/CD and produces seamless high quality maps of any part of the world in a range of scales from 1:250,000 to 1:150,000,000. The database contains some 50,000 names of places and features, fifteen classifications of rivers, eleven classifications of inland water bodies, eight coastline classifications, four coastal feature classifications, four transportation classifications and six classifications of national and disputed national boundaries. In addition to the complete database there are ten subsets or combinations of subsets of the main database. Another feature of the system is its ability to export map data into other software programs such as geographical information systems like SPANS and ARC/INFO, or into databases like dBASE III.

On-line Databases for Geographers, Including Bibliographic and Numerical Databases

Name: Aqualine Abstracts
Type: Bibliographical references from 1960 onwards
Owner/producer: Water Research Centre
Host: Pergamon Orbit Infoline
Printed version: *Aqualine Abstracts*
Subject coverage: Worldwide coverage of literature on commercial, technical and scientific aspects of water and wastewater industries, including legislation, management, monitoring water quality and water pollution.

Name: Bartholomew Digital Maps
Type: Digital map data
Owner/producer: Bartholomews
Host: University/Polytechnic Computing Services
Subject coverage: Great Britain data at 1:250,000 scale; includes all topographic information currently found on Bartholomew printed maps. Great Britain contour data at 1:250,000 scale. Europe at 1:1,000,000 scale, including Eastern Europe. World data at 1:14,000,000 scale, available in 1991. These data are being made available as part of a CHEST deal to provide the ARC/INFO geographic information system with digital map data.

Name: Central Statistical Office Databank
Type: Numeric
Owner/producer: Central Statistical Office (CSO)
Host: Scrimgeour-Vickers, Wefa Ltd, Datastream, DRI and IP Sharp
Printed version: *CSO Statistical Publications*
Subject coverage: UK statistics, including population census, economic and social data.

Name: A Computerised London Information System Online (ACOMPLINE)
Type: Bibliographical references with abstracts from 1973 onwards
Owner/producer: London Research Centre (Research Library)
Host: ESA–IRS
Printed version: *Urban abstracts*
Subject coverage: Urban and local government issues, including urban planning and transportation, urban renewal, housing issues, leisure and recreation, environmental design and architecture, and environmental conservation.

Name: Environmental Information and Documentation Centres (ENDOC)
Type: Textual/numeric
Owner/producer: Commission of the European Communities. UK representative: Department of the Environment

Name: Environmental Research Projects (ENREP)
Type: Textual/numeric
Owner/producer: Commission for the European Communities. UK representative: Department of the Environment
Host: Echo
Subject coverage: Information on environmental research projects within the European Community, including name of research establishment, type of research, costs, duration and references to publications.

Name: ESRC Data Archive
Type: Machine-readable data sets, from 1967 onwards
Owner/producer: Economic and Social Science Research Council/University of Essex
Host: University of Essex Computing Service
Printed version: ESRC Data Archive Catalogue: vol. I, *Guide and Indexes*; vol. II, *Study Descriptions*
Subject coverage: The largest British repository of accessible computer-readable data relating to social and economic affairs. The archive holds over 2500 data sets across the full range of the social sciences from UK and foreign sources, including the UK census data 1966, 1971, 1981, British Crime Survey, National Dwelling and Housing Survey, UK Rural Areas Survey, US Census Data.

Name: Geobase
Type: Bibliographical references with abstracts, from 1908 onwards
Owner/producer: Elsevier Publications
Host: Dialog
Printed version: *Geographical Abstracts*; *Geological Abstracts/Mineralogical Abstracts*; *Ecological Abstracts*; *International Development Abstracts*
Subject coverage: All aspects of human and physical geography, geology and ecology, and related subjects including cartography, climatology, demography, the environment, remote sensing, international development, natural resources, rural and urban planning, regional studies and transportation.

Name: Geoarchive
Type: Bibliographical references from 1969 onwards; also a referral service
Owner/producer: Geosystems
Host: Dialog
Printed version: Geotitles
Subject coverage: Geosciences, energy, petroleum, and minerals and mineral processing. Index of maps from the British Geological Survey.

Name: Geomechanics Abstracts (GMA)
Type: Bibliographical references, from 1977 onwards
Owner/producer: Rock Mechanics Information Service
Host: Pergamon Orbit Infoline
Printed version: *Geomechanics Abstracts*
Subject coverage: Worldwide coverage of rock mechanics, soil mechanics and engineering geology, including mining, tunnelling and support, foundation engineering, slope stability and soil improvement techniques.

Name: Information on Mining, Metallurgy, and Geological Exploration (IMMGE)
Type: Bibliographical references with abstracts, from 1979 onwards (with some earlier coverage)
Owner/producer: The Institute of Mining and Metallurgy, Library and Information Services
Host: Pergamon Orbit Infoline
Printed version: *IMM Abstracts*
Subject coverage: Worldwide literature coverage on economic geology, mining and extraction technology, mineral economics and management, health and safety, and environmental aspects of mining.

Name: Meteorological Office Library Accession Retrieval System (MOLARS)
Type: Bibliographical references, from 1971 onwards
Owner/producer: National Meteorological Library
Host: ESA–IRS
Printed version: *Monthly Accession List*
Subject coverage: Meteorology, climatology, hydrometeorology, surface oceanography and planetary atmospheres.

Name: National Online Manpower Information Service (NOMIS)
Type: Numeric, from 1978 onwards
Owner/producer: Durham University/Manpower Services Commission
Host: Durham University
Subject coverage: There are two main classes of data: the historical data set of UK population projections 1981–2001, employment, unemployment trends; and the current on-line series, including migration, employment and population data from the 1981 Census, and projections 1985–2011.

Name: Noble Denton Weather Services
Type: Numeric
Owner/producer: Noble Denton Weather Services Ltd

Host: Prestel (I.P. World Viewdata Services)
Reuters Ltd (Reuters Monitor)
British Telecom Gold (Petroleum Monitor)
Subject coverage: On Prestel: daily weather forecasts for 20 UK regions; weather and sea forecasts for waters around the United Kingdom, updated three times a day. Three-day UK national weather outlook; long-range outlook and special events forecasts; weather forecasts for European cities, updated daily.
Reuters Monitor service: daily weather forecasts for thirty European cities.
On Telecom Gold: five-day weather and sea state forecasts for four areas of the North Sea, updated daily; five-day weather forecasts for London and Aberdeen. Statistical weather and sea state information for the North Sea updated monthly.

Name: Planning Exchange Database (PLANEX)
Type: Bibliographical references with abstracts, from 1981 onwards
Owner/producer: The Planning Exchange
Host: Pergamon Orbit Infoline
Subject coverage: Literature on physical and economic planning, housing and the environment, including urban affairs, rural development, conservation, tourism and leisure, energy conservation, transport, development control and regional planning.

Name: SASPAC91
Type: Statistical
Owner/producer: Office Population Censuses and Statistics
Host: University and Polytechnic Computing Services
Subject coverage: The OPCS and local authorities are producing the small-area boundaries for the 1991 Census. The basic geographic unit for the small-area level in England and Wales is the Enumeration District (ED) – an area that contains between 175 and 225 households. In Scotland the system is based on postcodes. For the first time geography departments will have the opportunity to access small-area census data using SASPAC91 programs via their university or polytechnic computing services. The SASPAC91 is a software package for the processing of the Small Area and Local Based Statistics of the 1991 Census data.

Name: Urban Information Online (URBALINE)
Type: Bibliographical references with summaries, from 1981 onwards
Host: ESA–IRS (as part of ACOMPLINE)
Owner/producer: London Research Centre (Research Library)
Subject coverage: Urban and local government issues taken from daily and weekly news items, reviews, correspondence and central and local government press notices. Covering environmental sciences, local economic policies, traffic and transportation, housing and social studies, crime and race relations.

Name: Wildscape
Type: Bibliographical references with abstracts
Owner/producer: Nature Conservancy Council
Host: Profile Information
Subject coverage: Mainly UK coverage of nature conservation, wildlife management, farming and wildlife, and landscape/country planning.

V.7
Funding for Travel and Exploration in the United Kingdom

Shane Winser

There is little doubt that the majority of organizations who help finance undergraduates to travel either independently or as part of an expedition (plate V.7.1). do so to help further the education of the individuals concerned. This requires an immense act of faith on the part of these organizations, be they charitable trusts such as those listed below or the directors of commercial companies. Thus the candidates who are most successful in their applications for funding are those who show both enthusiasm and commitment to the ideas and projects they wish to undertake. Of course, clear concise documentation will help to give an air of professionalism but in most cases it will be you, the individual involved, who will be assessed in the context of your project appraisals. Therefore give the person to whom you are applying for funds every opportunity to understand why you wish to do this work, and how it fits in with your previous experience and future plans.

If environmental monitoring or geographical fieldwork is your *raison d'être* make sure you can demonstrate, on paper if necessary, that you have read the appropriate literature, consulted the relevant authorities both in your own and the host country and are conversant

Plate V.7.1 An undergraduate expedition.

with the methodologies you will use. Put the work in context and explain how it will be of benefit to yourselves and a wider, perhaps academic, community.

For those travelling for pleasure, or even adventure, rather than scientific investigation, fund-raising will be harder but by scouring the *Directory of Grant-Making Trusts* (Charities Aid Foundation, Tonbridge) and *The Grants Register* (Macmillan, London) and enquiring through your local library, possible sources of funding should become apparent. Start with local Trusts and work on from there looking for obvious links, perhaps a local company who trades with others in the country you intend to visit.

It will be very hard work. You might just consider it easier, to save a great deal of disappointment, to just try to find evening or holiday work to finance your trip, in which case you will be beholden to no one except for the moral responsibilities you have to your family and travelling companions.

In accepting grants or sponsorship monies from companies, you will be agreeing to give something in return. At the very least a detailed report of your activities will be required, and if you promise photographs do make sure these are sent as soon as possible after your return. For it is the quality of your experience and way in which you communicate it to others which will ease the path of future explorers and travellers seeking funds.

Travel and Exploration: A Directory of Funding for Geography Students

Adrian Ashby-Smith Memorial Trust
c/o Mr Jan Ivan-Duke, 39 Sutherland Drive, Newcastle-under-Lyme, Staffordshire ST5 3NZ

Established in memory of a young man killed during an eruption on Mount Sangay, Ecuador, and given to those taking part in their first expedition. Three awards are given each year in the following categories: member of scientific/exploratory expedition; handicapped member of expedition; member from an underprivileged background. Preference is given to candidates under 40 years of age. Application forms will be sent on receipt of a stamped addressed envelope. Closing date is 1 May with results out in June.

BBC/RGS Mick Burke Award Expedition Film Competition
c/o Wide World Series, BBC Television, Kensington House, Richmond Way, London W14 OAX

A competition for amateur camera persons taking part in an expedition between 1 May and 1 December of each year who would like to make a film of the expedition. All applicants must complete a form obtainable from the BBC which among other questions asks for a sample billing in the *Radio Times*. These forms must be returned by the end of February. The judges then sift through the forms and draw up a short-list of candidates for interview in March. Four teams are then selected to enter the competition which starts with a three-day training course usually in early April, the loan of equipment and film, together with a cash grant of £1000, which is a contribution towards post-production expenses as each camera-person is expected to edit their own film.

The Biological Council
Mrs B. Cavilla, Institute of Biology, 20 Queensbury Place, London SW7 2DZ

The Biological Council gives grants for student expeditions overseas with a specific biological aim or where there is a clearly defined biological component in a multidisciplinary project. The funds are to support or pay for the travel expenses of undergraduate or post-graduate students in the biological sciences. A statement of approval of the aims, proposed methods and planning of the expedition is required from the Head of the applicant's department. On their return from the expedition, successful applicants will be expected to provide the Council with a report within six months of its end. Awards of up to £250 are made in two parts: firstly with a letter of support to assist in further fund-raising and secondly on completion of the expedition to assist in the production of the report.

British Association for the Advancement of Science
Arthur Haydock Bequest and Bernard Hobson Fund, Fortress House, 23 Saville Row, London W1X 1AB

Between two and six grants are made each year for geological research which is part of an expedition or fieldwork project. The value of these grants varies from £10 to £250 each.

British Ecological Society
c/o Dr D. J. L. Harding, School of Applied Sciences, The Polytechnic, Wolverhampton WV1 1LY

Detailed guidelines are available on request but the intention of these awards is to give groups of students of sixth-form level and above the chance to widen their experience by participating in ecological fieldwork overseas. This is more important to the panel than the ability of the group to produce publishable data. About twenty awards of between £50 and £1000 are made each year. Letters of application and supporting documents must arrive before 31 January.

British Sub-aqua Jubilee Trust
16 Upper Woburn Place, London WC1H 0QW

Both amateur and professional divers may apply and the grant may be used for scientific or non-scientific projects as long as the work done involves diving on the aqua-lung. All applications must be made at least four months before the project starts and all applications will be considered by the Trustees on their individual merit. Grants are usually below £750.

Carnegie Trust for the Universities of Scotland
c/o Merchants Hall, 22 Hanover Street, Edinburgh EH2 2EN

Grants for Scottish University undergraduates doing supervised field research expeditions accompanied by a member of staff from a Scottish University. Apply early as decisions are made by March. Up to ten grants of a maximum of £1000.

Winston Churchill Memorial Trust
15 Queen's Gate Terrace, London SW7 5PR

Not really for expeditions or academic research but to broaden the experience of candidates in such a way that they will become more useful members of the community. Categories change each year but explorers, young sports men and women and travellers have qualified under a variety of headings in the past. Obtain an application form and list of categories, and then read the small print carefully. Forms must be returned by mid-October, so beware, this is much earlier than most other grant-giving organizations. Don't be put off if you think you only just qualify. People of all ages and backgrounds have been selected for interview and received these substantial awards (100 recipients of £3000 plus). Becoming a Churchill Fellow carries much kudos and opens many new doors.

Conservation Foundation Young Scientists for Tropical Rain Forests Award

1 Kensington Gore, London SW7 2AR (071-823-8842)

David Bellamy's organization which helps in all matters concerned with conservation from pond building to board games. However, this fund is particularly for ethno-medical and ethno-botanical studies in tropical forests and is worth up to £10,000. But be careful. This fashionable research topic is extremely difficult to do well and will certainly require the full co-operation of the host country.

Eagle Ski Club Georgina Travers Award
c/o Graham Elson, 25 The Mallard, Langstone, Havant, Hampshire PO9 1SS

Two grants of £130 are available each year for ski mountaineering expeditions and training courses. Application forms must be submitted by 12 December and candidates may be called for interview before a decision is reached in early February.

Edinburgh Trust No. 2
The Duke of Edinburgh's Office, Buckingham Palace, London SW1

Small grants are made to eight expeditions each year, and it will help if you have the backing of a recognized society such as the Royal Geographical Society. Closing date 31 January.

Explorers' Club Youth Activity Fund
Explorers' Club, 46 East 70th Street, New York, NY 10021, USA

Although primarily intended for American high school and college students, British undergraduates have succeeded in obtaining funds to enable them to participate in natural sciences fieldwork projects under the supervision of a qualified scientist. Application forms must be submitted by 15 April. Awards vary between US$100 and US$1000.

French Protestant Church of London
The Clerk to the Trustees, Queen Anne's Chambers, 3 Dean Farrar Street, London SW1H 9LG

For those under the age of 25 years of age for help with individual projects at home or abroad, preferably, but not necessarily, in connection with France. Grants will not be given for the sole purpose of learning a foreign language. Preference is given to those who are or whose parents are of the French Protestant Church of London, and to persons of French Protestant descent. However, monies have been forthcoming to a wide variety of applicants. Write before Easter enclosing a budget and a letter of support from an independent referee. The grant will not exceed 50 per cent of the total cost of the venture and has a maximum value of £300.

Ghar Parau Foundation
c/o Secretary: David Judson, Rowlands House, Summerseat, Bury, Lancashire BL9 5NF

For the original exploration, photography and survey of caves; scientific/speleological studies in caves, cave areas or associated features preferably in little-known, little-studied or remote areas. Evidence of experience, ability and research will be needed to complete the application forms which must be in by 31 January. Ghar Parau also administers the Sports Council International Grant for overseas caving expeditions with similar criteria and application procedures. Up to £5000 per application is possible, but under £1000 is more likely.

Gilchrist Educational Trust
229 Great Portland Street, London W1N 5HD

All applicants from British University-sponsored expeditions are asked for a detailed list of information and must submit this by 31 March. About thirty-five to forty awards are given annually and the average grant is approximately £375.

Norman Hart Memorial Fund
43 Northumberland Place, London W2 5AS

Applicants must be between the ages of 18 and 25 and planning short periods of study or study-related work projects that are concerned with further co-operation between Western European countries. Grants will not be made for university or other higher educational studies.

ICBPFFPS Conservation Expedition Competition
International Council for Bird Preservation/Flora and Fauna Preservation Society, 32 Cambridge Road, Girton, Cambridge CB3 ODL (0223-277318)

An enterprising competition which not only gives money but also encourages applicants to adopt a highly professional approach by giving a suggested format for the project proposal with the competition details. These have to be submitted by 31 January. The competition is for university or other ornithological teams planning an expedition with a conservation objective to work in a foreign country. Projects must be feasible and involve local students or other counterparts, have clearance from the host government and local institutes, and have at least one member with an academic background in life sciences. Expeditions within Europe and North America are not eligible. There are four prizes in two categories: birds (£1000 and £800) and all other animals and plants (£1000 and £800).

Jeff Jefferson Research Fund
British Cave Research Association, BCM/BCRA, London WC1N 3XX

All applicants must be members of the British Cave Research Association and must complete the form provided. Grants are made for specific scientific projects in any field of speleology and are worth between £50 and £200. There is no specific closing date and requests are usually considered within twelve weeks.

Manchester Geographical Society
274 Corn Exchange Buildings, Manchester M4 3EY

Grants for academic, educational or research purposes in the field of geographical or environmental enquiry, whether of the past or present, in areas local or distant. Preference is given to applications from the northwest of England. Forms should arrive by the first Tuesday in February but please note that grants are not normally made to individual undergraduates doing research for dissertations.

Mount Everest Foundation
Honorary Secretary: Bill Ruthven, Gowrie, Cardwell Close, Warton, Preston, Lancashire PR4 1SH

Don't think that the MEF is just for climbers, it isn't. Research in high mountain regions is equally important to the Trustees and can attract grants of up to £1500. There are two dates by which application forms must be submitted: 31 August and 31 December. Because of the origins of the Foundation only teams with the majority of their members from Britain and/or New Zealand should apply.

Albert Reckitt Charitable Trust
Southwark Towers, 32 London Bridge Street, London SE1 9SY

Stalwart and regular supporters of expeditions that are endorsed by their university or are Royal Geographical Society approved. There are no application forms: just write a good letter laying out your case before the end of January. Three to five grants of approximately £500 are given each year.

Royal Geographical Society
The Director's Office, Royal Geographical Society, 1 Kensington Gore, London SW7 2AR

The one organization all scientific undergraduate expeditions should approach for both approval and grant aid, as Royal Geographical Society (RGS) approval is highly regarded by other charitable bodies and commercial financiers of expeditions. The application forms are renowned for their length and complexity, but battle on. They should be submitted by 31 January (or 31 August for those going in the winter months). About one in three expeditions is selected for interview by an experienced screening panel before the decision is made to grant 'RGS approval' or not. Only then can the money be allocated to appropriate expeditions from one of the many funds and bequests administered by the Society. These include the Barclays Bank Award, Margaret Busk Fund, Augustine Courtauld Trust, Geographical Magazine Trust, Miss G. M. Gordon Award, Gough Island Fund, Penruddock Park Lander Fund, Barling Fisher Bequest, Sir Henry Rawlinson Fund, H.R. Mill Trust Fund, Percy Appleyard Bequest, Stephens Bequest, Margary Expedition Fund, RTZ Expedition Award, WEXAS International, and The Goldsmiths' Company.

Royal Scottish Geographical Society
10 Randolph Crescent, Edinburgh EH3 7TU

You must have a qualification or training in geography to apply for these awards, and the expedition must have a Scottish base or Scottish membership. The objectives of the expedition must be geographical and all recipients are required to submit a report of findings and a financial statement. Application forms are available from the Secretary and must be submitted by 31 January.

Scientific Exploration Society
c/o Phyllis Angliss, Waterpark, Frogmore, Kingsbridge, Devon TQ7 2NR

The Scientific Exploration Society (SES) Chairman, Colonel John Blashford-Snell, takes a keen interest in these awards. Submit expedition prospectus with covering letter by 31 March and it will be considered at the next quarterly Council meeting. Preference is given to expeditions organized by SES members with scientific objectives. Approval and recognition plus a small grant in the region of £100 is given in exchange for a report of the expedition.

Frederick Soddy Trust
9 The Drive, Hove, Sussex BN3 3JS

Grants for the study of the people and whole community of a particular area. Traditionally many university and college expeditions have applied but should include a staff or graduate member. Remember grants are given to groups only and not to individuals. No application forms are required and you can expectly a fairly prompt reply. Grants of between £200 and £300.

Springboard Charitable Trust
38 Suffolk Street, Helensburgh G84 9PD

Another charity which requires strong Scottish connections. The emphasis is on individual or team effort towards the attainment of specific expedition obejctives. Science or other research is not essential but a post-expedition report is required in consideration of the grant. There are no application forms and no formal closing date but letters of support are required from the sponsoring educational or charitable body. Applications will not be considered during June, July and August.

Rob Thompson Memorial Fund
c/o Steve Newton, Division of Ecology, School of Natural Sciences, University of Stirling, Stirling FK9 4LA

To assist young people under the age of 25 to do fieldwork on expeditions. Prospectus and application forms to be returned by 31 March for summer expeditions. Similar applications later in the year will also be considered for Christmas/New Year or Easter expeditions. A small number of grants of between £50 and £100 are available most years.

Trans-Antarctic Association
c/o British Antarctic Survey, High Cross, Madingley Road, Cambridge CB3 0ET

For fieldwork in Antarctica and associated research by nationals of the United Kingdom, South Africa, Australia and New Zealand. Apply by 1 February. Up to £1200 is available per group.

Twenty-seven Foundation
c/o Messrs Hays Allan, Southampton House, 317 High Holborn, London WC1V 7NL

For expeditions of scientific, humanitarian or conservational value supported by an organization having either exempt or registered charitable status which is willing to accept a cheque on behalf of the expedition if the application is successful. There are no forms to fill in but the covering letter must give particulars of destination, purpose, names and qualifications of the participants and a prospectus, to arrive by 1 March. Average grant £200–£300.

Paul Vander-Molen Foundation
Jack Vander-Molen, The Model Farm House, Church End, Hendon, London, NW4 4JS

To provide opportunities for people with disabilities to enjoy and participate in adventurous activities. Open to individuals, schools, clubs, expeditions etc. Write to Jack to explain your situation and he will advise on the best method of application.

Gino Watkins Memorial Fund and *Edward Wilson Fund*
c/o The Director, Scott Polar Research Institute, Lensfield Road, Cambridge CB2 1ER

Two funds for expeditions to polar regions only. Each Trust requires separate forms both of which should be returned by 1 March. Grants of approximately £400 are available from each.

V.8
Major Institutions
Andrew S. Goudie

Britain

In Britain there is a wide range of geographical institutions, though most of them now co-operate through the recently formed Council for British Geography (COBRIG).

Perhaps the senior institution (founded in 1830) is the Royal Geographical Society (RGS) (plate V.8.1). This has imposing premises (including a library, map room and lecture theatres) in the Kensington area of London, and the map room is open to the general public. Fellows, Associate Members and Educational Corporate Members can use all the facilities and attend lectures and technical meetings. The RGS mounts major overseas research projects of its own, but also gives grants and advice (through the Expeditions Advisory Centre) to undergraduate expeditions (see chapter V.7). The RGS is often regarded as being the prime centre co-ordinating British exploration activities, but in reality its remit is very much broader, and it maintains important links with business, politics and other learned societies, besides publishing the *Geographical Journal*.

Plate V.8.1 The Royal Geographical Society building, Kensington Gore, London.

Housed at the RGS is the Institute of British Geographers (IBG) which was founded in 1933. It is essentially the professional body for British academic geographers, and it holds a major annual conference (usually in deepest winter), has a wide range of study groups (see table V.8.1) representing the diverse facets of geography and publishes two journals (*Area* and the *Transactions*). It encourages postgraduate membership. It maintains a watching brief with respect to the interests of tertiary education.

The interests of the secondary and primary sectors of education, on the other hand, are represented by the Geographical Association (GA), which was founded in Oxford in 1893 and now has its headquarters in Sheffield. In recent years it has been notably successful in influencing government policy with respect to curricular changes in schools. The GA has numerous local branches and produces many publications including fieldwork guides, *Geography, Teaching Geography* and *Primary Geographer.*

Table V.8.1 Study groups of the Institute of British Geographers (1989)

Study groups
British Geomorphological Research Group
Biogeography Study Group
Developing Areas Research Group
Higher Education Study Group
Historical Geography Study Group
History and Philosophy of Geography Study Group
Industrial Activity and Area Development Study Group
Medical Geography Study Group
Planning and Environment Study Group
Political Geography Study Group
Population Geography Study Group
Postgraduate Forum Study Group
Quantitative Methods Study Group
Rural Geography Study Group
Social and Cultural Geography Study Group
Transport Geography Study Group
Urban Geography Study Group
Women and Geography Study Group

Other important geographical institutions in the United Kingdom include the Royal Scottish Geographical Society (which has its headquarters in Edinburgh and was founded in 1884) and Section E of the British Association for the Advancement of Science (which holds summer meetings and field trips, normally at provincial centres).

United States

There is also a diversity of geographic institutions in the United States.

The National Geographic Society, founded in 1888, and located in Washington, DC, produces its world-famous *National Geographic Magazine* for an essentially lay audience. However, it is increasingly concerned with raising awareness of the subject in the United

Table V.8.2 The specialty groups of the Association of American Geographers

Africa	Hazards
Aging	Historical
Applied	Industrial
Asian	Latin American
Bible	Mathematical Models and Quantitative Methods
Biogeography	Medical
Canadian Geography	Microcomputers
Cartography	Native Americans
China	Political
Climate	Population
Coastal and Marine	Recreation, Tourism and Sport
Contemporary Agriculture and Rural Land Use	Regional Development and Planning
Cultural	Remote Sensing
Cultural Ecology	Rural development
Energy and Environment	Socialist
Environmental Perception	Soviet and East European
Geographic Information Systems	Student Geographers
Geographic Perspectives on Women	Transportation
Geography in Higher Education	Urban
Geomorphology	Water Resources

States and promoting its teaching in schools. It also has a large research fund to support geographical research, and publishes the results in *National Geographic Research.*

Rather comparable with the IBG is the Association of American Geographers (AAG, founded in 1904), which holds major annual conferences for professional geographers, has its own 'specialty groups' (see table V.8.2) and publishes various journals, most notably *Professional Geographer* and the *Annals.*

Another venerable US institution (founded in 1851) is the American Geographical Society (AGS). This body, which has a library housed at the University of Wisconsin in Milwaukee and publishes amongst other things the *Geographical Review*, has, like many learned institutions, had some problems in recent decades. The AGS is involved in contemporary academic research, in fostering exploration and in publishing.

Also in the United States is the National Council for Geographic Education, which was organized in 1914 'to promote and improve the effectiveness of education in geography'. Its official publication is the *Journal of Geography*, and it is located at the Indiana University of Pennsylvania.

In Britain and the United States, therefore, geographical institutions are of three types: those concerned with exploration and the popularization of geography; those concerned with academic geography at a professional level; and those concerned with

Table V.8.3 Selected English-language geographical institutions outside the United Kingdom and United States

Country	Institution	Founded	Aims
Canada	Canadian Association of Geographers	1951	The exchange of ideas among geographers in Canada, the fostering of scholarly research in geography and the improvement of teaching in geography. Publishes *The Canadian Geographer*
Australia	Institute of Australian Geographers	1959	
New Zealand	New Zealand Geographical Society	1944	To promote and stimulate the study of geography, particularly the geography of New Zealand and the SW Pacific. Publishes *New Zealand Journal of Geography* (mainly for teachers) and *New Zealand Geographer*
South Africa	South African Geographical Society	1917	Publishes *South African Geographical Journal*
Ireland	Geographical Society of Ireland	1934	Publishes *Irish Geography*
Zimbabwe	Geographical Association of Zimbabwe	1966	Publishes *Geographical Journal of Zimbabwe*
India	Indian Geographical Society	1926	Publishes *Indian Geographical Journal*

geographical education. This is a pattern that is repeated to a varying extent in other countries. It would be impossible to give full details of all such institutions in all countries. However, a selection is given in table V.8.3.

Finally, it is worth making reference to the international body which co-ordinates the discipline on a global basis. This is the International Geographical Union (IGU), a component member of the International Council of Scientific Unions. It holds a major congress every four years (e.g. Sydney, 1988; Washington, DC, 1992).

Part VI
What Next?

What happens when you've finished your geography course and must go out into the wide world? The other parts of this volume should have convinced you that geography is not only worth doing, but is a valuable training for a citizen of the world in the last decade of the century. Geography, however, is not like law or medicine, which seem to lead the graduate into obvious career destinations. Here, David Chamberlain and Joseph M. Cirrincione offer some comments on the employability of geographers, while Tony Binns makes a special plea for more students to return to the subject as teachers. Anyone who has completed a geography course will probably have a desire to travel and see the places they have spent so much time trying to know from books. Nick Middleton suggests some of the ways you can turn this curiosity to your advantage. Five other chapters explain how you can further your geographical interest by doing postgraduate research in various countries.

VI.1
Careers for Geographers in the United Kingdom

David Chamberlain

With some subjects, like electrical engineering, it is possible to state the national demand and to specify the likely employers of graduates. This is not true of geography: the picture is far more varied. There have been considerable advances in the discipline, especially in such matters as analytical and quantitative techniques, so that geographers have come to fill vacancies in fields which were formerly the preserve of scientists. Nevertheless, whether the bias in their course has been towards science or social subjects, the majority of graduates take posts which appear not to have too much immediate connection with their studies. Such is the record of success that this should not be seen as discouraging but rather as confirmation of the wide range of careers open to geographers and the fascination of the topic. In this chapter we examine those areas of employment, explain something of the patterns of success and describe how those emerging from higher education approach the choice of a career.

The relationship between the subject of degree study and employment is complex, and misconceptions abound. Perhaps it was easier in the 1960s when almost two-thirds of British geography graduates went into planning or teaching. Both of these remain substantial recruiters on a national level but at a much reduced rate. In the case of the first, the expansion in local government in the 1970s involved strong recruitment, which was inevitably followed by a dearth of openings for planners as local authorities could not maintain the level of recruitment. Now we are into a new phase, with more encouragement to the prospective planner than in the last decade, although not to the same extent as fifteen years ago. Environmental issues are more prominent and some background knowledge of planning can be a useful acquisition.

Some 335 geographers are selected for postgraduate teacher training each year in Britain, and most of these take up teaching vacancies in secondary schools or colleges at the end of their one-year course. A somewhat smaller number train to teach in primary schools, not necessarily losing sight of their geography but essentially preparing for more general classroom teaching. The demand for teachers fluctuates but will rise again substantially when the secondary school rolls begin to rise again from 1991–2. At primary level there is already a strong need, although the variation depends rather on location: the prosperous areas of the southeast with their high cost of living have more vacancies than the north.

In contrast, the shortage of lectureships in higher education is well documented. Teaching at this level is an option only for those who, after an outstanding record on their first degree course, go on to take a further degree, probably a doctorate. Such academic posts arise infrequently, but this problem is not one exclusively associated with geography.

Readers will find elsewhere descriptions of geographical studies and some of the more scientific and technical topics. These developments have certainly enhanced the range of jobs which geographers can contemplate. There are opportunities in civil engineering and consultancy, surveying and cartography, landscape architecture and all kinds of environment planning for those who have taken degree courses associated with field experience of data gathering and problem-solving, information retrieval, systems, map analyses and design, automated cartography, remote sensing and environmental monitoring – or perhaps have taken a first degree in geography and have gone on to do further study in such topics. It is always a sound principle to choose options carefully, ensuring that the relevant ones are taken. Equally important may be the early identification of fields of interest and the gaining of related work experience during the vacation. There are not necessarily sufficient openings in these technical fields in a given year, nor do all students wish to pursue these lines. However, the skills acquired, notably in computer and systems work, are keenly sought by a range of employers. Thus, although they may not be employed as geographers, such graduates are making direct use of the skills acquired during their higher education studies.

A healthy minority of graduates, then, use their geography as their major qualification in choosing their career, especially those who have concentrated on the physical and scientific aspects and can develop its technical applications, while those in the educational field are ensuring the future of the subject. How are the others employed? For these graduates, the process of their study is more significant than the subject matter. This statement implies no challenge to the value of geography as an education: indeed, there are advantages over some other disciplines, whose graduates have traditionally entered varied occupations which have little obvious connection with their courses of study.

The first of these advantages results from the fact that geography bridges the natural and the social sciences, encouraging in its graduates the better qualities of both disciplines. The range of skills which can be acquired is unique. Some courses insist on a pass at Advanced Level in mathematics and many others require an ability in that subject. Numeracy is a much prized quality, frequently mentioned by employers who otherwise consider graduates from any subject background. Not only are quantitative techniques so well practised but most courses develop powers of statistical analysis and an ability to interpret and present complex data. The presentation and analysis of data in graphs, charts and diagrams can be invaluable skills. Most of all, the use of computers, not only for graphics but also for data gathering, analysis and information retrieval, will be an eye-catching item on job applications, as mentioned above. There remains a wealth of opportunities in computer programming and systems analysis, and any experience incorporated into a degree course must give useful pointers to employers about likely potential in that area of work.

The methodical approach may not be unique to geography, but it is nonetheless a virtue. Likewise, an articulate nature should be encouraged and developed in tutorials and associated exercises. Nor should geographers lack literacy, since essays, seminar work and projects or dissertations are common features. In short, the geography graduate should command the four skills of communication: the various facets of numeracy, graphicacy, articulateness and literacy.

It is important to look for these aspects of courses and to ensure that the selected one has the right balance. There is another extension to higher education, however, which in career terms can be equally important. It is striking how some departments, apparently teaching the same subject, seem to give added value to the product of their course. Living and studying among intelligent contemporaries can be such a maturing process. The opportunities for extracurricular activities, clubs and societies of endless variety are also part of the general educational process. Not only do they bring enjoyment and memorable experiences but they help to develop personal, social and transferable skills, especially if students have taken an active role in the planning, promotion and organization of such activities.

More and more secondary schools in Britain give help and some practice in preparing profiles and records of achievement. Young people will in future, it is hoped, have developed more ability in self-analysis and in defining their skills. For most people, this is the first stage in the process of career decision-making. Then comes the matching of these personal characteristics to the requirements of various careers. The next stage is the investigation of job vacancies within the preferred areas. Finally, the application procedures must be followed. In some cases, this will involve the compilation of a curriculum vitae, a skeletal outline of personal details, qualifications, aptitudes, interests, so that in the letter and at interview the applicant can expand in a positive manner on how skills and personal characteristics listed fit the profile given in the job description. For some students, this can be a new exercise, even an intrusion into study time. Careers advisers are available for help, but the wise student embarks early on the process, even if the actual jobs-search is deferred. Most large organizations have their own application form but the self-analysis and the careful choice of the manner of self-promotion are still important exercises in the preparation for approaching employers.

And who are these employers? Geographers have the flexibility to respond to the constantly changing demands. In recent years, the financial sector has offered most vacancies. The recruitment into chartered accountancy has risen steadily, so that it has become one of the most regular introductions into business management. The other branches of accountancy – management, certified, public finance – also have a wealth of vacancies. The last one, for example, has given geography graduates, among others, a professional qualification and an insight into the financial control of local authorities and some central government departments too. The training schemes for the clearing banks are well known but the banks have diversified their services and consequently the functions into which they recruit. By the same token, the building societies have become more prominent and have begun to make serious impact on graduate recruitment.

Actuarial vacancies are restricted to those with strong mathematics: the occasional geographer who especially enjoys the statistical side has been welcomed. Insurance has additionally various roles which graduates can fill; the most likely are claims and underwriting, but geographers may find appealing the opportunities in surveying. Chartered surveying has been mentioned above but in this section come the increasing opportunities in the valuation and management of property. Estate agencies are now less given to insisting on a cognate degree, as they too develop a wider range of services.

Geography graduates are certainly among those selected for the management training schemes of the nationally known retail groups, but they have also been attracted by the distribution function in manufacturing industry. Transport has always been closely associated with geography and there are some good schemes of training with the organizers of rail, road, air and sea transport, both on the operational side and in such roles as sales, finance, personnel and management services.

Aptitude in computer programming will again be a key quality in the last-mentioned function, which is also an entry into many areas of work – from local government to manufacturing industry, from information work to banks. Geographers more readily see themselves in market research, as they feel familiar with the methods of social survey; a few make the grade, but it is not generally recognized how comparatively few openings there are in any given year. In contrast, the leisure and recreation field continues to develop strongly. Traditionally more associated with temporary and seasonal jobs, it now begins to offer more chance of career positions and progression. Relevant vacation experience, as in so many other work areas, helps to give the graduate applicant credibility.

Some graduates have shown particular enterprise in setting up their own business soon after the completion of their studies, not least in the provision of leisure-time activities. Others with a commercial outlook, perhaps after the Graduate Enterprise Programme, have begun to produce, market and sell innovatory items, responding to customer demand in our affluent society. Most large corporations have graduate training schemes but rarely do they recruit into general management at this level, but rather into a specific function such as sales, marketing, production, distribution or management services – all of these are open to the geographer.

Of the Civil Service departments, probably Environment and Transport have most immediate appeal, but the Department of Trade and Industry, the Foreign and Commonwealth Office and the Ministry of Agriculture, Fisheries and Food could be equally attractive. In the competitions for entry as Administrative Trainee and Executive Officer, there are other departments with comparable opportunities: indeed, the Deputy Director of the Civil Service Selection Board is a graduate in geography.

It would be impossible to list all those employers likely to be open to those who complete a course of higher education in geography but this is a selection of some of the most regular choices.

Geography is not a profession, but many geography graduates do become true professionals in the educational and scientific worlds. Unemployment rates from higher education courses in the subject are low; indeed, from some institutions, very low. Characteristics which make graduates employable may not have been determined precisely by the subject of study but more probably by the process of study, although the bridging of the arts and science, the enhancement of skills in numeracy and statistical analysis and the use of computers can give geographers an advantage over many others. As with most other disciplines, a significant minority take some form of further qualification: finance and law are nowadays as common as the more traditionally preferred teaching. Britain still sets high value on the potential of graduates and will employ them in the widest range of functions. Thus, about 40 per cent of jobs notified to graduate careers services are open to those of any subject background, and geographers are well represented among the successful applicants. The possibilities for career development are then limitless.

In short, a geography degree is a valuable qualification in the contemporary labour market, with personal characteristics being just as important in the competition for most entries. Choosing the right career is probably one of the most difficult decisions in life but insights into personality which come with the wider experience of life afforded by higher education help graduates to relate to the multitude of available opportunities.

VI.2
At the Chalk Face: Teaching Geography in the United Kingdom

Tony Binns

Teacher Supply

Britain needs good (geography) teachers!
(plate VI.2.1)

In fact the present need is greater than for many years – why? Geography has been designated a foundation subject in the National Curriculum, which means that for the first time all children in state schools will be required to study geography for the period of compulsory schooling between the ages of 5 and 16. So more children will have to be taught geography. However, in most primary schools there is no specialist geography teacher and in many secondary schools geography is taught by teachers with qualifications in other subjects. The situation will get worse as rising pupil rolls in the primary sector feed into secondary schools in the 1990s at a time when the full National Curriculum is being implemented.

Looking at the supply side, the number of geography teachers being trained has declined steadily since 1972. In that year some 717 geography graduates were admitted to Postgraduate Certificate in Education (PGCE) courses, whereas the

Plate VI.2.1 Geography teacher leading a geography field trip.

figure for 1984 was only 346, a reduction of more than 50 per cent. The other route into teaching, the Bachelor of Education degree (BEd), has also suffered major cuts from a high point in 1972. Much of this decline in the training of geography teachers has been due to the retirement or redeployment of specialist staff in institutions of higher education. There seems to have been no nationally co-ordinated reduction in the number and location of training places. To make matters even worse, the number and calibre of students applying for these fewer places seems to have declined in recent years. It seems that geography graduates are now very employable in a wide range of other jobs and careers, and the days when half or more graduates went into teaching are no more (Briggs, 1988). One can only conclude from this somewhat depressing survey that there is an urgent need for an injection of good quality geography specialists into all sectors of the teaching profession.

Is Teaching for Me?

But, one may ask, is teaching really for me, and in any case is it a good time to enter the profession? If you read the newspapers or watch the television, teaching seems to be surrounded by gloom and doom! Low pay, low morale, low public esteem, coupled with a barrage of inadequately resourced educational innovations, make many think twice about a career in teaching. Such a climate in schools, one could argue, might serve to put off all but the most dedicated from entering the profession. But is it really as bad as the media portray? Is teaching still a worthwhile career? These questions need to be answered before you go further and apply for a teacher training course.

One thing in teaching that has not changed over the years is the children! They may be more sophisticated than twenty years ago, but the majority are still fresh-faced, impressionable and very eager to learn. Teaching more than anything enables you to play a vital formative role in the development of young people's personalities and intellects. There can be few, if any, other careers where this is possible. Enthusiasm for one's subject should not be the only reason for wanting to train as a teacher, but it is nevertheless a great asset. Perhaps of greatest importance is an interest in working with young people and an ability to motivate and communicate with them. You should be able to empathize with children – put yourself in their shoes. You should be adaptable, versatile, have common sense, an ability to organize, a sense of humour and if possible a bit of charisma! Does anyone have *all* these qualities, you might ask? In truth the answer is probably no, and for many teachers it takes years of practice and experience to acquire some or all of them.

You may want to become a teacher because you enjoyed school and did well yourself. But you should remember that you are one of the successes of the education system. Many children experience difficulties, lack motivation and are likely not to have their sights set on A-level and university. If you think you would like to become a teacher, arrange to visit a school, or preferably several, and talk to teachers. Ask them about their work and see them teach. How do they relate to and motivate children? What are the good and bad points about being a teacher? Try to weigh up the pros and cons of, say, discipline problems and marking on the one hand with reasonably long holidays and opportunities to organize and take part in school trips and fieldwork on the other.

You should try to become familiar with some of the current educational innovations in schools, such as the National Curriculum and other aspects of the 1988 Education Reform Act, the most wide-ranging piece of educational legislation since 1944. Schools are currently considering or introducing such topics as profiling, appraisal, new methods of assessment and modular courses. Some are heavily involved with the TVEI (Technical and Vocational Education Initiative and CPVE (the Certificate of Pre-Vocational Education). The General Certificate of Secondary Education (GCSE) is well under way, replacing both GCE O-level (General Certification of Education) and CSE (Certificate of Secondary Education) at 16-plus. It is likely that A-level will have to change so that it becomes a more natural progression from GCSE rather than the now defunct O-level/CSE. Should A-level students take more courses to give them greater breadth of knowledge or is depth more important? Already some schools have introduced shorter AS-level (Advanced Supplementary) courses alongside more traditional A-levels. Some teachers would suggest that a complete re-formulation of post-16 work in schools is long overdue. The pace of change in our schools is indeed very rapid and the new initiatives, as with so much in the educational world, are full of jargon. But don't be put off! There is a wonderful opportunity for young teachers to play a major role in formulating and implementing many of these innovations.

Another important consideration for prospective teachers is the age group you would like to teach – infant, junior, secondary, sixth form, tertiary or further. You may wish to work with the full primary (5–11 years) or secondary (11–18 years) ranges, or alternatively specialize with a narrower age range such as infant (5–7 years) or sixth form (16–19 years). There are many possibilities, but again you should talk to teachers and visit a number of different schools and colleges. If you are particularly keen to teach your subject for most of the time, then secondary or tertiary work is more appropriate. In a primary school you are likely to be responsible for a class, to which you will teach a range of subjects and topics across the curriculum. In all schools you may be asked to teach subjects other than your specialism and you will be required to spend a considerable part of each day engaged in pastoral work, marking and a range of administrative duties. You should be prepared on a day-to-day basis to relate in different ways to pupils, teacher colleagues, heads and deputies, inspectors, advisers, parents and governors.

Geography in Schools

For those who really want to get on in teaching there are a great many different career paths, and incentive allowances have been introduced over and above the main professional pay scale to reward those with extra responsibilities. Because of the eclectic nature of geography as a discipline, geographers are seen as being versatile and in possession of a wide range of skills. It is quite possible to be a head of department in a large comprehensive school before you reach 30, a deputy head by the age of 35 and a head by 40. There are many schools where senior positions of heads and deputies are occupied by geographers. However, it should be pointed out that the teaching profession overall has a 'flat' career structure with about half of all employees in state schools on the main professional grade with no incentive allowances.

Of all the subjects in the school curriculum, geography must be one of the most exciting and dynamic. Many would say that the subject has the most interesting and innovative textbooks and other resources such as computer programs, video and fieldwork materials. In the case of the latter, geography teachers have developed considerable expertise in field investigation over many years and fieldwork is a compulsory yet popular element in virtually all syllabuses. The subject has benefited from having three curriculum projects funded by the former Schools Council. These are 'Geography for the Young School Leaver' or the Avery Hill Project (1971–86); 'Geography 14–18', often known as the Bristol Project (1970–5 and 1978–81); and 'Geography 16–19', an A-level project based at the Institute of Education at the University of London. These projects have produced some excellent syllabuses and resources which are the envy of many other school subjects.

The main subject association, The Geographical Association (GA), founded in 1893 and with a current membership of over 10,000, mainly teachers, produces three journals – *Geography, Teaching Geography* and *Primary Geographer* – and has an impressive array of up-to-date publications relating to the teaching of geography at different levels. The prospective geography teacher could gain much from becoming a student member of the GA and getting involved in local branch activities. The GA has worked hard to ensure the vitality of the subject in schools and played a key role in getting geography recognized as a foundation subject in the National Curriculum.

Geography is a highly popular subject in schools, such that in the first GCSE examination of June 1988, geography was the fourth most popular subject after mathematics, English language and English literature. With some 295,163 candidates, geography was in fact the most popular of the non-compulsory subjects. A-level entries and applications to study geography at universities, polytechnics and colleges are equally impressive. The subject's popularity is undoubtedly due partly to the way it is taught in schools and the resources available for this. But the employability of those with a qualification in geography, together with the increasing concern for environmental issues which are seen as being within the domain of the subject in schools, must be other good reasons for the buoyancy and popularity of the discipline – all the more important to ensure an adequate and well-trained body of good teachers for the future.

Taking it Further: Choosing a Course

If the idea of becoming a teacher appeals to you after you have visited some schools and talked with teachers, then you should identify which course seems most appropriate for you. The so-called *NATFHE Handbook* is a useful resource for considering the range of different initial teacher training courses (NATFHE, 1989). There are two routes into teaching: a three- or four-year B Ed course or a one-year PGCE taken after you have obtained a first degree, normally a BA or BSc. The advantage of the B Ed route is that in each of the three or four years you will have an opportunity to gain experience in schools. The total time spent in schools will be very much greater than for students taking a 36-week PGCE. One possible shortcoming of the BEd is that it is very much geared to producing teachers, and if you should wish to change career at some later stage then a BA or BSc might be regarded as more versatile and acceptable by potential employers. Furthermore, the depth of knowledge obtained in a specific academic discipline is likely to be greater from a three-year BA or BSc course, which could put you in a stronger position for teaching at sixth-form level and above.

If you already possess a first degree (usually BA or BSc) then a PGCE course is the most appropriate route into teaching. No particular class of degree is formally required, but it is one of a range of factors taken into account by those involved in the selection process within individual university or college departments. A degree in geography is preferred, but not essential, for those wishing to train as geography teachers. Degrees in environmental science or development studies, for example, may be regarded as acceptable by individual departments. In addition to a degree, applicants for PGCE are required to have obtained a grade C or better in English language and mathematics at GCE O-level, GCSE (or equivalent) level (e.g. CSE grade 1; CEE grades 1–3; SCE). You should write for an application form and course booklet to the Graduate Teacher Training Registry (GTTR), which acts as a clearing house and sends your application to the four institutions that you have listed in order of preference.

In your search for an appropriate course you should also write to individual institutions asking them to send course prospectuses. From detailed study of these documents you will see that PGCE courses vary considerably. You should look for evidence in the prospectus on the balance between theory and practice and how this is organized. How much time will you spend in school? Will this be a continuous school experience on a two or three day a week basis throughout the course, or two or more block practices of perhaps five weeks spent away from the college or university? Increasingly popular with the teaching profession, the Department of Education and Science and Her Majesty's Inspectorate (HMI) are courses where theory and practice run alongside each other, with students spending typically three days in a school and two days each week in the college or university. These 'school-based' courses enable students to obtain a good amount of classroom experience over a longer period of time and may culminate with a block practice (perhaps three or four weeks) in the school. Students are encouraged to use the two days spent each week in college or university to relate their practical experience to educational theory and broader issues and to reflect upon their schoolwork. Such a close and continuous juxtaposition of theory and practice in school and college or university seems preferable to having a term divided into, for example, a discrete five-week block of college or university work followed by a five-week block of school experience. A common complaint of students on PGCE courses, particularly of the block practice type, is that there is too much educational theory which is not related to classroom experience.

Other important aspects of the course which might be gleaned from prospectuses include the manner in which the PGCE is assessed. Does the course have written examinations? What role do practising teachers play in evaluating students' work? You might also look for evidence of close interaction between school and college or university teachers. For instance, do school teachers contribute to teaching in the college or university? You should enquire whether there is a published HMI report on the course that provides an independent evaluation.

Application and Interview

To have a reasonable chance of securing a place in the institution of your first choice you will need to submit your application during October of the year before proposed entry. The application form should be completed in as much detail as possible and particular care should be taken in listing experience of working with young people. In the section of the form headed 'Additional Information' you should give a clear and carefully worded statement on your reasons for wanting to enter the teaching profession. You should approach two persons who know you well to act as referees and can report on your academic background and your suitability for teaching. They should be ready to write their references as soon as requested by the appropriate institution, usually your first choice. Delays in submitting references can slow down the selection process and may jeopardize your chances. If for some reason your first choice of institution rejects you, then GTTR will send your application to the other institutions in your order of preference. If you are unsuccessful in all four choices, GTTR usually continues to send your application to other institutions which still have vacancies in your field of interest.

Before offering places on PGCE courses most institutions will want to see applicants for interview. The interview is usually fairly informal and focuses mainly on practical issues relating to teaching. You may be asked some simple questions about developments in education and it is useful to have followed the educational columns in a newspaper such as *The Times*, *The Guardian* or *The Independent*. The weekly *Times Educational Supplement* is particularly recommended. An experienced school teacher who is familiar with the course may be one of the two or three people conducting the interview. You will have a chance to ask questions about the course and the schools used for school experience and to look around the institution and evaluate facilities such as the library, bookshop, accommodation etc.

When you have been offered a place you should try to become more familiar with developments in the educational world and perhaps arrange to visit a school on a regular basis. Before starting the course you will almost certainly have to undertake a period of observation in a school. The PGCE course is rewarding but hard work. Students often find the work in school most enjoyable, but soon realize that this requires considerable physical stamina!

Concluding Thoughts

It should be remembered that both the B Ed and PGCE are *initial* training courses. A considerable amount of on-the-job training will follow during the probationary period, when many schools will give newly appointed teachers a lighter timetable to enable them to undertake further training and settle into teaching. Help and advice are always close at hand during this early phase. There is always more to learn, and if you are to grow in your job you should adopt a positive attitude to in-service training (INSET). Many teachers after a few years in the profession enrol for specialized courses, perhaps leading to a diploma or Master's degree.

Teaching is a challenging and rewarding profession where every day brings a different set of experiences and interactions. Geography is a dynamic, forward-looking and topical school subject with first-class resources and support from organizations such as the GA. 'Geography is going places' in more senses than one, but the continuing success of the subject depends on a regular influx of high calibre teachers. Are *you* still interested?

Background Reading

Bailey, P. and Binns, J.A. (eds) 1987: *A Case for Geography*. Sheffield: Geographical Association.

Boardman, D. (ed.) 1986: *Handbook for Geography Teachers*. Sheffield: Geographical Association.

Briggs, J. 1988: Jobs for geographers: some hard evidence. *Geography*, 73, 2, 137–40.

Department of Education and Science (DES) 1987: *Quality in Schools: the initial training of teachers – an HMI survey*. London: HMSO.

Department of Education and Science (DES) 1988: *The New Teacher in School: a survey of HM Inspectors in England and Wales, 1987.* London: HMSO.

Her Majesty's Inspectorate (HMI) 1985: *Good Teachers*, Education Observed Series No. 3. London: Department of Education and Science.

Her Majesty's Inspectorate (HMI) 1988: *Initial Teacher Training in Universities in England, Northern Ireland and Wales*, Education Observed Series No. 7. London: Department of Education and Science.

Lee, R. 1985: Teaching geography: the dialectic of structure and agency. In D. Boardman (ed.), *New Directions in Geographical Education*, London: Falmer Press, 199–216.

Mills, D. (ed.) 1988: *Geographical Work in Primary and Middle Schools.* Sheffield: Geographical Association.

Molyneux, F. and Tolley, H. 1987: *Teaching Geography: a teaching skills workbook.* London: Macmillan.

NATFHE (National Association of Teachers in Further and Higher Education) 1989: *The NATFHE Handbook – the handbook of initial teacher training, other degrees and advanced courses in Institutes/Colleges of Higher Education, Polytechnics, University Departments of Education in England and Wales.* Mansfield: Linneys ESL.

Richardson, R. 1983: Daring to be a teacher; In J. Huckle (ed.), *Geographical Education: reflection and action*, Oxford: Oxford University Press.

Wiegand, P. (ed.) 1989: *Managing the Geography Department.* Sheffield: Geographical Association.

Useful Addresses

AGCAS (Association of Graduate Careers Advisory Services). Various publications on careers in teaching and advice on applications and interviews are obtainable from Careers Advisory Services in universities, polytechnics etc. or directly from their publishers: Central Services Unit, Crawford House, Precinct Centre, Manchester M13 9EP.

DES (Department of Education and Science) publications, including HMI reports, are available from DES Publications Despatch Centre, Honeypot Lane, Canons Park, Stanmore, Middlesex HA7 1AZ.

GA (The Geographical Association). Membership details and publications are available from the Association's headquarters at 343 Fulwood Road, Sheffield S10 3BP.

GTTR (Graduate Teacher Training Registry). PGCE application forms from GTTR, 3 Crawford Place, London W1H 2BN.

HMSO (Her Majesty's Stationery Office). Various publications on educational matters from HMSO bookshops or by mail from PO Box 276, London SW8 5DT.

NATFHE Handbook is available in libraries or direct from the printers: Linneys ESL, Newgate Lane, Mansfield, Nottinghamshire NG18 2PA.

TASC (Teaching as a Career). Various leaflets on teaching are available from TASC Publicity Unit, DES, Elizabeth House, York Road, London SE1 7PH.

VI.3

Careers and Employability of Geographers in the United States

Joseph M. Cirrincione

Someone in the United States contemplating a career as a geographer faces a challenge from the very beginning. First, there is no clear-cut category called 'geographer' which can provide data as to demand or employability. Second, geographers in the United States often face the problem of defining their professional field to the public. The current concern about 'geographic illiteracy' in the United States has renewed interest in geography. However, the interest is in 'place-name geography', or knowing where major political and physical features are located. This narrow definition or view of the subject poses a problem with the general public and potential employers concerning geography as a career or professional field.

This in part reflects the comfort people generally feel in defining a field or profession by a specific content orientation or set of activities, such as electrical engineer, chemist or teacher. Someone contemplating a career in geography will not be allowed the luxury of simply saying 'I'm a geographer' and assuming that employers or the general public see a clearly defined role. Rather, you as a geographer must define yourself, your skills and the contribution you are capable of making to a business, an agency or an organization.

To select geography for career training is to accept the fact that your chosen field is best identified by the type of questions asked, the approach used, the data generated and organized and the necessary skills to deal effectively with the challenges posed. The various chapters in this volume give some indication of what those questions might be.

Rather than list careers commonly associated with geography, this chapter will present a basic approach that any student can adopt in developing or making decisions about career training in geography. It will also identify basic attributes associated with three career categories commonly used in the United States.

The usual undergraduate geography curriculum in the United States is organized to serve both the liberal arts and the career goals of a degree programe. When considering a career, any programme should be viewed in terms of flexibility and structure. A student should make a critical assessment of their strengths, interests and temperament. The business world emphasizes economic questions, and a concern for profitability and productivity. By contrast a career with a government agency such as the Geological Survey will require quite different specializations and temperaments. The programme chosen should be flexible enough to allow for the development of such qualities.

Most degree programmes in geography provide a structure based on work in geography, related fields of study and geographical techniques. The geography component provides the opportunity to develop a geographical perspective, to acquire a sound knowledge base and to analyse problems from both physical and human perspectives. It also allows for topical specialization which may be supported by course work in related fields. The selection of these fields may also be related to a specific career description. For example, a specialization in physical geography and historical geography, with recreation as a related field, can be a good preparation for a career with the National Park Service.

The development of skills associated with geographical techniques is also relevant. Familiarity with cartography is necessary for any geographical analysis, while the growing use and importance of geographical information systems in business, government and planning agencies has increased the need for computer literacy. The explosion of data from satellite imagery has also extended the demand for remote sensing.

The term geographer is rarely used either in the title or the description of a career position. Three areas that have a demand for people trained in geography and have provided employment opportunities are business, government and planning. Although the boundaries between these three fields are not impermeable, their goals differ and they are associated with selected career attributes and characteristics.

Business

The world of business offers a wide range of opportunities for geographers (plate VI.3.1). The titles commonly associated with geographers in business reflect this diversity: demographer, transportation analyst, cartographer, geographic information specialist, health care planner, market researcher, industrial planner, travel agent, land-use consultant, real estate developer, environmental manager, recreation planner, location research analyst, resource analyst, and so the list goes on. The only limits are set by the geographer's imagination and training. That geography deals with questions of location, where things are, is in fact the major asset allowing the development of challenging careers.

There are characteristics of the business world with which a person must feel comfortable and which they must see as challenges. One must be able to function in a highly competitive atmosphere. The pressure of competition is invariably coupled with the pressure of deadlines, with many projects operating on a short-term 'end result' basis. A 'staff' position may involve providing advice and recommendations based on research and analysis. A 'line' position may place you in the position of having to make the decisions yourself and assume responsibility. In either case the end result of business is to make a profit and business people are expected to share this goal.

The geographer brings to the business world unique skills of being able to combine natural and social science knowledge and research. The training allows you to work with a diversity of topics and data, and to cross boundaries when solving problems. Yet, the geographer is not a generalist. Your training operates from a disciplinary perspective and with a disciplinary rigour. It also provides you with effective geographical information systems, quantitative and presentation skills, giving you a competitive edge at the entry level. Training in supporting fields such as business, economics, transportation and environmental sciences strengthens the topical dimension of your business career choice and enhances your communication in that area.

According to the Department of Labor the demand for people with training in geography will keep pace with overall employment increases. Continued emphasis on monitoring rapid changes in the environment and in urban areas has generated the greatest demand for

Plate VI.3.1 The Manhattan skyline.

environmental and urban management, geographical information systems, resource planning and health planning. The need for impact studies, either for business projections or governmental regulations, has given additional impetus to the need for location analysts, with many firms looking to consulting services for this advice.

Government

Government employment has been a fruitful area for many geographers, although the title is seldom used. They have found employment in a wide range of governmental agencies at the Federal, state and local levels. The agencies employing geographers vary significantly. At the Federal level, for example, geographers work for the Bureau of the Census, the US Geological Survey, the State Department, the Central Intelligence Agency, the Environment Protection Agency, Defense Mapping, the Energy Department, the Corps of Engineers, the Transportation Department, the Library of Congress, and the Archives. Comparable positions are found at state and local levels with a greater emphasis on economic development, natural resources, transportation and planning. Some positions are comparable with the business 'staff' position, concentrating on research, while others are 'line' positions involving the enforcement of regulations or the provision of governmental consulting services.

Government agencies operate on mandates which define their objectives and scope of activity, and they recruit geographers with topical specializations in line with their mission. The opportunities available in an agency will vary as government priorities shift. Currently there is a high demand for geographers with training in cartography, physical geography, remote sensing and computing, while a strong background in mathematics and statistics is also demanded. This reflects the expanding use of technology to monitor the environment and the use of geographical information systems. The increased need for environmental impact studies has expanded the opportunities for geographers with training in both the physical and human sides of the discipline. Those with foreign area specialization and training in foreign languages have opportunities with the State Department and intelligence agencies. The continued expansion of the global economy has created openings for geographers interested in international trade, economic development and location analysis. A training in computer mapping continues to provide entry-level advantages at a wide range of agencies.

Geographers interested in a career in government must realize that they will be working in an environment that tends to be more regulated than is found in the business world. It is also a world where goals tend to be more complex and not defined by the generic term 'profit'. It requires a person who can set and achieve goals based on the objectives of their agency. A personal motivation must be based on completion of projects and providing a public service, in addition to the basic career aims of financial security and continued personal growth and development. As in business, government agencies are increasingly concerned with the productivity and the development of their personnel, requiring people who continually take advantage of opportunities within the regulated world of government work.

Planning

Planning is a future-oriented problem-solving profession. It not only deals with the 'what is' but also has a legitimate concern with the question of 'what should be'. Planners attempt to meet the social, economic and even physical needs of a society by devising strategies or plans of action. The range of areas of planning is varied. Geographers often think of land-use planning as the major area having a strong affinity with their training. Planners trained in geography are also involved in transportation planning, community development, health and human services planning, emergency planning, historical preservation, economic development, regional development and Third World development. Specialists in physical geography are sought in planning agencies dealing with resource development, physical planning and natural hazard planning.

Employment openings for planners are primarily at the local level with city, county and regional planning agencies dominating. Planning at the state level is an important function, with the range of positions being extensive, though the numbers are limited. Careers in planning at the national level are usually found within major Federal agencies. Some, such as the Department of Housing and Urban Development or the Agency for International Development, would obviously have a higher demand for planners. Finally, more and more planners are finding opportunities in private industry, especially in consulting firms involved in providing public services such as utilities and banks.

Geographers contemplating a career in planning should realize the attributes necessary to function effectively. A planner operates primarily in the public forum. They must be able to examine issues from diverse, and at times conflicting, points of view. They must be effective listeners and enjoy working with a variety of people, both professionals and the general public. Planners are advocates of change who must not only devise effective plans for implementing change but also communicate and educate clients or potentially hostile patrons on the need for change.

Regardless of the specific role, a technical analyst, researcher or someone dealing directly with the public, one must be able to deal competently with public scrutiny and evaluation.

Conclusion

Contemplating a career in geography does require a person who is willing to develop a clear picture of their strengths and interests. It requires research on current and future career openings and the type of training required to obtain and grow with the position. It requires the selection of a programme allowing the effective integration of geography, geographical skills, and work in supporting fields of study.

Career opportunities in the United States for people with training in geography are extensive and expanding, and can accommodate a wide range of interests and strengths. The current emphasis given to applied geography attests to the increased role of geographers outside academic settings. Geographical professional organizations in the United States (see chapter V.8) are actively involved in expanding career chances by providing information on career training and positions. Geography is clearly not viewed by the profession as a purely academic or school subject. The only limitation is the imagination and drive of the geographer, whether a beginning undergraduate or one undertaking graduate training.

VI.4
Postgraduate Studies in Australia

Michael Taylor

Undertaking postgraduate research can be very valuable and rewarding, and moving to another country to undertake that work adds significantly to the depth and breadth of a researcher's understanding. To do PhD research in Australia you will need a good first degree – a first class or a good upper second class honours degree.

There is a continent of research opportunities in Australia but a prerequisite is that you must be sufficiently committed to the topic you choose to carry you through two or three years of specialized work. You cannot do research if your motives are those of a tourist.

Not all universities in Australia offer opportunities for postgraduate study in geography. There are eighteen principal geography departments in Australia, but geographical research can also be undertaken in the social science, applied science and environmental studies schools of a number of newer and recently designated universities.

Research Topics

Australia offers a staggering array of geographical research opportunities (plate VI.4.1). Its physical environments stretch from the tropics through deserts to the temperate latitudes, and there is a strong research commitment in Antarctica. The ocean environments surrounding the continent are varied and unique and the issues of climate and, more recently, climate change are pressing.

As a highly urbanized, multicultural society that continues to be one of the world's major migrant destinations, demographic and social issues are also very prominent in Australia. Equally, as the country endeavours to restructure its economy, to diversify away from primary production, to strip away the barriers of economic protection and to integrate more fully into the international economy, an array of significant economic, regional and social issues is emerging which need urgently to be addressed. Australia is also shifting its orientation towards Asia, and the process of identifying Australia's role and growing integration into Asia offers major opportunities for geographical research of all types.

Scholarships and the Money Question

Postgraduate research is expensive and requires money to cover students' fees, living expenses and research costs (especially the costs of fieldwork). As in all countries, overseas student fees in Australia are substantial, virtually ruling out privately funded postgraduate research. For geography, overseas student fees currently range from $A12,000 to $A15,000 a year depending on whether the research is social science in nature or laboratory based. Almost invariably, therefore, overseas students will need to secure a scholarship, and

Plate VI.4.1 Limestone cliffs, Napier Range, in the Kimberleys, NW Australia.

these are available from a range of sources.

Australian Postgraduate Research Awards are available from the Australian Government for permanent residents of Australia. However, the Australian Government also makes scholarships available to specific groups of intending overseas students. The Equity and Merit Scholarship (EMS) scheme is available to students from a list of developing countries in Africa, South, Southeast and East Asia, and the Pacific Islands and is a source of support for undergraduate as well as postgraduate study. Graduates from eligible countries need to lodge 'expressions of interest' with the Australian Diplomatic Mission in their home country by about April of each year.

For intending postgraduate students from developed countries, the Australian Government makes available each year approximately 110 scholarships as part of the Overseas Postgraduate Research Scholarship scheme (OPRS). Ten additional scholarships are available under this scheme for students from developing countries not covered by the EMS scheme. These scholarships are awarded to students intending to work in areas of research priority established by each Australian university.

Commonwealth Scholarships are also a valuable source of support for graduates with First class degrees. These scholarships cover fees and travel and incorporate a living allowance.

Most universities also offer a number of scholarships for full-time graduate research, and these are available to Australian and overseas graduates on a competitive basis. The terms and values of these scholarships vary from university to university and it is important for intending students to explore the availability of this type of scholarship support with the Australian university they would like to study at. In addition, many universities offer small numbers of fee waivers and additional living allowances and, as you would expect, competition for this extra funding is very keen.

Individual geography departments may also offer small amounts each year to at least partially cover postgraduate students' research expenses. These funds are in short supply in the current university funding environment in Australia. Additional funds may also be available from research grants held by potential supervisors in geography departments. To gain access to these funds, however, intending postgraduates need to be in close touch with individual Australian departments and particular academics.

This last point cannot be stressed enough. The structure of funding in Australia makes it vitally important for any overseas student wanting to undertake postgraduate work in geography to begin discussing plans early with not just a particular department but an individual staff member within a department. This is the only realistic way to secure funding.

Applying for Admission

There are two major bureaucratic steps involved in undertaking postgraduate research in Australia. You must

1 obtain a student visa and meet the requirements of the Department of Immigration and
2 apply for and gain admission to an Australian university.

These two requirements are quite independent of each other. A vital first step for all intending students is therefore to enquire at the nearest Australian Diplomatic Mission (Embassy or High Commission) about visa and immigration requirements. At the same time, it is important to apply directly to the university you have chosen for admission to undertake postgraduate study.

Obviously, you take neither of these steps until you have had lengthy discussions and correspondence with a staff member and a department in an Australian university to secure funding and supervision and to settle on an appropriate and achievable topic.

It is important to begin discussing opportunities for postgraduate research in Australian universities with departments and possible supervisors as early as possible. Below is a list of the principal geography departments in Australia with their postal addresses and telephone numbers.

Addresses of the Principal Geography Departments in Australia

University of Adelaide
Department of Geography
GPO Box 498
Adelaide SA 5001
Telephone: (08) 228 5643

Australian Defense Force Academy
Department of Geography and Oceanography
Campbell ACT 2600
Telephone: (06) 268 8294

Australian National University
GPO Box 4
Canberra ACT 2601

1 Department of Geography
The Faculties
Telephone: (06) 249 2745
2 Department of Human Geography
Research School of Pacific Studies
Telephone: (06) 249 2234
3 Department of Biogeography and Geomorphology
Research School of Pacific Studies
Telephone: (06) 249 4340

Flinders University of South Australia
School of Social Sciences
Bedford Park SA 5042
Telephone: (08) 275 2166

James Cook University of Queensland
Department of Geography
PO Box 999
Townsville QLD 4811

Telephone: (077) 81 4521

Macquarie University
School of Earth Sciences
Sydney NSW 2109
Telephone: (02) 805 8387 and
(02) 805 8428

The University of Melbourne
Department of Geography
Parkville VIC 3052
Telephone: (03) 344 6339/6340

Monash University
Department of Geography and Environmental Science
Clayton VIC 3168
Telephone: (03) 565 2910

University of Newcastle
Department of Geography
Rankin Drive
Newcastle NSW 2308
Telephone: (049) 68 5654

University of New England
Department of Geography and Planning
Armidale NSW 2351
Telephone: (067) 72 2761

University of New South Wales
Department of Geography
PO Box 1
Kensington NSW 2033
Telephone: (02) 697 4386

University of Queensland
Department of Geographical Sciences
St Lucia QLD 4072
Telephone: (07) 377 2060

University of Sydney
Department of Geography
Sydney NSW 2006
Telephone: (02) 692 2886

University of Tasmania
Department of Geography and Environmental Studies
GPO Box 252C
Hobart TAS 7001
Telephone: (002) 20 2463

The University of Western Australia
Department of Geography
Nedlands WA 6009
Telephone: (09) 380 2697

University of Wollongong
Department of Geography
Northfields Avenue
PO Box 1144
Wollongong NSW 2500
Telephone: (042) 27 0721

Postgraduate research in geography can also be undertaken at a range of newer and recently designated universities including Curtin University, the University of Canberra and Griffith University.

VI.5

Postgraduate Studies in Canada

David Ley and Matt Sparke

The university geography department in Canada still occupies its historic position midway between British and American models of education. The Canadian department is typically larger than its American counterpart; its staff include a number of Britons and ex-Britons (often educated in the United States), as well as a smaller peripatetic band born elsewhere in the Commonwealth; and physical geography is a strong element in research and teaching, involving between a quarter and a third of instructors in most departments. At the same time the Canadian university follows the American in the pursuit of mass education; a common source of postgraduate funding comes from teaching assistantships, requiring eight to twelve hours of work a week with large undergraduate classes; and postgraduate degrees include both a coursework and a thesis component, in approximately equal measure over the period of a two-year MA or MSc programme or a four-year PhD programme.

There are twenty-one universities in Canada which offer postgraduate degrees in geography, with slightly fewer at the doctoral level (table VI.5.1). An additional five francophone universities in Quebec offer advanced degrees. Like any national system of departments they include a wide range of specialties. An important first step for a would-be applicant is to secure a copy of the annual *Directory* of the Canadian Association of Geographers, which lists every department, its faculty and their research interests and publications over the previous twelve months (available from Helen Rowlands, Department of Geography, McGill University). Somewhat less detailed information on Canadian departments is provided by the *Guide to Departments of Geography in the United States and Canada*, published annually by the Association of American Geographers. A list of addresses is appended at the end of this chapter.

Table VI.5.1 English-speaking geography departments and postgraduate enrolments, 1987–1988

University	Graduate enrolment	
	MA/MSc	PhD
Alberta	39	19
British Columbia	19	20
Calgary	19	9
Carleton	13	3
Guelph	18	
Manitoba	15	7
McGill	22	17
McMaster	12	16
Memorial	8	
Ottawa	19	12
Queens	35	13
Regina	n.a.	
Saskatchewan		6
Simon Fraser	18	19
Toronto	59	25
Victoria	17	8
Waterloo	49	32
Western Ontario	14	13
Wilfred Laurier	20	
Windsor	13	
York	18	13

Source: Canadian Association of Geographers, *Directory*, compiled and edited by William Barr (CAG, Montreal, 1988)

With these directories serving as a

general guide, more systematic enquiries might then be sent to selected universities who will provide their own advertising material. The choice of a department for graduate study is an important decision, as advanced degrees require a concentrated programme of study in close contact with several faculty. Compatibility of research interest (and perhaps paradigm) should be assessed carefully, either through a visit or communication with present or past graduates of the department, and through reading the research publications of faculty who would form a potential research committee. This process may usefully be followed up with a direct letter to a faculty member who might act as the student's research supervisor. Like any locational decision a range of criteria may direct decision-making, including the environmental amenity of different regions (the aesthetic migrant) (plate VI.5.1) and the funding arrangements that are presented to research students (the optimizing migrant). For serious research students, however, the opportunity to work with the best faculty members in the relevant field should take precedence over other considerations, including in most instances the size of the financial package which a department is able to offer.

It is not possible here to comment on the quality of specialties of the twenty-one departments offering advanced degrees. While most observers would identify (in alphabetical order) British Columbia, McGill, McMaster, Queens and Toronto as the five most competitive departments across the board, other departments have specialized programmes of high quality. Most aspiring postgraduates send three to five applications to a range of departments which they perceive to have varying admission thresholds. The lower threshold for admission in most instances is a middle to high B grade (72–76 per cent, or an Upper Second degree standing in the United Kingdom). Departments expect applications to include a record of courses taken and marks awarded (a transcript), letters of reference from undergraduate instructors, and a short statement outlining research interests and possible thesis areas. This one- or two-page document should be carefully composed, as it is often perceived to provide a measure of research maturity by a graduate selection committee.

The deadline for a completed application file is mid-February in the majority of departments, although late submissions may be successful. In many departments, however, it is well worth working to an earlier deadline of, say, Christmas as this makes applicants eligible for university fellowships which have a deadline considerably before mid-February. University fellowships are the ideal means of funding a postgraduate programme, as they usually have no citizenship restrictions and entail considerable independence for awardees. But they are competitive and are won by only a minority of applicants. Other scholarships, such as those awarded by the Social Sciences and Humanities Research Council, require Canadian citizenship and a separate application by students. Others again, including Commonwealth Scholarships and, in the United States, National Science Foundation Scholarships, are transferable to Canadian departments. A second funding source, mentioned earlier, is teaching assistantships which require some supplementary teaching and marking in undergraduate courses taught principally by faculty. While scholarships offer generally an adequate if modest income, teaching assistantships may supply inadequate funds to carry over the summer. Students should enquire about supplementary summer funding, perhaps through a research assistantship, before accepting a place in a graduate programme, or they may find themselves resorting to savings

Plate VI.5.1 Canada provides a wide range of landscapes and environments in which to study. These include the great mountains of the West, where natural splendours predominate. In other parts of Canada, as in Quebec, humans have achieved major landscape transformations, harnessing the power of the great rivers.

or loans to carry them over the summer period. Another important financial question is to clarify whether graduate fees are to be paid by the student or are waived as part of the offer of admission.

Postgraduate programmes are stimulating, but also challenging, and require a high degree of commitment and motivation. The potential graduate student should share a love of knowledge and intellectual problem-solving. While he or she will acquire a set of skills which may well be relevant in the marketplace, the desire for a job certificate in the absence of intellectual curiosity is unlikely to lead to a rewarding outcome in the pursuit of an advanced degree.

These observations are derived from the experience of more than ten years on a graduate selection committee in a Canadian university. But, if practical, they are also rather formal and present the story from a single perspective. In the second part of this chapter we present a second perspective, and reproduce the deliberations of a postgraduate applicant before and after admission to a Canadian geography department.

Deciding to Come to University in Canada: a Personal Account

When a number of friends and advisers in Oxford first began to say that the United Kingdom was a bit dead and that I should think about making transatlantic graduate school applications I was both excited and sceptical: excited by that naive 'I'd love to live in the new world' type of feeling; and sceptical because of the aura of general 'far-awayness'. With finals looming and a dissertation to complete I hastily asked tutors for advice, made 'party-rate' phone calls to North America for application forms, struggled through entering for the Graduate Record Examination (far more tricky than the exam itself), and duly made the pilgrimage to the Fulbright Commission. In all these efforts there seems to be no clear success-bound path. Filling every form in sight may seem sensible but is costly and means you cannot spend enough of that so-significant 'care' on the one form that matters. Equally just applying for one scholarship and one place may leave you disappointed. Some places come with scholarships (that are quasi-attached) provided by the university which are open to foreign student application; but most places are granted independent of funding which must therefore be sought in the form of a scholarship, or teaching assistantship employment and fee-waivers. In my case the strategy was simplified in so far as I ended up applying to just two universities and four funding sources. This choice (of the Geography Department at the University of British Columbia (UBC) in Canada and the Department of Architecture and Urban Planning at the University of California at Los Angeles (UCLA), United States was made after I had looked carefully at the lists of resident academics and assessed them in the light of my own interests and commitments to critical theory and urban/social geography. It was further consolidated when, after writing and informing people of my interests, I received some generous replies and friendly invitations.

Whilst UCLA's application fee was not too high, UBC was completely free. (I have since discovered that this is because with a holdover of positivist predictive prowess UBC seems to have designated the United Kingdom a Third World country). Busy months passed but in the end I heard that I would receive funding and a place from UBC and so quickly acknowledged my acceptance. Thus as the weeks before finals sped on I felt the confidence of having at least something 'in the bag' that was not dependent on my having to get a 'First'.

Later, with the exams over, the excitement growing and the scepticism fading fast, friends asked me whether I would mind having to do courses; whether I thought it would be a waste of time to plod through a Master's degree when in the United Kingdom I could get cracking right away with a PhD; and whether I agreed that the idea of getting course grades was rather crazy. But I had already thought about these things and had at least justified them in my own mind. Doing a Master's with people whose work I respected and liked would surely involve simply reading literature that I would want to look at anyway. It seemed that the Master's thesis would be the ideal way of testing the waters for the PhD research. Equally, doing courses which would involve discussion over material that people had actually read (rather than having learned the summary dismissal off pat) seemed a good way of working through problems in a text. As for grades, they could be ignored just like all other marks (except of course at the next round of scholarship applications).

In the event many of these justifications I had repeated to myself have been confirmed. I have found my expectations more than simply met and have felt really challenged. The reading is not so exhaustive that I cannot do my own and nearly always consists of something I would have wanted to look at anyway. Similarly I am sure that the thesis I will undertake will provide a good basis for beginning the PhD. Most of all, the free and constant discussion amongst fellow graduates and faculty persistently reaffirms my original decision. I am convinced that this facility to chat with a large number of people about topics we are all reading together provides me with a great benefit, and is a major strength of the postgraduate education I have encountered in Canada.

Useful Addresses in Canada

University of Alberta
Department of Geography
Edmonton, Alberta
TG6 2H4

University of British Columbia
Department of Geography
Vancouver, British Columbia
V6T 1W5

University of Calgary
Department of Geography
Calgary, Alberta
T2N 1N4

Carleton University
Department of Geography
Ottawa, Ontario
K1S 5B6

University of Guelph
Department of Geography
Guelph, Ontario
N1G 2W1

University of Manitoba
Department of Geography
Winnipeg, Manitoba
R3T 2N2

McGill University
Department of Geography
Montreal, Quebec
H3A 2K6

McMaster University
Department of Geography
Hamilton, Ontario
L8S 4K1

Memorial University
Department of Geography
St John's, Newfoundland
A1B 3X9

University of Ottawa
Department of Geography
Ottawa, Ontario
K1N 6N5

Queen's University
Department of Geography
Kingston, Ontario
K7L 3N6

University of Regina
Department of Geography
Regina, Saskatchewan
S4S 0A2

University of Saskatchewan
Department of Geography
Saskatoon, Saskatchewan
S7N 0W0

Simon Fraser University
Department of Geography
Burnaby, British Columbia
V5A 1S6

University of Toronto
Department of Geography
Toronto, Ontario
M5S 1A1

University of Victoria
Department of Geography
Victoria, British Columbia
V8W 2Y2

University of Waterloo
Department of Geography
Waterloo, Ontario
N2L 3G1

University of Western Ontario
Department of Geography
London, Ontario
N6A 5C2

Wilfrid Laurier University
Department of Geography
Waterloo, Ontario
N2L 3C5

University of Windsor
Department of Geography
Windsor, Ontario
N9B 3P4

York University
Department of Geography
North York, Ontario
M3J 1P3

VI.6
Postgraduate Studies in New Zealand
Eric Pawson

There is one very good reason for doing postgraduate work overseas: you can see the world and earn a degree at the same time! And that is a path that increasing numbers of geography graduates are following, particularly to North America and Australasia. Of course, you will have to want to do postgraduate work in the first place. Commitment to an extra three or four years study beyond undergraduate years is no mean thing, but it is an exciting option if your interest in geography has been aroused to the point where you want to explore some aspect of the subject further.

It is also an increasingly realistic option if you earn a first degree of high quality. Even then, you will not start or finish a PhD unless you have real interest. But having finished, you will expect to be able to use it. The prospects of using a PhD for entry into a university career are brighter in the 1990s throughout the English-speaking world than at any time for two decades. The reason for this is that the 'bulge' of university staff appointed during the expansionary years of the 1960s will be reaching retirement over the next ten years or so, creating more university job vacancies than since the early 1970s. In addition, PhDs in key areas such as environmental management, geographical information systems and policy analysis are in demand in government consultancy and the private sector.

So do not be put off considering a PhD. It can enable you to do something you really want, and it can give you an increasingly valuable job qualification.

Why New Zealand?

New Zealand is perhaps not the most immediately obvious place to consider. It is a long way from home. They do things differently there too, but not so differently that anyone from the English-speaking countries would feel alienated; yet sufficiently different, if you have any interest in new places, to provide a wealth of novel experiences.

There are six geography departments amongst its seven universities. Some of these enjoy international reputations and geography itself is a subject of high standing both on and beyond the campus, largely because it has always had a keen focus on applied research problems. But the New Zealand landscape provides far more in the way of research issues than its own geographers can ever hope to exhaust: you will never be short of a research topic of some originality!

In fact, it has a physical geography that actually looks like those diagrams in textbooks, with alpine and coastal scenery that is generally very accessible. Its human geography engages all the issues of urban development, economic restructuring and, in the relations between the indigenous Maori and Europeans, bicultural management. New Zealand has undergone radical change, social and economic, since the election of the Labour government in 1984, which has thrown up hosts of issues of importance in geographical research. You can get a taste of these from recent publications by the New Zealand Geographical Society: *Southern Approaches*, edited by P. G. Holland and W. B. Johnston in 1987, an 'Alternative Perspectives' issue of the *New Zealand Geographer* (1987) and a volume on restructuring, *Changing Places*, edited by S. Britton, R. Le Heron and E. Pawson (1992).

New Zealand Universities

The largest geography departments are at Auckland (fifty or so graduate students) and Canterbury (in Christchurch, about thirty students), with equally long established departments at Otago and Victoria (in Dunedin and Wellington respectively) and newer smaller units at Massey and Waikato (in Palmerston North and Hamilton). Geography graduates are also to be found in tourism (Otago), regional planning (Otago and Massey) and environmental science (Auckland) – all courses associated with geography, as well as in parks and recreation, resource management and landscape architecture at Lincoln (the seventh campus, outside Christchurch).

New Zealand undergraduates do a three- or four-year general degree, there being no tight time restriction. In order to earn honours in geography (or any other subject) they must take an additional year of graduate coursework or, for a Master's, an additional two years including a thesis. Most graduate students are therefore doing honours or Master's courses, or degrees or diplomas in the other subjects mentioned above. For a New Zealander, an honours or Master's degree is the route to a PhD.

However, as an overseas student with a three-year honours degree, you can go straight into a PhD. And there will be no formal coursework requirement as in Canada or the United States. The largest doctoral programmes are at Auckland and Canterbury (five to ten PhD students each), but importantly you will also have the company of all those other graduate students. In the other departments, with fewer PhDs, you will still have this wider convivial network of people your own age, and this really matters when you are engaged in a programme of research on your own. Having plenty of people to share work ideas with, and to socialize with, makes a great difference.

Scholarships and Supervision

All countries charge overseas students fees. In New Zealand, these are in the range of NZ$20,000–24,000 per annum for 1992. A privately funded PhD is therefore not feasible and you will need scholarship support to pay both fees and a living allowance. The best option is a Commonwealth Scholarship, which is awarded to holders of First class and sometimes Upper Second degrees. New Zealand universities do not generally have funds for this purpose, but it is worth enquiring as circumstances are always changing.

Commonwealth Scholarships pay a living allowance. This can usually be supplemented with some paid teaching and sometimes by free, or discounted, living in a hall of residence in exchange for sub-warden duties. In addition, it is usual for a department to seek extra support from granting agencies to cover individual PhD research expenses.

In order to gain initial acceptance by a department, it is necessary to indicate a general topic for research. After some weeks, or months, in the country, you will be able to write a firm research proposal. A supervisor is appointed to oversee your research once the university has accepted your proposal. You can expect to develop a close working relationship with your supervisor who, in turn, is responsible to the university for ensuring that you progress. Supervision is, if anything, often tighter than elsewhere. Consequently geography PhDs in New Zealand have a very high completion rate (approaching 100 per cent) and would normally have to be completed in three to four years, although six is the allowable maximum.

Topics and Places

New Zealand geography departments tend to specialize in particular themes within the discipline, and hence it is important that you select one with staff expertise in your chosen field. For example, alpine research is strong at Otago and Canterbury (plate VI.6.1), coastal studies at Canterbury and Auckland, biogeography at Massey and Otago, hazards research in almost all cases. Waikato has no physical geography but does have a separate earth sciences school. It is also notable for feminist and bicultural studies. Development studies, focusing on the Pacific and Southeast Asia, are available at Auckland, Victoria and Canterbury, historical and cultural geography is well represented at Otago, Canterbury and Auckland and there are specialists in restructuring at Massey, Victoria and most other departments.

The only way to get up-to-date information on departmental interests is to write for the graduate handbook in each case. Some departments are particularly well equipped and may be able to offer better research and working facilities than others. The availability of ancillary resources on each campus is important too. For example, the existence of an engineering school with expertise in hydrology, or a sociology department with urban affairs interests, may make a lot of difference to the viability and richness of your research experience.

Plate VI.6.1 The Southern Alps of New Zealand provide a magnificent environment in which to undertake all types of geographical research, including the study of mass movements, tectonics, glaciation and periglacial processes.

Jobs

In the 1980s, thirty-six geography PhDs were completed in New Zealand. Of these, ten now have university posts in New Zealand, twelve have university jobs overseas and the remainder are employed by research agencies in New Zealand and overseas. On completing your PhD you may decide to stay or to join the 'international circuit', taking a post in another overseas university, or you may return home older, wiser and with a qualification every bit as good as that available in most other countries.

The basic prerequisites for a successful and happy postgraduate path overseas are a good honours degree, a real interest in a particular aspect of geography and a willingness to experience the culture of another country. And you cannot avoid making many new friends.

Addresses

For publications of the New Zealand Geographical Society, go to your university library or write to:

The Secretary
New Zealand Geographical Society
Department of Geography
University of Canterbury
Christchurch 1
New Zealand

For information on departments, write to the head of the Department of Geography:

University of Otago
PO Box 56
Dunedin

University of Canterbury
Christchurch 1

Victoria University of Wellington
Private Bag
Wellington

Massey University
Palmerston North

Waikato University
Private Bag
Hamilton

Auckland University
Private Bag
Auckland

VI.7

Postgraduate Studies in the UK

Peter Jackson

While most of the *Companion* has been concerned with studying geography at undergraduate level, this section provides an introduction to postgraduate studies in geography in the United Kingdom. It gives a brief description of graduate life, the range of courses available, the principal sources of funding and methods of application.

Postgraduate Life

As an undergraduate you will probably only encounter postgraduate students as 'demonstrators' in laboratory or statistics practicals, assisting on field classes, taking tutorials or giving the occasional lecture. These are some of the ways that postgraduates supplement their meagre incomes and gain useful professional experience. They spend the rest of their time attending courses, conducting research (in the field, library or laboratory) and writing their thesis or dissertation.

Postgraduates can be divided into two basic sorts: those pursuing a taught course (usually for a Diploma or Master's degree) and those doing original research (for a DPhil or PhD). Taught courses include various professional qualifications such as the Postgraduate Certificate of Education (PGCE) (see chapter VI.2 for more details) or planning courses that lead to membership of the Royal Town Planning Institute. They include area studies courses (on Latin America or Africa for example), technical courses (on remote sensing or geographical information systems), courses on subjects related to geography (such as conservation or environmental studies) and even, occasionally, Master's degrees in geography itself.

Such courses are usually assessed by a combination of examination and project work (generally including a dissertation) after one or two years of full-time study. By contrast, most PhD degrees (in Britain at least) are assessed entirely in terms of a thesis or dissertation, written towards the end of three or more years of full-time study and examined through an oral defence or 'viva'. There is considerable variation on this pattern as American-style PhD programmes are gradually being introduced in Britain, with greater emphasis on technical training (usually through taught courses) and relatively less emphasis on the dissertation. American PhDs also generally require the successful completion of a 'comprehensive' examination showing all-round ability in the chosen subject, before the students go on to research for a particular dissertation. In Britian, most PhD students register initially for a Master's degree (MA, MSc or MPhil), transferring to PhD status after a year or two when they have made sufficient progress.

Postgraduate studies can be lonely and stressful in comparison with the more structured and sociable life of the average geography undergraduate. However, they offer the potential for extending your knowledge of those aspects of geography that you most enjoy and even, in the case of PhD research, the chance to contribute in an original way to the development of the discipline.

Supervision

Life as a postgraduate will also very significantly depending on whether you study in a large or a small department (both have their attractions) and on whether you are registered full-time or part-time (in order to take paid employment, because of domestic responsibilities etc.). But probably the most important significant factor in the lives of most postgraduates is their relationship with their supervisor. Most postgraduate students choose or are assigned to a single supervisor. They may have no choice in the matter if only one member of the department has the necessary expertise or if the funding agency specifies a particular supervisor. But it is increasingly common for departments to appoint a supervisory committee to offer an alternative point of view to the main supervisor, an 'escape route' in cases of personal disagreement, or simply to supplement the range of expertise available to the student. It may also be possible to transfer to another supervisor if a personality clash occurs.

How to Apply

Master's courses in Britain are often advertised in the newspapers and in specialist journals such as *New Scientist* or the *Geographical Magazine*. The *Times Higher Education Supplement* is also useful, as are the noticeboards in your undergraduate department. Most colleges and universities have careers departments which include postgraduate study within their remit. But the best starting point is the member of staff in your undergraduate department with whom you get on best, your current tutor or the person who is most knowledgeable about the field in which you are interested. Don't be shy of asking for advice; tutors usually respond enthusiastically to enquiries about postgraduate courses and research degrees. They are in a good position to advise about which departments and individuals have the best reputations in a particular field of study. You can then write to those individuals and departments that seem to offer the best prospects in your field, requesting further information. The *Student's Guide to Graduate Studies in the UK 1990* (Careers Research and Advisory Council, Hobson, Cambridge) may also provide useful starting information.

Unlike the undergraduate Universities Central Council on Admissions (UCCA) system, there is no centralized postgraduate admissions procedure in the United Kingdom. The onus is on the prospective graduate student to make enquiries and complete the necessary application forms. For Master's degrees and other taught courses you can usually obtain an information pamphlet and application forms from the institution concerned. Most departments also now produce publicity brochures for prospective graduate students. In the case of PhD research, however, there is no substitute for contacting the individuals with whom you would most like to work, outlining your own interests and background. You will probably be asked to fill in an application form and you may be called for interview. If you are asked to outline your proposed research, you should take this very seriously, consulting with your undergraduate tutors. Writing a convincing proposal, properly researched and fully referenced, is often the most telling part of the application process. If you are invited for interview, try not to be too passive. Ask questions about available facilities (for computing and word-processing, libraries and laboratories); ask about supervisory practices and completion rates; ask whether you will have your own desk or office space; whether you will be able to do any teaching (and, if so, how much you will be paid); and be sure to enquire what financial support is available for attending conferences or conducting overseas fieldwork. Conditions vary enormously between departments and it is worth making full enquiries even at this early stage.

Finances

Getting accepted for a postgraduate course or research degree is relatively straightforward, especially if you have a First or Upper Second class degree. The main problem is funding. If you are self-financing you will need to find sufficient money to cover tuition fees and living expenses for the duration of the course. Most postgraduate grants are assessed at the same rate as the standard undergraduate grant, extended over 52 weeks and allowing about six weeks vacation. There are few opportunities as a postgraduate for supplementing your grant with vacation work but other sources of income may be available (see below). Fees vary widely between courses and institutions, and are much higher for non-UK and non-EEC students. It is relatively cheap to take a part-time Master's course, but much more expensive to study full-time for a PhD. Accommodation and living expenses add further to these costs and should be borne in mind when estimating the amount required to pursue any particular course.

Some Master's courses have a limited number of ear-marked grants and you should enquire about these when you apply. Others are awarded on a competitive basis across a range of courses. The main sources of funding for PhD degrees are the Research Councils (listed below) which offer postgraduate awards for specific topics and a limited number of 'open' awards for topics chosen by the student, allocated on a competitive basis. You can enquire about these awards directly from the research councils or from the departments which are successful in attracting research council funding. You can also keep your eye on newspapers and journals which advertise the awards as they become available.

The Economic and Social Research Council (ESRC) funds research in human geography, planning and related areas. Their address is:

Postgraduate Training Division
Polaris House
North Star Avenue
Swindon SN2 1UJ
Telephone: 0793-413000

The Natural Environment Research Council (NERC) funds research in geomorphology, meteorology, hydrology, ecology and conservation. Their address is:

Polaris House
North Star Avenue
Swindon SN2 1EU
Telephone: 0793-40101

A limited amount of geographical research is funded by the Medical Research Council (MRC), the Agriculture and Food Research Council (AFRC) and the Science and Engineering Research Council (SERC). The British Academy, the British Council, the Royal Society and the Department of Education and Science (DES) generally do not fund postgraduate research in geography although they do administer small supplementary funds (such as the Twentieth IGC Fund and the Dudley Stamp Award) for specific purposes. A different system applies in Northern Ireland where postgraduate studies are funded by the Department of Education for Northern Ireland:

Pathgael House
Balloo Road

Bangor
County Down BT19 2PR
Telephone: Bangor 66311

Besides the Research Councils, there are a whole range of governmental and private agencies (including educational charities) who fund postgraduate research. It is worth searching carefully for these as they are often undersubscribed. Useful sources include the *Directory of Grant Making Trusts* (Charities Aid Foundation, Tonbridge), the *Association of Commonwealth Universities' Scholarship Guide* and *The Grants Register* (Macmillan, London). Some universities, departments and colleges, particularly in Oxford and Cambridge, have their own funds for postgraduate study. Look carefully through departmental literature for any sources not mentioned here.

Going Overseas

The option of undertaking postgraduate study abroad is proving increasingly popular with British students and is certainly worth considering.

Information on postgraduate studies in the United States can be obtained from the Educational Advisor at (and see chapter VI.8)

The Fulbright Commission
6 Porter Street
London W1M 2HR
Telephone: 071-486 1098

Information on Commonwealth Universities can be obtained from (and see chapters VI.4, 5, 6)

Association for Commonwealth Universities
36 Gordon Square
London WC1H OPF
Telephone: 071-387 8572

who also publish the *Commonwealth Universities Yearbook* and the *Scholarships Guide for Commonwealth Postgraduate Students*. There are also a range of scholarships for postgraduate studies in the United States, including the Fulbright, Harkness and Thouron Awards which are advertised on an annual basis.

Support Groups

In order to reduce the potential isolation among research students, a range of support groups for postgraduate students in geography have developed over the last few years. BEARINGS provides a regular informal forum for geography postgraduates in London and the southeast with a lively programme of seminars and workshops. The Institute of British Geographers (IBG) has a Postgraduate Forum which publishes a newsletter called *PRAXIS*, campaigns on a variety of postgraduate issues and provides a network of support and information to postgraduate students. The IBG has special rates for postgraduate members and encourages them to participate in its study groups and conferences. The Royal Geographical Society welcomes postgraduate members and has a Young Members Committee to represent their interests. Postgraduate students are also eligible for membership of the National Union of Students and can apply for associate membership of the Association of University Teachers (AUT) or the National Association of Teachers in Further and Higher Education (NATFHE).

Careers

Finally, it is worth thinking about the kind of job opportunities that are open to people with a postgraduate qualification in geography or related subjects. In the case of taught courses or professional qualifications, such as planning or teaching, this may be obvious. In other cases, an academic career as a university or polytechnic lecturer may be possible although the opportunities in Britain are now relatively few and salaries are poor in relation to potential earnings in other fields. Short-term contracts as a Research Assistant or Junior Research Fellow may be available but seldom offer the prospect of secure long-term employment.

If your research project has been financed in whole or in part by a private company or government department, they may provide a career opening. In general, though, you will have to be flexible and exploit the range of skills you have acquired as a postgraduate student. Geographers are still unusual in the variety of abilities they possess: they are not restricted to literary or scientific skills, but also have the ability to organize an independent project, to conduct fieldwork at home or overseas, to design and implement a questionnaire survey, to do a library or archive search and to synthesize the findings. Many people can do one or two of these things; a well-trained geographer can usually do all or most of them. Further information on careers for geographers can be obtained from the Institute of British Geographers or from the Royal Geographical Society, both at 1 Kensington Gore, London SW7 2AR (and see Chapter VI.1). Useful short courses for second-year postgraduates are organized by the Careers Research and Advisory Council (CRAC).

As this section has shown, postgraduate studies in geography cover a wide range of different experiences and it is hard to generalize about them all. But if you research the opportunities carefully and pick the right course and department, the prospects for postgraduate geographers are exciting and varied.

Overseas Students

All British higher education institutions are anxious to recruit overseas students for postgraduate studies in both research and taught courses. There is no centralized system of application and no examination system (although you will be expected to show proficiency in the English language), since each institution sets its own standards and requirements. You should therefore apply to the department(s) of your choice directly. A list of courses offered can be found in *The Student's Guide to Graduate Studies in the UK* available from the Careers Research and Advisory Centre, Sheraton House, Castle Park, Cambridge CB3 0AX. However, given the British system of a close relationship between supervisor and postgraduate it is important that you enquire whether or not there will be specialists in your area of interest before committing yourself.

Like many other countries, Britain is an expensive place in which to do a degree; in recent years, fees charged to overseas students have increased considerably. There are some centrally administered scholarship schemes, notably the Commonwealth Scholarship, of which 1000 are offered for all subjects each year. To be eligible you must be a citizen of a Commonwealth country, under 35 years of age, and intending to return to your home country. These scholarships cover the full cost of tuition, fares and living expenses. Information on them can be found in your country's British Council office or Embassy or High Commission, or in *The Scholarships Guide for Commonwealth Postgraduate Students* available from the Association of Commonwealth Universities, 36 Gordon Square, London WC1 0PF. For students from outside the European Community there is an Overseas Research Student award scheme funded by the British government. These do not pay full fees, but enable the student to pay the equivalent of home students. The award holders are nominated by the institution on the basis of academic quality rather than financial need. You should therefore apply to your chosen institution directly.

Most institutions have their own scholarships award schemes and fee waivers. You should inquire about them when you make your initial approach, bearing in mind that the competition will be tough.

VI.8
Postgraduate Studies in the United States

Brian W. Blouet

Over the past decade the opportunities to pursue higher degrees in some countries have been limited by financial problems. In North America entry into Master's and doctoral degree programmes is straightforward for those possessing good qualifications. Funding in the form of graduate teaching assistantships is the normal way of financing work towards a higher degree. Assistantships usually require the recipient to instruct in introductory human or physical geography classes. Many universities have fellowships to support graduate work. The competition for fellowships is more intense than for assistantships. Apply for both, but do not count on more than an assistantship.

Many geographers who have earned first degrees in Britain, Australia and New Zealand have gone to North America for higher degree work. It is important to recognize that there are differences between the systems.

In most American universities graduate degrees at the Master's and doctoral levels require coursework in addition to the completion of a Master's thesis and a doctoral dissertation. Usually students are required to complete a Master's degree before proceeding to the doctoral degree. It is difficult to complete the two degrees in less than five years.

The application procedures are time consuming and require documentation that is unfamiliar to applicants from abroad. Applicants must provide official *transcripts* of all coursework completed. Your registrar's office will be able to provide this on institutional notepaper embossed with the official university stamp, which validates the document at the American end. Transcripts are sent directly, by the home registrar, to the graduate college of the university to which you are applying. Transcripts are not sent to the department to which entry is sought unless a specific instruction is given.

Most universities in the United States require that applicants provide Graduate Record Examination (GRE) scores for the quantitative and verbal sections of that test. The GRE can be taken at centres in Britain and elsewhere, but it is necessary to plan well ahead. It is seen as a nuisance by well-qualified applicants from outside the United States, but the scores are often used not only to determine entry but to decide who will receive the better financial awards to support graduate work. The *GRE Information Bulletin* is available free from

Graduate Record Examinations
Educational Testing Service
PO Box 6000
Princeton, NJ 08541-6000
USA

The test is offered a few times a year. It needs to be taken early in the year of application if the scores are to be available to support applications for fellowships and assistantships.

Determining which offer of financial aid to accept is not easy. The value of awards varies. Living costs differ markedly from one part of the continent to another, and some universities wholly or partly remit the cost of tuition with the award of an assistantship or fellowship; others do not. An assistantship of apparently modest value from a midwestern university may be easier to live on than an apparently more valuable award from a California institution.

A *Guide to Departments of Geography in the United States and Canada* is published annually by the Association of American Geographers, 1710 Sixteenth Street NW, Washington, DC 20009-3198. The publication costs $10 for prospective students plus postage. It is a useful source on the nature of programmes. However, departments buy space in the directory and write their own entries, under specified heads. There is no external review of entries. A department with one physical geographer may claim physical geography as a speciality alongside another department with three or four physical geographers. The entries do not always reveal the specific requirements of graduate programmes offered.

It is important to recognize that there may be two sets of requirements for a higher degree at a university. One set of requirements will be fully spelt out in the graduate college catalogue that will be supplied to all applicants. A second set of requirements may be imposed by the department. It is essential to ask, in the enquiry phase, what requirements a department has in addition to those listed in the graduate college catalogue. Some departments have pamphlets that spell out departmental policies for higher degrees. For example, the Department of Geography at the University of Kansas has a thirty-page booklet that details application procedures, financial aid, the degree options available, the courses all students must take in the options, language and research skills that students must demonstrate and the areas in which the department is prepared to supervise higher degrees. It is essential to understand these matters before agreeing to join a department, otherwise you may be forced to meet requirements that are uncongenial. Some departments have few requirements other than those spelt out in the graduate college catalogue; other departments have far too many rules of their own and should be approached with care.

In order to enter the United States for higher degree work a visa is required. Most graduate students enter on a student visa. Advice on the necessary documentation, and how to apply for a visa, is usually provided by the university you decide to accept.

Start the enquiry and application process in September or October of the year previous to the desired date of entry into graduate school. A later start can exclude you from some of the financial awards. Write to many departments. Apply to five or six universities.

Below is a list of the departments in the United States which offer the Master's and doctoral degree in geography. There are many other excellent departments, not listed here, which offer only the Master's degree.

Addresses of Geography Departments Offering Master's and Doctoral Degrees in the United States

Arizona State University
Department of Geography
Tempe, AZ 85287

University of Arizona
Department of Geography and Regional Development
Tucson, AZ 85721

Boston University
Department of Geography
Boston, MA 02215

University of California at Berkeley
Department of Geography
Berkeley, CA 94720

University of California, Davis
Department of Geography
Davis, CA 95616

University of California, Los Angeles
Department of Geography
Los Angeles, CA 90024

University of California, Riverside
Department of Earth Sciences
Riverside, CA 92521

University of California, Santa Barbara
Department of Geography
Santa Barbara, CA 93106

University of Chicago
Committee on Geographical Studies
Chicago, IL 60637

University of Cincinnati
Department of Geography
Cincinnati, OH 45221

Clark University
Graduate School of Geography
Worcester, MA 01610

University of Colorado
Department of Geography
Boulder, CO 80309

University of Delaware
Department of Geography
Newark, DE 19716
(The PhD degree offered is in climatology)

University of Denver
Department of Geography
Denver, CO 80208

University of Florida
Department of Geography
Gainesville, FL 32611

University of Georgia
Department of Geography
Athens, GA 30602

University of Hawaii, Manoa
Department of Geography
Honolulu, HA 96822

University of Illinois
Department of Geography
Urbana, IL 61801

Indiana State University
Department of Geography and Geology
Terre Haute, IN 47809

Indiana University
Department of Geography
Bloomington, IN 47405

University of Iowa
Department of Geography
Iowa City, IA 52242

Johns Hopkins University
Department of Geography and Environmental Engineering
Baltimore MD 21218

University of Kansas
Department of Geography
Lawrence, KS 66045

Kent State University
Department of Geography
Kent, OH 44242

University of Kentucky
Department of Geography
Lexington, KY 40506

Louisiana State University
Department of Geography and Anthropology
Baton Rouge, LA 70803

University of Maryland
Department of Geography
College Park, MD 20742

Michigan State University
Department of Geography
East Lansing, MI 48824

University of Minnesota
Department of Geography
Minneapolis, MN 55455

University of Nebraska
Department of Geography
Lincoln, NE 68588

University of North Carolina
Chapel Hill
Department of Geography
Chapel Hill, NC 27599

Ohio State University
Department of Geography
Columbus, OH 43210

University of Oklahoma
Department of Geography
Norman, OK 73019

Oregon State University
Department of Geography
Corvallis, OR 97331

University of Oregon
Department of Geography
Eugene, OR 97403

Pennsylvania State University
Department of Geography
University Park, PA 16802

Rutgers University
Department of Geography
New Brunswick, NJ 08903

University of South Carolina
Department of Geography
Columbia, SC 29208

Southern Illinois University
Department of Geography
Carbondale, IL 62901

State University of New York at Buffalo
Department of Geography
Amherst, NY 14260

Syracuse University
Department of Geography
Syracuse, NY 13244

University of Tennessee
Department of Geography
Knoxville, TN 37996

Texas A & M University
Department of Geography
College Station, TX 77843

University of Texas at Austin
Department of Geography
Austin, TX 78712

University of Utah
Department of Geography
Salt Lake City, UT 84112

University of Washington
Department of Geography
Seattle, WA 98195

University of Wisconsin
Department of Geography
Madison, WI 53706

University of Wisconsin
Milwaukee
Department of Geography
Milwaukee, WI 53201

VI.9

Prospecting for Money for Graduate Work in North America

Ronald F. Abler

If you need money to underwrite your graduate course or your research, this chapter will help you find it. If you are reasonably energetic and talented, you have a good chance to win support for your education and research if you investigate diligently, ask carefully and apply well in advance of need.

Assistantships and Fellowships

You will have little chance of obtaining the funds you need unless you find a match between your needs and a donor's programme. Three-quarters of success consists of making that match. Once you have identified places that would serve your needs, you should apply promptly.

Over 200 geography departments in North America offer graduate assistantships and fellowships. Most departments award financial assistance solely on the basis of academic qualifications, without regard to nationality of applicants. Institutions normally require that applications be submitted in January for assistantships and fellowships that begin the following Fall.

How do you decide where to apply? First, decide what you need. What specializations do you wish to pursue in your graduate work? Do you seek a Master's degree or the doctorate, or are you unsure whether you want to continue beyond the Master's? Be able to answer these questions before you start searching for a graduate department. If you do not, you will search aimlessly.

Once you know what specialty you wish to pursue, find out who does the best work in that specialty. Who is the acknowledged leader? Who is the rising young star on the region or topic? 'Best' is always relative to some criterion. A seasoned scholar may publish extensively and lead a field. He or she may also be too busy leading to work closely with graduate students, and may be coasting on somewhat dated ideas. Someone who is just starting to make a name in your specialty may be more accessible and might provide you with training that has greater staying power. The individuals and departments that offer the best work in your specialty constitute your targets. Read the literature in your selected specialty and consult your mentors for advice. Use the guide to graduate programmes published by the Association of American Geographers (AAG) (details are given in chapter VI.8).

Select two to four places that seem to offer a good match with your needs. Write to each of them, preferably in September of the year before you wish to begin your graduate work. Give a brief summary of your background and interests. Be sure to include in your letter at least a paragraph explaining why you want to study at each of the institutions you have selected. Tailor that paragraph to each of the schools to which you apply. Conclude your letter by requesting application forms for assistantships and fellowships, and any other pertinent information.

When you receive the forms and information you have requested, begin to craft your application. Provide all the information requested in precisely the form requested. You may be required to take the Graduate Record Examination (see chapter VI.8), and, if English is not your native tongue, the Test of English as a Foreign Language (TOEFL). The need to have your scores on such examinations in the hands of the departments to which you will apply is one reason why you should begin your enquiries in September or even August. Finding out what tests you must take, arranging to take them, and having your scores sent to your designated institutions will take months. You will not meet the January deadlines unless you start early.

You will also be asked to arrange for letters of reference. Select referees who know you and your work well. Ask them to submit their letters well in advance of the application due date. If you have not previously discussed your plans for graduate work with the individuals you select, provide each with a copy of your academic record and a brief statement of your educational and career goals. You are responsible for providing your referees with the raw materials they need to write convincing letters on your behalf. Ask your referees to notify you when they have sent their letters, and keep track of those that are outstanding. Don't be hesitant to issue gentle reminders if the deadline for receipt of materials is near and some of your mentors have not yet written.

You should be informed of the result of your application in February or March. If you receive only one offer, you have a simple choice. If you are lucky enough to receive more than one, you must choose which you prefer. If at all possible, visit the department or departments you are considering, even if the expense is considerable. You are making one of the largest investments of your life in deciding where to obtain a graduate degree. A few days and a few hundred dollars spent on making sure which is the right place for you will be time and money well spent, even if you have to borrow the money. (Visiting departments *before* you apply is an excellent idea if it is possible). If you receive more than one offer, compare them carefully in all respects. Every offer will have pluses and minuses that you should take into account.

Don't be dazzled by dollar signs. Consider the entire financial package. Some universities offer high stipends but subtract a good chunk of it for tuition. Others have smaller stipends with tuition waivers. Find out how much money you will be able to put in your pocket after your fees are paid, but don't rank money ahead of programme quality. You will live a life of genteel poverty during your graduate career regardless of the stipend, but what you will learn will shape the rest of your career. Choose the most exciting and innovative programme you are offered.

Be aware also that virtually all universities in the United States and Canada subscribe to a convention regarding offers of financial assistance for graduate work that specifies certain ground rules with respect to offers and acceptances. You have until 15 April to accept an offer, or to change your mind about an offer you previously accepted. Conversely, if your acceptance of an offer remains in force on 15 April you may not accept another offer and are obligated to meet your commitment. If you are pressured to give a commitment before 15 April, remind the person with whom you are in negotiation of the Council of Graduate Schools agreement. The text of the agreement is printed in the AAG's *Guide to Departments of Geography in the United States and Canada*.

Don't be abashed about informing a department that has made you an offer about other choices you are considering. You have no obligation to do so, but neither should you feel that you have anything to hide. Talented graduate students are scarce and the competition for them is keen, a fact of which all graduate departments are well aware. It is not unknown for a department to sweeten an offer if it becomes aware that a student it wants has other options.

Research and Project Support

As a graduate student or new faculty member you may need backing for the projects you wish to undertake. External support for Master's-level research is unusual, but many doctoral candidates receive partial support for their dissertation research, either by working on a funded project in their departments or by obtaining their own support from an agency such as the National Science Foundation (NSF). Mounting a major research effort in a technical field may require tens or hundreds of thousands of dollars. The process of obtaining those funds is similar in kind to getting assistantships and fellowships.

Again, your first task is to decide what *you* want to do. You must be able to state your need clearly, concisely and in an appealing manner. You stand little chance of convincing someone else of the merits of your endeavour unless you have first convinced yourself.

A fundable research problem consists of a sharply defined question that you can answer. When you can state your question or describe the likely outcome of your project in a single sentence that makes sense to specialists in your field, you are ready to start looking for money. In addition to stating the problem, of course, you must be able to specify the data that bear upon the question, know what analytical techniques to apply and know what constitutes a satisfactory answer to the question.

Avoid ambiguity and uncertainty when formulating your project. Don't try to convince anyone to underwrite your work until you have formulated a well-structured plan of attack that lacks only resources to ensure its satisfactory completion. No matter how interesting or important the question you wish to address, unless you have a good chance of resolving it you will not be likely to be given money. Fascinating research topics abound: fundable projects are scarce. Translating interesting topics into viable research projects demands rigorous thought.

When you know what you want to do, start looking for the help you need. Consult foundation directories and your colleagues, and examine lists of previously funded projects. Every organization welcomes some topics and ignores others. Learn the culture and history of promising organizations. Make sure your needs match an organization's interests before you apply. Don't waste time on sources that have little interest in your topic.

The NSF, for example, supports basic research, and it responds to investigator-initiated proposals. If your research will improve geographic theory, NSF might welcome your proposal. If your project's strongest justification is better policies or more effective applications of known principles, you are unlikely to succeed at NSF. Never ask an NSF programme officer what topics he or she wants to fund. You should decide what you want to do, and then convince your NSF programme officer that your project is worthy of Foundation support.

On the other hand, some funding agencies and institutions are prescriptive, and may not entertain unsolicited proposals. When approaching that kind of research sponsor, you should ask what kinds of projects the agency or foundation wishes to support. Obtaining money from a source that does not welcome unsolicited proposals requires that you persuade someone to issue a request for proposal (RFP) on the topic that interests you. Often, such persuasion occurs in what seems like casual discussion.

Never – *ever* – submit a proposal to a funding agency before making personal contact with one of its programme officers. Never – *ever* – contact a programme officer unless you can explain in five minutes exactly what you want to do, why it is worth doing and about how much money you need. Unless the programme officer knows your specialty well, pitch your presentation to a generalist.

Attend carefully to the programme officer's reaction, for the response you receive may be ambiguous. If you detect encouragement, ask about ways your project could be made more attractive and about weaknesses or ambiguities that should be excised form your formal submission. If you sense discouragement, ask about other programmes or agencies that fund the kind of work you want to do. If you intend to submit a proposal, make sure you understand exactly when and how your proposal should be submitted.

A complete research proposal consists of an answerable question, a plan for answering that question and a research budget. Each agency or foundation wants research proposals submitted in its own format. Follow instructions scrupulously. Do it *their* way, even if you don't understand why.

At the very outset, clearly identify the question you want to answer or the problem you want to solve. Don't lead up to it with lots of background and context. State the problem and then explain why it is important. State your problem in as compelling a way as you know how. Then describe how you will answer the question or solve the problem, by presenting your research plan. Relate what data you will use. If you propose survey work, submit your questionnaire as an appendix. If you intend to use data from government or commercial sources, identify them definitively. Whether your evidence is qualitative or quantitative, you are somehow going to move from evidence to an answer to your question. Describe the sequential steps of your reasoning from raw evidence through to the conclusions you hope to derive. Specify your analytical techniques and discuss candidly any weaknesses in your chain of analysis and reasoning. Cleanse your proposal of ambiguity.

Within reason, let the project govern the size of your request. Your pre-proposal discussion with the programme officer will have yielded some guidelines regarding allowable expenses and how much the agency or sponsor will award. Don't include expenditures that are not absolutely required in order to do the research. Explain every expenditure in a narrative that documents what the money will purchase and why the item is needed. Answer all questions an intelligent reviewer or research administrator might ask about your request.

Winning proposals are self-contained, technically perfect and exciting. A proposal that identifies unavoidable weaknesses and allays the reservations they engender bespeaks a meticulous dedicated scholar. Demonstrate that you have considered all possible outcomes of your research. Ask a trusted colleague to read your proposal sceptically, and incorporate the resulting suggestions. You will greatly diminish your chances of obtaining funds by submitting a shoddy proposal. Spelling and grammatical and arithmetical errors will raise doubts about your competence in the minds of reviewers. Present a proposal that is neat and technically perfect. Write your narrative in a way that conveys commitment to your work. Write clearly and elegantly. How you make your case will affect reviewer evaluations. Polish the narrative by repeatedly rewriting the draft until it presents your request in clear and vivid language that will capture the attention and interest of those who read it.

Your proposal will be reviewed by experts chosen to advise the organization to which you submit it. (You may be asked to suggest qualified reviewers when you submit the proposal.) They will evaluate your request in relation to the criteria the agency has established. In one way or another, they will judge whether you are capable of doing what you say you will do, whether the task is intrinsically worth doing and whether anyone will use the results of your project. In addition to evaluation by individuals, some agencies and foundations employ panels of experts to recommend which proposals should be supported. At most funding organizations, final decisions regarding funding rest with the programme officers.

The pre-proposal discussion should have given you some idea of when you will be informed of your proposal's fate. If you have not heard from the agency two or three weeks after that date, call or write and ask whether a decision has been made. Do not call on the exact date that decisions are due. Give the programme officer some latitude. How much you will learn about why your proposal was funded or declined will vary from sponsor to sponsor. Some render only their decision with little or no explanation. Others provide verbatim copies of reviewer comments and a summary of the advisory panel's discussion. Most provide some explanation.

An award is cause for celebration. You can be proud that your peers judged your project worthy of support. But you should pay careful heed to what your peers have said about your proposal before starting your project. Your proposal was probably read by a number of experts in your field, and their comments may contain valuable recommendations that will strengthen your research plan. Use their advice. Think also about how you will share the results of your work with your peers. The best way to get a second award is to publicize the results of your first grant in a way that convinces the sponsor that it received good value for its investment.

A rejection is no disgrace, but deep disappointment inevitably results. No matter how keen the competition, you hoped to win. Your colleagues will probably be aware of your application, and you will have to tell them you've been turned down. If you receive reviewer comments, it may be clear that one or more did not understand your proposal. Once the original pain has subsided, examine reviewer comments carefully and objectively, for you must decide whether to resubmit the proposal in the next round of competition. If the tenor of most reviews is negative, you may have an inherently weak proposal that should be written off in favour of pursuing other projects you wish to do. If most of the criticism are constructive, talk to the programme officer about whether, in his or her opinion, submission of a revised proposal is warranted. Revised proposals that respond forthrightly to reviewer suggestions normally succeed more often than first-time submissions.

Remember, above all, that most proposals submitted to most programmes by most applicants are declined. Few agencies or foundations fund more than one in three of the proposals they receive. Few scholars consistently receive funding for more than one in five of the proposals they submit. You must be willing to accept failure if you hope to succeed in your quest for external support. Unless you are truly exceptional, you must be willing to be declined four times for every time your projects are supported. Scholars who consistently win funding fail often, but they submit enough proposals so that their occasional wins keep them going.

Remember also that obtaining external

research support is a learnable skill, and it is a skill that improves with practice. Ideally, you should receive instruction and practice in crafting proposals during your graduate training. If you did not, you will have to teach yourself and obtain advice from colleagues who have been successful at winning awards. Proposal writing workshops can offer good advice, and academic institutions will often pay the fees of junior faculty who wish to attend such programmes. Search for advice and assistance before you start your search for money.

If you are or intend to be a practising professional, the principles outlined above apply to the projects you will wish to pursue within your agency or firm. You will make your mark and advance in the organization by proposing new programmes. To obtain assent and support for them, you will have to convince your superiors that you have formulated a worthwhile feasible project that will augment the organization's efforts. You may be required to submit a formal proposal. More probably, you will be asked to prepare an oral presentation or briefing. Alternatively, your proposal might be presented in a casual conversation over lunch. Whatever the forum and format, you can increase your chances of approval by following the steps noted above. You must have a clear and precise idea of what you want to accomplish, and a cohesive plan for marshalling the resources you are requesting in pursuit of an achievable and worthwhile objective.

Summary

Start your search for external funds with a viable project to which you are intellectually and emotionally committed. Never submit a proposal before you have consulted someone at the target agency and received confirmation that you are not wasting your time and everyone else's by addressing your request to that organization. Make every effort and take every pain to craft a self-contained elegant proposal that is free of contradictions and ambiguities. If you succeed, rejoice. If you do not, be persistent. Don't plan to fail, but remember that failure is the price of success in prospecting for money.

Further Reading

Abler, R. F. 1989: Becoming a grant swinger: how to get extra-mural research funds. In M. S. Kenzer (ed.), *On Becoming a Professional Geographer*, Columbus, OH: Merrill.

Abler, R.F. and Baerwald, T. J. 1989: How to plunge into the research funding pool. *The Professional Geographer*, 41, 1, 6–10.

Levine, F. J., 1986: Social and behavioral science support at NSF: an insider's view. In S.D. Quarles (ed.), *Guide to Federal Funding for Social Sciences*, New York: Russell Sage Foundation, 46–63.

VI.10
Overseas Work and Independent Travel

Nick J. Middleton

The call of overseas work and travel is one that all geographers have felt at some time or another. Many people from all walks of life travel abroad for holidays and relaxation, but the desire to cross the English Channel in search of 'real travel' or employment is a slightly different kettle of fish. If you are considering overseas independent travel or work you should start by working out exactly why it is you want to go. Are you attracted by a sense of excitement and adventure, or better weather, or are you running away from a bad situation at home or in search of a different culture abroad? If you manage to identify your motivation then it will be easier to narrow down the sorts of options you might consider.

Working Abroad

Many of the general skills of geography graduates, such as the ability to collate and summarize large amounts of information, computer skills and having a broad perspective of a range of views and issues, are in demand from all sorts of potential employers who operate or who have offices overseas. Large companies in areas such as banking, finance, marketing, trade and insurance all operate internationally and therefore may offer overseas postings after initial training periods. For a more directly geographical job some consultancies, such as Environmental Resources Limited (ERL), a London-based international consulting group, have offices in a number of foreign countries. If you are interested in a career in these sorts of fields then be sure to check out the possibilities of postings abroad while surveying your potential employers.

Voluntary Services Overseas

Voluntary Services Overseas (VSO) is an organization that attempts to combat world poverty not by sending food and material aid but by sending people. VSO recruits and sends more than 650 people each year to Africa, Asia, the Caribbean and the Pacific (plate VI.10.1). They include teachers, doctors, nurses, engineers, foresters, social workers, builders and agriculturalists. Their job is to live and work in a community in a developing country and to pass on their skills and knowledge to local people.

Most of the people that VSO sends abroad have several years of practical experience in their field, but they also send younger less experienced volunteers (their recruiting age range is 20–65 years). Although you will be paid this is not a job for getting rich. The minimum period of work is two years, and so the most important quality you must have is commitment. If you are interested in volunteering for VSO write to

Enquiries Unit
VSO
9 Belgrave Square
London SW1X 8PW

Plate VI.10.1 Voluntary Service Overseas (VSO) is one of the major bodies that recruits volunteers to work in developing countries such as the Sudan.

The British Council

The British Council, an independent body funded by the British government, exists to promote Britain abroad. This promotion includes British ideas, skills and experience in education, the English language, the arts, sciences and technology. Its 4000 staff are based in the United Kingdom and in eighty-two countries around the world. An overseas job with the British Council demands administration skills and diplomacy. The nature of the jobs can be as wide as the British Council's brief. You could be in charge of visiting British students, running an English language library in a foreign city, organizing tours of British arts or musicians, or selecting suitable applicants to visit the United Kingdom for training in a particular skill.

If you are interested in a career with the British Council write to

Tony Lockhart
Head of Recruitment
Personnel Management Department
The British Council
65 Davies Street
London W1Y 2AA

Independent Travel and Work

The idea of being a foreign correspondent or a travel writer is an appealing one to many geography graduates. Although getting such jobs is not impossible it is not easy either. Essentially, if you want it badly enough, and you have the capability, it is possible, but the key to these sorts of independent overseas/travel jobs is that it is very much up to you to create your own opening. Not often do you see such jobs advertised in the newspapers.

Before setting off it is worth your while ringing around some of the publications you have targeted to ask whether they are interested in pieces from the country you will be visiting. It is also worth trying some of the many guidebook publishers, since they are continually in need of updated information and they may pay for it. Some worth trying include Rough Guides, Lonely Planet (based in Australia) and Vacation Work. In most cases editors will tell you to contact them on your return. You must be very lucky or have good contacts to get a commission to write something from a foreign trip.

Your next step is to go abroad. Outside Europe it is certainly worth trying to get some sort of opening with an English language newspaper in the capital, writing about your impressions of the country or a piece about Britain perhaps. If you have done some student journalism this will improve your chances, so take copies of your work. If the publications you are aiming to write for also publish photographs with stories you should take lots of film. You should shoot transparencies for colour reproduction, prints for black and white, although quite good black and white prints can be made from colour transparencies nowadays.

Conversely, you can try to get something published on your return. In these circumstances you often come up against the Catch-22 situation of an editor demanding to see copies of your other published work. If you do not have any you can only write a piece and send it in. In almost every case, your writing will not be news, it will be a feature. Make sure you are familiar with the publication you want to write for, their style, usual length, types of subject etc. Send your manuscript, typed double spaced on one side of the paper, to someone by name (find out the name of the editor or editor of the section that you are aiming for), and follow this up with a phone call soon after it arrives. Editors are busy people and they receive dozens of unsolicited manuscripts. Do not be surprised or put off if they are brusque on the phone, but putting a voice to an anonymous piece of paper on their desk puts you above the other anonymous pieces of paper.

There are numerous publications to try for: newspapers and magazines of all kinds have regular travel slots and other geographical-type pages such as environment or developing country issues. Almost all use some freelance contributions. It pays to research the sort of publications you aim to try for before setting off on your travels. Geographical magazines include *Geographical Magazine, World* and *The Adventurers.* An invaluable guide to freelance writing of all kinds is *The Writers and Artists Yearbook* (A & C Black, London).

If you just want to go off and work your way around a continent or two there are often certain types of job you can get, such as teaching English, which will help fund your wanderings, although not many that directly exercise your geographical expertise. A useful book to guide you through the possibilities is *Working Your Way Around the World* (Vacation Work International, Oxford).

Index

Note: Alphabetical arrangement is word by word. Abbreviations (e.g. BBC) are found at the beginning of each letter sequence. Organizations are entered under their name, but US government departments etc. are to be found under United States.